Geomagnetics for Aeronautical Safety

# NATO Security through Science Series

This Series presents the results of scientific meetings supported under the NATO Programme for Security through Science (STS).

Meetings supported by the NATO STS Programme are in security-related priority areas of Defence Against Terrorism or Countering Other Threats to Security. The types of meeting supported are generally "Advanced Study Institutes" and "Advanced Research Workshops". The NATO STS Series collects together the results of these meetings. The meetings are co-organized by scientists from NATO countries and scientists from NATO's "Partner" or "Mediterranean Dialogue" countries. The observations and recommendations made at the meetings, as well as the contents of the volumes in the Series, reflect those of participants and contributors only; they should not necessarily be regarded as reflecting NATO views or policy.

**Advanced Study Institutes (ASI)** are high-level tutorial courses to convey the latest developments in a subject to an advanced-level audience

**Advanced Research Workshops (ARW)** are expert meetings where an intense but informal exchange of views at the frontiers of a subject aims at identifying directions for future action

Following a transformation of the programme in 2004 the Series has been re-named and re-organised. Recent volumes on topics not related to security, which result from meetings supported under the programme earlier, may be found in the NATO Science Series.

The Series is published by IOS Press, Amsterdam, and Springer, Dordrecht, in conjunction with the NATO Public Diplomacy Division.

**Sub-Series**

| | |
|---|---|
| A. Chemistry and Biology | Springer |
| B. Physics and Biophysics | Springer |
| C. Environmental Security | Springer |
| D. Information and Communication Security | IOS Press |
| E. Human and Societal Dynamics | IOS Press |

http://www.nato.int/science
http://www.springer.com
http://www.iospress.nl

**Series C: Environmental Security**

# Geomagnetics for Aeronautical Safety

## A Case Study in and around the Balkans


edited by

Jean L. Rasson

Royal Meteorological Institute, Centre de Physique du Globe,
Dourbes, Belgium

and

Todor Delipetrov

Faculty of Mining and Geology,
Štip, Republic of Macedonia

Published in cooperation with NATO Public Diplomacy Division

Proceedings of the NATO Advanced Research Workshop on
New Data for the Magnetic Field in the former Yugoslav Republic of Macedonia for Enhanced Flying and Airport Safety
Ohrid, the former Yugoslav Republic
18–22 May 2005

A C.I.P. Catalogue record for this book is available from the Library of Congress.

ISBN-10 1-4020-5024-0 (PB)
ISBN-13 978-1-4020-5024-4 (PB)
ISBN-10 1-4020-5023-2 (HB)
ISBN-13 978-1-4020-5023-7 (HB)
ISBN-10 1-4020-5025-9 (e-book)
ISBN-13 978-1-4020-5025-1 (e-book)

---

Published by Springer,
P.O. Box 17, 3300 AA Dordrecht, The Netherlands.

*www.springer.com*

*Printed on acid-free paper*

---



Printed in the Netherlands.

**TABLE OF CONTENTS**

# INTRODUCTORY ADDRESS[1]

Data on the geomagnetic field in the Balkan region and state borders were regarded as confidential information for a long time. Unfortunately this meant that geomagnetic field information was confidential information. The Republic of Macedonia was in a complicated situation because geomagnetic investigations were carried out by experts from Belgrade, Serbia and Montenegro. When Macedonia became an independent country, a team of experts from the Faculty of Mining and Geology, Department for Geology and Geophysics in Stip and Faculty of Natural Sciences and Mathematics, Institute of Physics in Skopje, started activities to establish a Geomagnetic Observatory in Macedonia. In the last four years, with the help of Dr. Jean Rasson from Institut Royal Météorologique, Centre du Physique du Globe in Dourbes, Belgium, a network of 15 repeat stations for measurement of the geomagnetic field in the Republic of Macedonia was created. For the first time since independence, all elements of geomagnetic field were determined.

Detailed measurement of the geomagnetic field is especially important at airports. Without information about the geomagnetic field there is real danger that aircraft compasses can not be calibrated at the airport. The magnetic compass is still the primary navigation device on aircraft. In case of failure of other electronic navigation devices (GPS, VOR) the magnetic compass will play an important backup role. The failure to correctly calibrate magnetic compasses represents a big threat to airport navigation systems.

Knowing the geomagnetic field elements is of interest in navigating airplanes. The most important geomagnetic element is declination. Precise values of declination make it possible to calculate mathematically exact geographic directions critical to navigation. Geographic north and magnetic north do not coincide. The difference between the two is the angle of declination. This is why there should be a correction made to the angle of the compass on the airplane.

Precise declination measurements must be made to increase airplane safety. There are special locations (compass certification pads) where airplanes can test the accuracy of their compass. These locations are free from magnetic contamination and have a minimal field gradient. In these locations precise directions of the geomagnetic field and geographic north are plotted so that when an airplane is at the site, its compass can be calibrated.

[1] Speech given at the inauguration ceremony of the NATO Advanced Research Workshop



J.L. Rasson and T. Delipetrov (eds.), *Geomagnetics for Aeronautical Safety*, 1–3.

Currently, this kind of certification is not done at Macedonian airports. This workshop will be useful to evaluate different solutions based on the experiences of participating countries. Also we will organise a round table for improving procedures of geomagnetic field measurements at airports.

Knowledge of the magnetic field distribution over the Republic of Macedonia also provides the means to produce magnetic charts of declination. Such maps are necessary for completing aeronautical charts and to compute the magnetic headings to be followed in order to navigate from one airport to another. The international collaboration proposed by this ARW is especially useful for this purpose.

Bearing in mind the central geographical position of the Republic of Macedonia, it is of special interest to use geomagnetic field data of neighboring countries (Bulgaria, Greece, Albania and Serbia and Montenegro).

Collaboration is important because the geomagnetic field depends on geological conditions and does not recognize state borders. Through an exchange of information at the workshop, these goals may be obtainable.

We must improve airplane safety and adopt procedures for measuring the geomagnetic field elements at airports. It is of special interest for airports in Macedonia and for all airports in the Balkan region.

The workshop will result in the transfer of knowledge, data and exchange of recent experiences, as well as the possibility to define new methods and procedures in observations of the geomagnetic field at airports for better safety of flying.

This workshop is motivated by recent geomagnetic measurements made in the Republic of Macedonia and the need to connect our data with data from neighboring countries and the presentation of this data to the public. On the other hand, a workshop like this, in the Balkan region with colleagues from EU countries and NATO members will be a contribution for better collaboration and understanding, which, unfortunately, in this region is not yet at the proper level.

The conclusions from the workshop will help to determine procedures for geomagnetic field measurements at the Macedonian airports. This should happen as soon as possible, to improve airport safety.

Basic scientific motive is the connection of the geomagnetic field in the territory of the Republic of Macedonia with neighbouring countries. This may help solve some border problems with interconnection and interpretation of the geomagnetic field. Very often measurements in border zones were impossible and extrapolations had to be made. Now that we can compare our data from both sides of the border, we have an opportunity to define the exact values for the geomagnetic field, and in some cases, possible common measurements to improve the data.

The basic motive in the presentation of experiences and discussions about procedures for geomagnetic measurements at the airports is the introduction of these procedures to the airports in the Republic of Macedonia and dissemination of this type of experiences from EU countries to countries in the Balkan region.

Another motive is that countries from the region may cooperate in advanced techniques, such as geomagnetic field measurements among themselves as well as with other developed countries, although in the Balkan region different destructive processes have taken place for a long time.

In general, the workshop should initiate collaboration between countries from the Balkan region and EU countries in the field of exploring and observing the geomagnetic field. The Workshop should contribute to implementation of the highest standards for measurements of geomagnetic field elements at the airports in Republic of Macedonia and make them safer.

The Workshop will also promote recent measurements in the Republic of Macedonia (carried out during 2002 – 2004).

**President of the Parliament of the Republic of Macedonia,**
**Dr Ljupco Jordanovski.**

# INTRODUCTION TO THE ADVANCED RESEARCH WORKSHOP: "NEW MAGNETIC FIELD DATA IN THE FORMER YUGOSLAV REPUBLIC OF MACEDONIA FOR ENHANCED FLYING AND AIRPORT SAFETY"

JEAN L. RASSON[2]
*Institut Royal Météorologique de Belgique*

## 1. Introduction

### 1.1. MACEDONIA, THE BALKANS AND SURROUNDINGS

The Republic of Macedonia, formerly part of Yugoslavia, has taken the initiative to call on the expertise of both scientific researchers in geomagnetism (modelers, cartographers, surveyors and, geophysicists) and aeronautical experts (pilots, aircraft operators, and airport managers) to improve aeronautical and airport safety.

This ARW will unite the two professional groups around a navigation instrument: the magnetic compass. During this workshop, we will review how the knowledge of geomagnetism and in particular magnetic declination, can be used to improve aeronautical safety.

The recent splitting up of Yugoslavia and subsequent political evolution of the Balkans, along with its rich scientific past (Nikola Tesla was born and lived here) contribute to the value of holding the workshop here.

### 1.2. MAGNETIC FIELD

The geomagnetic field is a vector quantity, and as such it is characterized either by its cartesian components X, Y and Z or by:

- two angles: declination, D, and inclination, I, expressed in degrees and
- an intensity, F (modulus or "Total Field"), expressed in nanoteslas.

[2]Address for correspondence: J Rasson, IRM/CPG, Rue de Fagnolle, 2 Dourbes, B-5670 Viroinval, Belgium ; e-mail : jr@oma.be



*J.L. Rasson and T. Delipetrov (eds.), Geomagnetics for Aeronautical Safety*, 5–12.

Magnetic observers strive to measure those quantities with an accuracy of one second of arc for D and I and one tenth of a nanotesla for F. This high accuracy is necessary to ensure correct extrapolations when forecasts are made, a procedure which tends to amplify the observation errors.

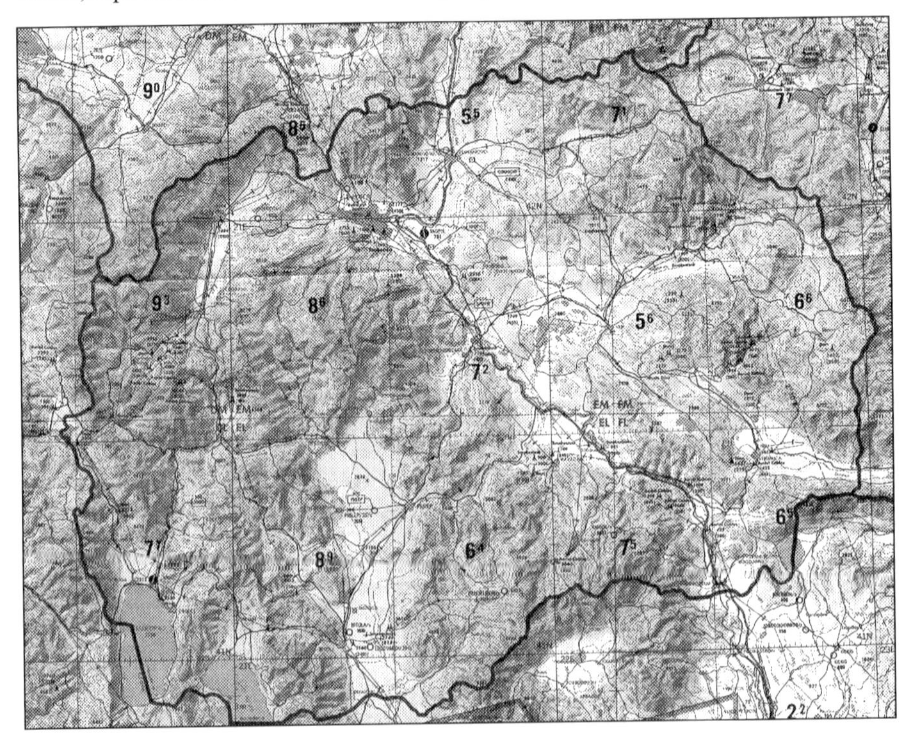


Figure 1. The Republic of Macedonia: an aeronautical map.

## 2. Aim of this ARW

The aim is to enhance the security in aircraft and airports throughout Macedonia, the Balkans, and the surrounding area.

How?

- By providing the correct value of the magnetic declination where and when it is needed:
    - At the airports now
    - At the airports in the future
    - Over the Balkan region's airspace
- By ensuring that aircraft magnetic compasses are working properly
    - Certify compass roses for the compass swinging procedure
    - Calibrate aircraft compasses

## 3. Aeronautical conditions in Macedonia

Figure 1 is a map of the Republic of Macedonia showing airport locations, elevation, and other physical features. The very mountainous terrain creates a difficulty for aircraft. The two major international airports of Ohrid and Skopje are indicated as well as other small airports like the ones in Bitola and Ponikva .

When consulting the web-pages of the Macedonian airports, we looked for magnetic declination information. Skopje Airport did not give this data while the airport in Ohrid gave data which is 15 years old (obsolete) as seen in Figure 2.

Ohrid Airport

| | |
|---|---|
| Airport: | OHRID |
| IATA Code: | OHD |
| ICAO Code: | LWOH |
| City: | OHRID |
| Referent Point: | 41°10'46"N 20°44'51"E |
| Site: | 1 275m BRG 018°GEO from RWY THR 02 |
| Distance and direction from city: | 10 km BRG 327°GEO from Center of Ohrid |
| Airport Referent Point (ARP) site: | Center of RWY |
| Elevation: | 705 m |
| AD REF Temperature: | 28°C AUGUST |
| Magnetic Variation: | 2°E (1990) |

Figure 2. Extract from the web-page of Ohrid airport (2005).

Enhanced safety in aeronautics can be obtained by using up-to-date geomagnetic measurements. An integral part of this workshop is to point out the necessity of using correct and up to date geomagnetic data.

## 4. What are our assets?

### 4.1. PROTECTED SITES

The geomagnetic community has at hand a set of about 150 magnetic observatories in Europe and worldwide, where the geomagnetic field is measured on a continual basis, often at a rate of one full vector of the field at 1 sample per minute or faster. These sites are carefully protected from magnetic perturbations, so as to ensure that they observe the natural magnetic field and not one perturbed by cultural or technical noise. This is the same unperturbed field which will be measured by the compass of an aircraft flying aloft. Many of these observatories belong to the INTERMAGNET network, offering their data in near real-time on the INTERNET. Figure 3 shows a protected magnetic observatory site. Note that the observatory buildings are made exclusively of non-magnetic materials like wood, copper, white sandstone, and nonmagnetic concrete. It is worth mentioning here that the Republic of Macedonia is contemplating the construction of a magnetic observatory soon.

Figure 3. A protected site: the geomagnetic Observatory of Dourbes, Belgium.

## 4.2. TRAINED AND DEDICATED PEOPLE

Geomagneticians are available at observatories as well as at specialized geophysical, geological, meteorological, university, or topography institutes, to observe and measure the Earth's geomagnetic field over time as a time series or in space as spatial variations.

These professionals are very keen to work for the aeronautical sector. The application of geomagnetics to aeronautics extends work into the commercial realm. In the past, geomagnetic studies have been mostly academic.

## 4.3. AVAILABILITY OF GLOBAL AND REGIONAL MODELS

A sizeable part of the geomagnetic community is very busy with the modelling of the geomagnetic field. Measurements of the Earth's magnetic field are incorporated into regional and/or global models. Special mathematical techniques, used to distribute data on spherical surfaces are used. Thanks to these techniques, aircraft pilots have a clear idea of how geomagnetic declination behaves on the Earths Globe and know precisely how they must interpret the bearings given by their magnetic compasses.

## 4.4. KNOW-HOW

Accurate measurement of the geomagnetic field is not easy, especially at the sub-nanotesla and second of arc level. Additionally, measurement difficulty stems from the fact that only the natural field is to be measured. The observer must be magnetically clean. The observatory buildings and surrounding underlying terrain must be free from magnetic contaminants.

Up to now only trained geomagneticians had the expertise to make accurate measurements. Now, some topographical and military institutes are capable.

## 4.5. UP-TO-DATE AND VAST GEOMAGNETIC DATA BASES

Magnetic observatories have existed for more than 500 years. A huge database of geomagnetic observations has been accumulated in the so called "World Data Centres". They can be reached with the links given in Table 1.

Table 1. List of the Geomagnetic World data Centers.

| LOCATION | URL or Email |
|---|---|
| KYOTO | http://swdcwww.kugi.kyoto-u.ac.jp/catmap/index.html |
| COPENHAGEN | http://web.dmi.dk/projects/wdcc1 |

| LOCATION | URL or Email |
|---|---|
| EDINBURGH | http://www.geomag.bgs.ac.uk/gifs/on_line_gifs.html |
| MUMBAI | abh@iigs.iigm.res.in |

Updates to the WDC's addresses can be found at http://www.ngdc.noaa.gov/wdc/list.shtml

Another database is offered by the INTERMAGNET consortium (www.intermagnet.org). INTERMAGNET provides high quality data at a sampling rate of 1minute and accessibility in near real time. This worldwide network presently offers data from about 100 magnetic observatories, as depicted on Figure 4.

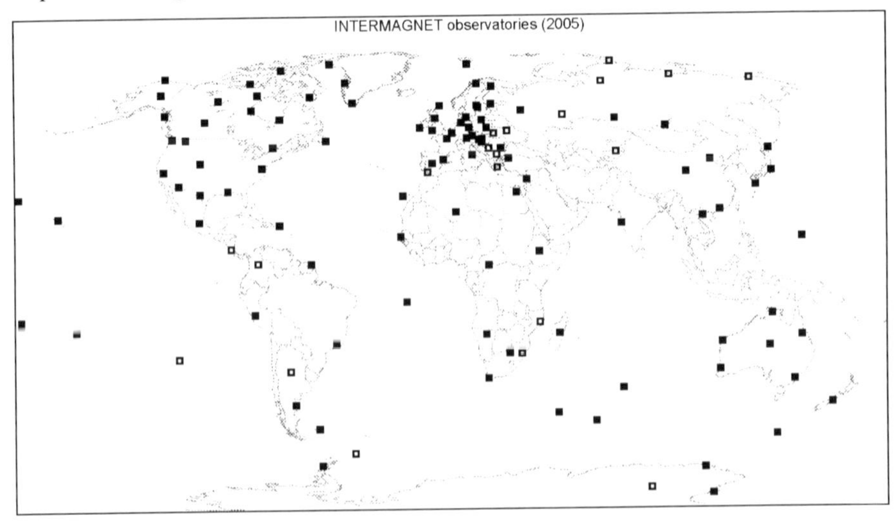


Figure 4. World map with the INTERMAGNET magnetic observatories as of 2005.

The timely availability of online or archived data is important for successful delivery of magnetic declination data to the aeronautical community.

## 4.6. QUASI MONOPOLY

Due to the specialized nature of the work, the costly infrastructure of geomagnetic observatories, and the relatively low demand from the commercial sector, one institution per country can provide geomagnetic information to interested parties. Therefore, a *de-facto* monopoly exists for the supply of these services and products. This is a favourable situation for those who intend to sell geomagnetic products and services.

## 5. How will this Workshop address its topics?

### 5.1. KNOW YOUR GEOLOGY

The local geology is of importance in an investigation of the geomagnetic features of a region. We will have a thorough review of the relevant geological features of Macedonia and of the Balkans in the paper by T Delipetrov. As Dean of the Faculty of Geology in Štip, Republic of Macedonia, he is one of the best informed scientists on the subject.

### 5.2. INVENTORY OF FIELD AND OBSERVATORY DATA

Among the participants of this Workshop there are key persons dealing with magnetic field measurements in the Balkans and surrounding areas. These scientists are going to show a detailed view of the geomagnetic data available for our purpose of compass navigation. Balkan countries, old and new and also the neighboring states have to get scientists together in order to rationalize and unify data, so that any discrepancies (at the borders for instance) can be normalized.

### 5.3. METROLOGY OF GEOMAGNETIC FIELD

Measuring the geomagnetic field depends very much on the availability of good instrumentation. Obtaining the required accuracy is a constant concern for the surveyor. Additionally, the reduction of staff observed during the last 20 years increasingly calls for faster measurements, protocols, and more user-friendly interfaces in order to carry on the ever increasing workload.

Therefore, advances in geomagnetic instrumentation frequently have to do with increasing the operator's comfort and reducing the operational tasks by taking advantage of automated procedures.

The methods for logging geomagnetic data in the observatory environment are not entirely satisfactory. We will have a few papers on how to make advances on that topic.

Finally, the geomagnetic community is a step closer to realizing its dream of having a fully automatic magnetic observatory operation as discussed in a paper on the automatic DIM.

### 5.4. SERVICES AND PRODUCTS FOR AERONAUTICS

A sizeable part of our workshop will be devoted to comparing methods used by the various attending experts. There has been a tendency toward individualism in the past as each group performing services and measurements has been working in relative isolation.

This situation surely is not satisfactory, and the workshop will be a unique opportunity, offering a discussion forum where all our procedures and experiences can be evaluated by the geomagnetic community.

### 5.5. MATHEMATICAL PROCEDURES

Processing measurements into useable form is critical when creating products for the aeronautics sector. As the time of the actual measurements always precedes the publication of maps, values list, or spherical model, by about a year, there is a need to forecast the data. Customers like to have declination values which apply to the time interval they will be working in.

Sophisticated mathematical procedures can help greatly. We will have presentations on the Spherical Cap Harmonic Analysis technique applied to the computation of aeronautical maps, the Chaos theory will be put to use demonstrating how accurate forecasting of the geomagnetic field values can be achieved.

# GEOMAGNETIC FIELD OF THE REPUBLIC OF MACEDONIA

TODOR DELIPETROV[3]
BLAGICA PANEVA
*Faculty of Mining and Geology*

**Abstract.** The paper presents geomagnetic investigations carried out in the territory of the Republic of Macedonia. It also gives the geologic and geographic location of the country on the Balkan Peninsula. The detailed description of the tectonic setting contains reference of the neotectonic distribution. The paper presents investigations that commenced in the 19th century and those carried out in 2003. Geomagnetic investigations have been separated from investigations of the normal magnetic field and those of the anomalous geomagnetic field and presented in a chronologic manner. Analysis of activities carried out during geomagnetic investigations indicate that the most intense were those of 1950s and 1960s for the discovery mineral deposits and those in 1970s for the investigation of normal magnetic field. Of note are investigations that have been carried out since 2002 in order to study the normal magnetic field in the territory. During this short period of time a grid of repeat stations was established and the site for the construction of the geomagnetic observatory in the country was selected.

**Keywords:** Geology, Republic of Macedonia, Geomagnetism, Balkans

## 1. Geotectonic position of the Republic of Macedonia

The territory of the Republic of Macedonia occupies the south central part of the Balkan Peninsula which geotectoniclly belong to the Alpine - Himalayan geosynclines area.

[3] To whom correspondence should be addressed at: Faculty of Mining and Geology, Department of Geology and Geophysics, 2000 Štip, Republic of Macedonia. Email: todor@rgf.ukim.edu.mk.



*J.L. Rasson and T. Delipetrov (eds.), Geomagnetics for Aeronautical Safety,* 13–42.

During the tectonic evolution, eight geotectonic units were differentiated in the Balkan Peninsula, which were later separated into eight younger units also important for the regional geotectonic setting of the area. The geotectonic units, from east to west include (see Figure 5):

- The Mesian plate
- The Carpatho - Balkanides
- The Rhodope mass - Serbo - Macedonian massif
- The Vardar Zone
- The Dinarides - Helenides
- The Adriatic massif (Apulian plate)

Of these, the following are present in the Republic of Macedonia; parts of the Carpatho - Balkanides (separated as Strumica zone), the Rhodope mass - Serbo Macedonian massif (separated as Macedonian massif), the Vardar zone, Dinaride - Helenides (separated as Serbo - Macedonian zone and Pelagonian massif) and the Adriatic massif (present as a small part at the border with Albania, separated as Korabides).

The Mesian platform represents the western part of the Ponto - Caspian table which belongs to the Russian platform. The Mesian platform includes part of northeastern Serbia, a large part of Romania and northern Bulgaria as far as the Black Sea.

The Carpatho - Balkanides is part of the northern border of the Alpian orogeny, which, in the eastern Alps, is an arc that continues through the Carpathian mountains in Romania and follows the Danube to eastern Serbia. From there it goes to Bulgaria, and continues east to the Black Sea.

The Rhodope mass - the Serbo Macedonian massif, as a first order geotectonic unit embraces the central and eastern part of Serbia, the eastern part of Macedonia, from where it continues to central and southeastern Bulgaria, eastern Greece, and continuing to Romania in the north. In the east and northeast the massif overthrusts the Carpatho - Balkanides, and in the west and southwest overlies the Vardar zone.

Starting from the Alps, the Dinarides - Helenides extend to the south through the terrain of Slovenia, Croatia, Bosnia and Herzegovina, the western parts of Serbia and Macedonia, Albania and Greece where they bend to the east continuing to Asia Minor.

According to their strike, the Dinarides are divided into Dinarides (proper), developed as far as the Skutari - Pec fault, to Helenides which include the western terrains of Macedonia (west of the Vardar zone), Albania, Greece, and to Taurides which are developed in the central and western parts of Asia Minor (Turkey).

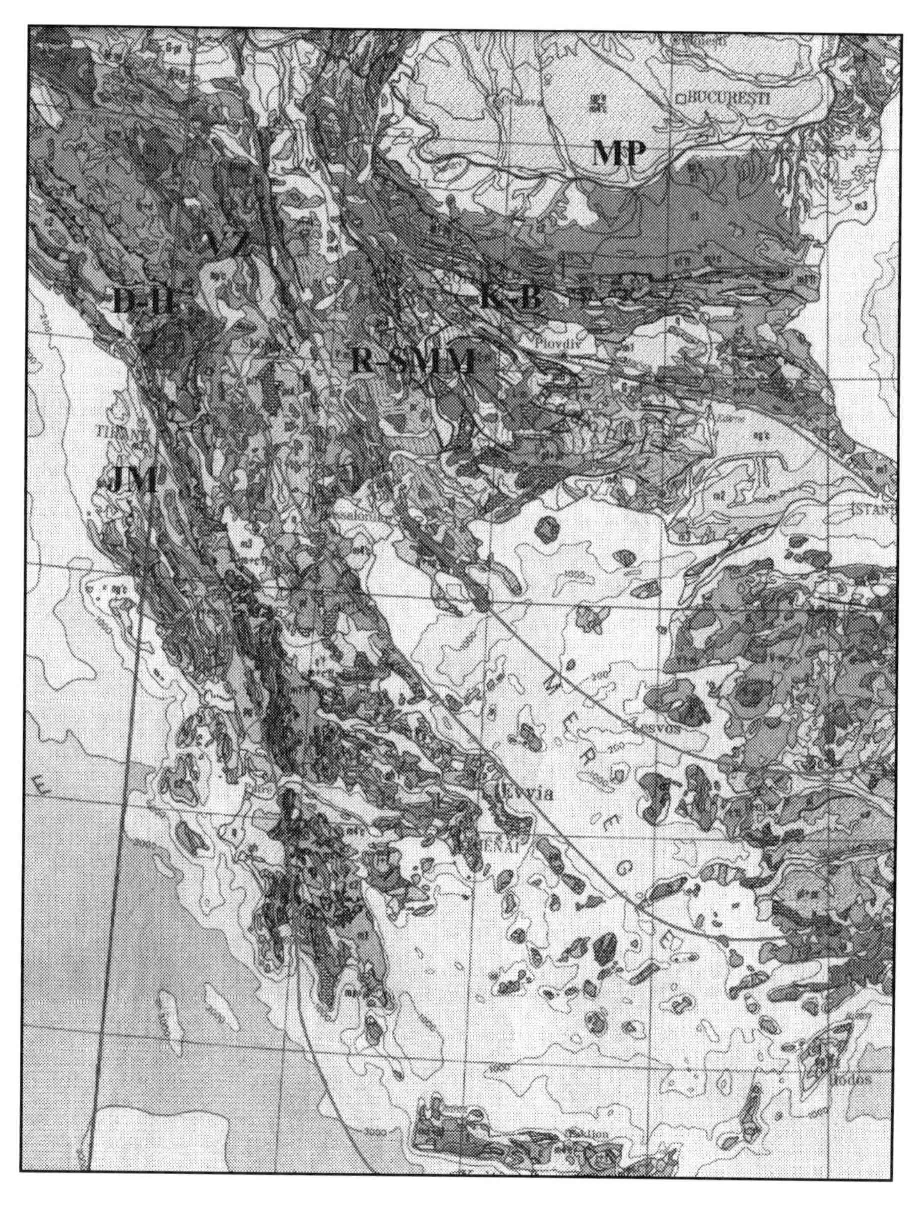


Figure 5. Map of Balkan Peninsula with some geotectonic units MP - Mesian plate, K-B - Karpatho-Balkanides, R-SMM - Serbian-Macedonian massif, VZ - Vardar zone, DH - Dinarides-Helenides, JM - Adriatic massif (Apulian plate).

The Vardar zone, an old geosyncline, was formed during the break up of the Grenville earth crust in the Riphean - Cambrian. Until the Triassic it had geosynclinal development. During the Jurassic, the opening of the continental crust in the area resulted in the formation of ocean type crust within the Vardar- Izmir- Ankara zone.

The Adriatic massif is situated between the Alpine - Apeninian and Dinaride - Helenide geosyncline system, representing part of the African platform. It played an important role during the formation of the Dinaride - Helenide tectonic structures in the east and the Apennines in the west. The massif is also known as the Apulian platform (part of the African platform), the Adriatic geosyncline area, and an intermountainous depression.

## 2. Regional tectonic setting

Information offered by many authors (Arsovski at al. 1975) helped to compile Figure 6 and give the regional tectonic setting of the Republic of Macedonia.

The Republic of Macedonia belongs to the Dinaride and the Rhodope system. The part of Macedonia west of the Presevo - Zletovo - Strumica - Dojran line belongs to the Dinaride system and east of the line is part of the Macedonian massif which, with the Ograzden massif, joins the Rhodope system.

The Macedonian massif in this part of the Alpine orogeny is a geological anticline zone, or mid - position massif that separates the Dinarides from the Carpatho - Balkanides. In the area bordering Bulgaria east of the Berovo - Pehcevo line, elements of the Carpatho - Balkanides have been forced as a wedge into the old Rhodope massif, known as Strumica zone (Kraistides) separating the Rhodope and the Eastern Macedonian massif.

The territory of Macedonia, west of the Presevo - Zletovo - Strumica - Dojran line belongs to the Dinaric system. Four zones have been distinguished: the Vardar zone, the Pelagonian horst - anticlinoriums, the western Macedonian zone, and the Serbo Macedonian massif.

These zones represent individual structural facies units with their own geological evolution.

The structural zones are characterized by their own geological evolutions which can be seen from various lithological complexes that differ in composition, age, and dislocation.

Different types of rocks (metamorphic, intermediary to igneous), from Precambrian to Cenozoic are present.

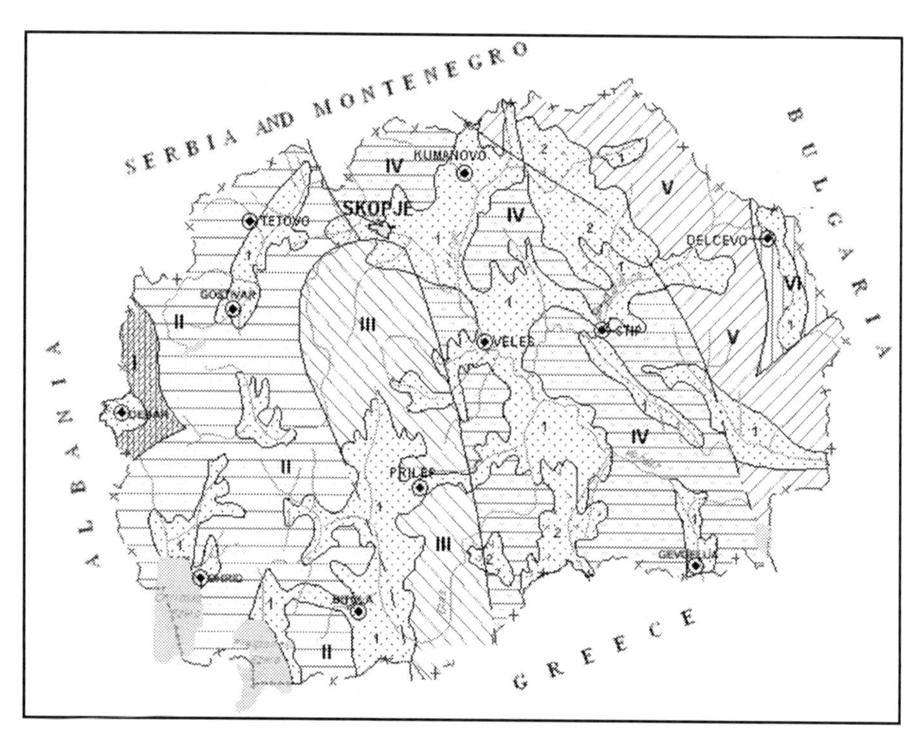


Figure 6. Regional tectonic setting of the Republic of Macedonia I - Cukali Krasta zone, II - Western Macedonian mass, III - Pelagonian horstanticlinorium, IV - Vardar zone, V - Serbo-Macedonian massif, VI - Kraistide zone.

## 2.1. NEOTECTONIC CHARACTERISTICS AND ZONATION OF THE TERRITORY OF THE REPUBLIC OF MACEDONIA

The geological evolution during the Neocene and Quaternary in all of Macedonia is characterized by continental development, uplift, overthrust, and subsidence. During this period, volcanic activity, the outflow of large andesite - dacite volcanic masses and tuffs occurred only in the area of Zletovo. Along reactivated deep faults there was outflow of volcanic material of some 1000 km$^3$. A similar volcanic mass also developed in the area of Kozuf - Vitacevo. According to data, the volcanic activity took place periodically, although it started earlier in the Zletovo area. In the area of Kozuf it continued to the beginning of the Quaternary.

Modern relief was formed in limnic basins due to active neotectonic processes. Terrigenous layers of molasse with interbeds of coal were deposited in the depressions. During the Pliocene the terrigenous material became coarse as a result of atomization caused by tectonic movements. These processes have continued to the present time and manifest themselves as earthquakes (Skopje in 1963, Valandovo in 1931 etc.). At the end of the Pliocene and at the beginning of the Quaternary the volcanic activity consisted of outflows of basalts near Nagoricani and some other localities. Today, only traces of this activity can be seen in the area of Ohrid (the village of Kosel) in the form of sulphatara - fumarola.

All geotectonic units mentioned, starting in the Neocene, developed as continental phase. During the first phase, peneplenisation of structures developed through orogenic processes (end of Paleocene – Oligocene). In the second phase, commencing in the Miocene, a neotectonic phase took place and basic structures seen today as modern relief were formed. Mountainous massifs formed as elements of uplift and depressions formed in areas of relative subsidence.

The neotectonic processes spurred the development of new structures and at the same time reactivated structures formed earlier. Many of the underthrust faults reactivated such as the Drim fault zone, some in the Vardar zone and other places.

Neotectonic zonation of the Republic of Macedonia took place (see Figure 7). In the western part, morpho-structures of uplift up to 2000 m in size formed. These structures of uplift are blocks elongated with meridian strike. Graben structures were formed with meridian strike as well.

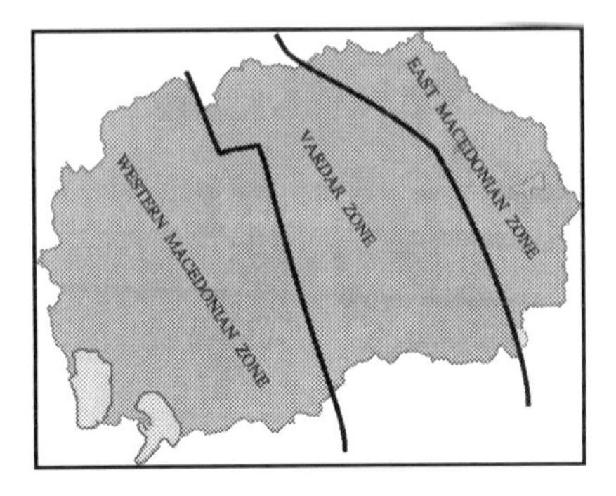


Figure 7. Neotectonic map of the Republic of Macedonia.

This indicates that during the neotectonic stage the pattern of general uplift was an east - west expansion. The morphostructures in the Vardar zone are characterized by mountainous massif morph structures of uplift of 1000 to 1500 m (500 m lower than those structures in western Macedonia).

In the Vardar zone, depressions were the dominant structures. Skopska I, Ovcepolska II, Tikveska III, are situated above the older structures and consist of complex shapes of 100 - 400 m height. From the intensity of vertical movements, whose impact can be seen in the modern relief, and based on higher order morph structures, it is concluded that the horizontal component of extension is of a different orientation in the zones.

Unlike the Vardar zone, the morphostructures of uplift in eastern Macedonia are present as mountainous massifs 1600 - 1800 m high and the depressions are present as grabens oriented east - west. The main strain is of vertical extension, whereas the axis of extension is of meridian strike.

The neotectonic zones of Western Macedonian, Vardar, and Eastern Macedonia are rather different.

## 3. Review of geomagnetic investigation carried out

Geomagnetic investigations can be divided into three periods: those started in the 19$^{th}$ century up to 1945, those carried out between 1945 and 1991 when Macedonia was a constitutional part of Yugoslavia, and those after the declaration of independence in 1991.

During the first period, due to unstable political conditions and the scientific backwardness of the Turkish Empire and the Kingdom of Serbs, Croats and Slovenians, scientists from European countries investigated the magnetic field of the Balkans including a small part of the territory of Macedonia only on rare occasions. These investigations are of historic importance but due to their sparcity did not offer any deep scientific understanding of the geomagnetic field in the area.

During the time of the Federal Republic of Yugoslavia, a geomagnetic observatory was established in Grocka, near Belgrade. They initiated and carried out geomagnetic investigations on an ongoing basis and occasionally acquired data from geomagnetic field stations across Yugoslavia. During this period field stations were established in Ohrid and Strumica. The observatory also initiated all field investigations in terms of defining areas of geomagnetic anomalies.

After the Second World War, particularly during the 1950,s and 1960's, intensive geomagnetic investigations were completed with the aim of discovering new deposits of mineral raw materials.

The following two sections of this paper discuss the geomagnetic investigations carried out so far. Section 3.1. discusses investigation and study of the geomagnetic field, and section 3.2. discusses geomagnetic investigations for the discovery of deposits of mineral raw materials.

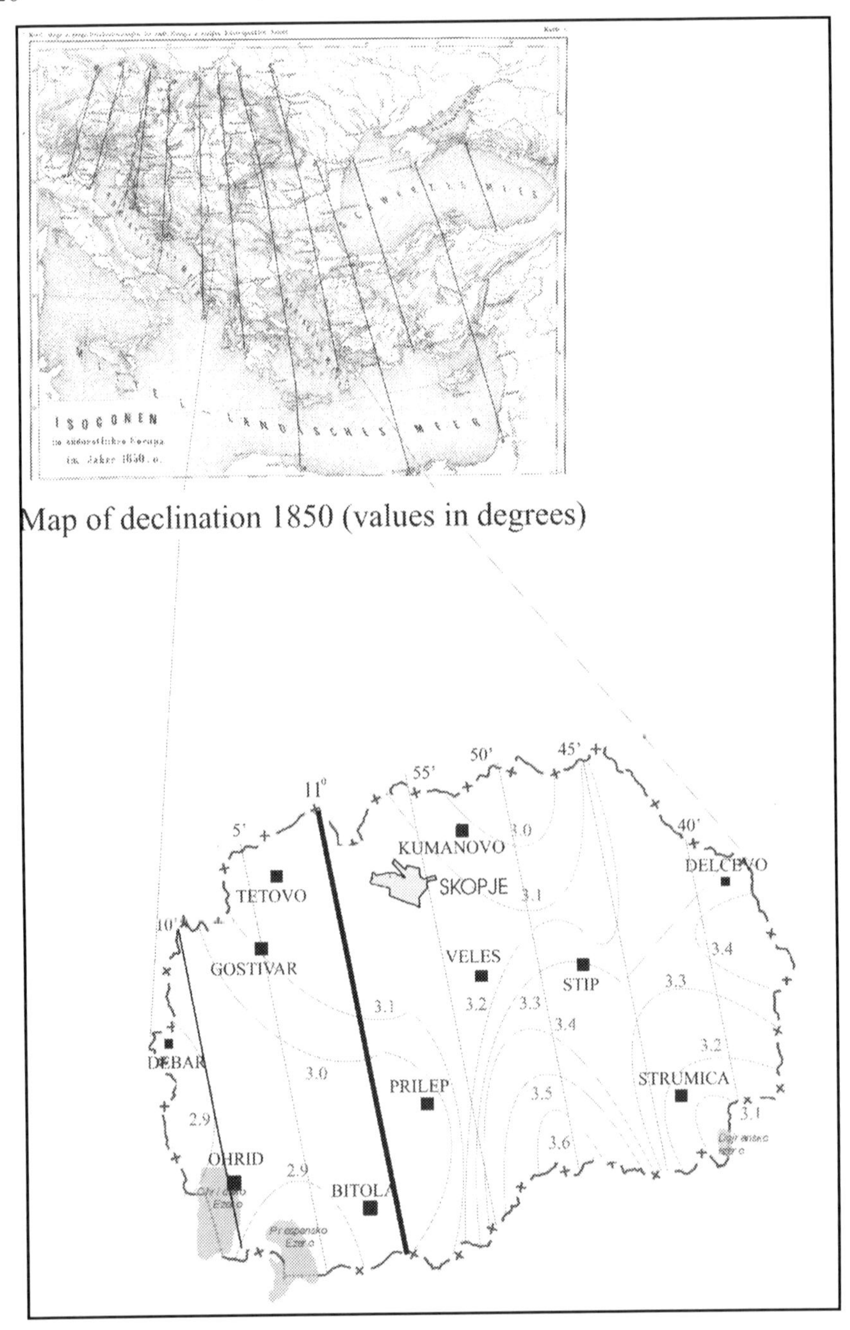


Figure 8. Map of declination in Macedonia 1850.0 and 2003.5.

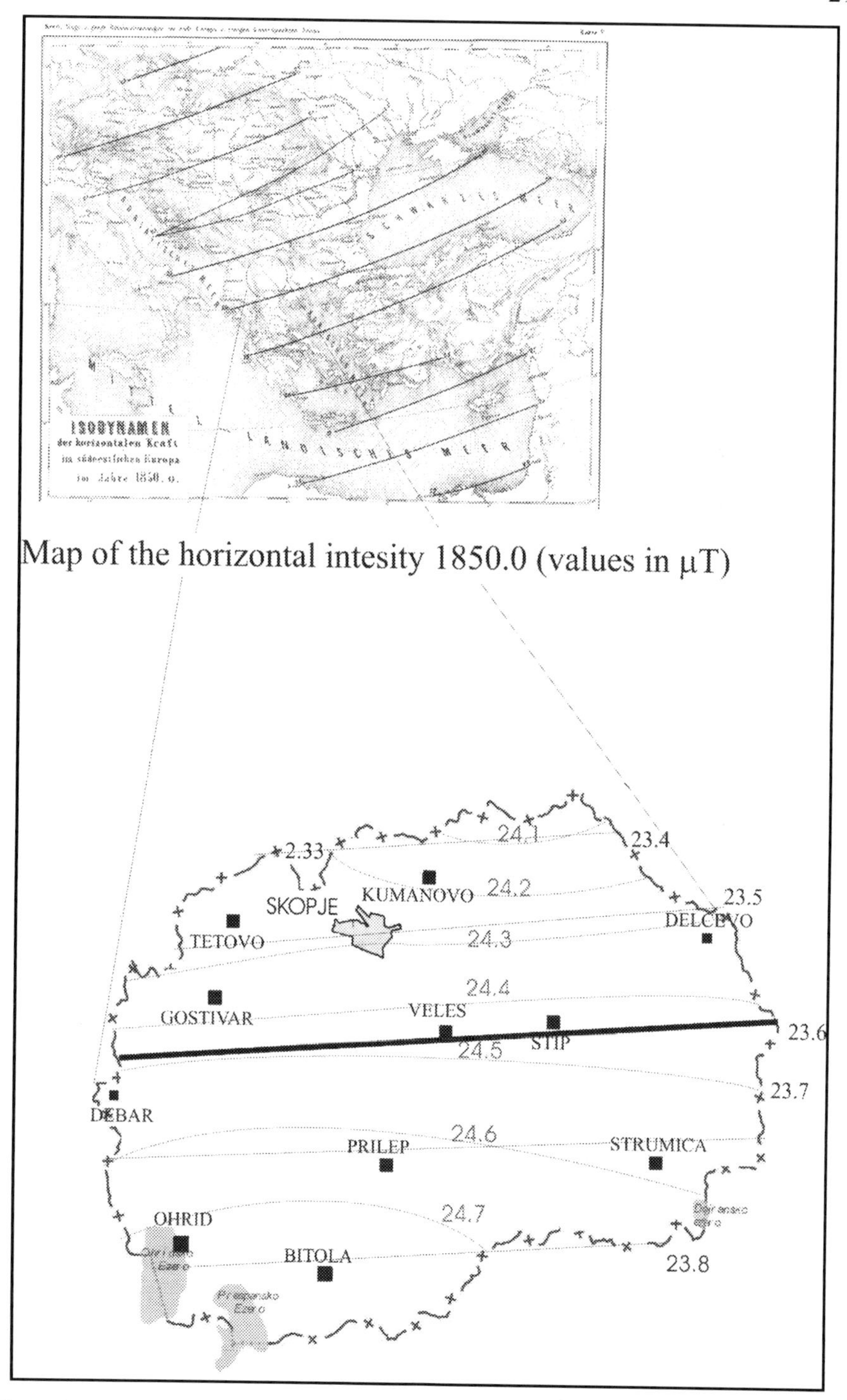


Figure 9. Map of horizontal intensity in Macedonia 1850.0 and 2003.5.

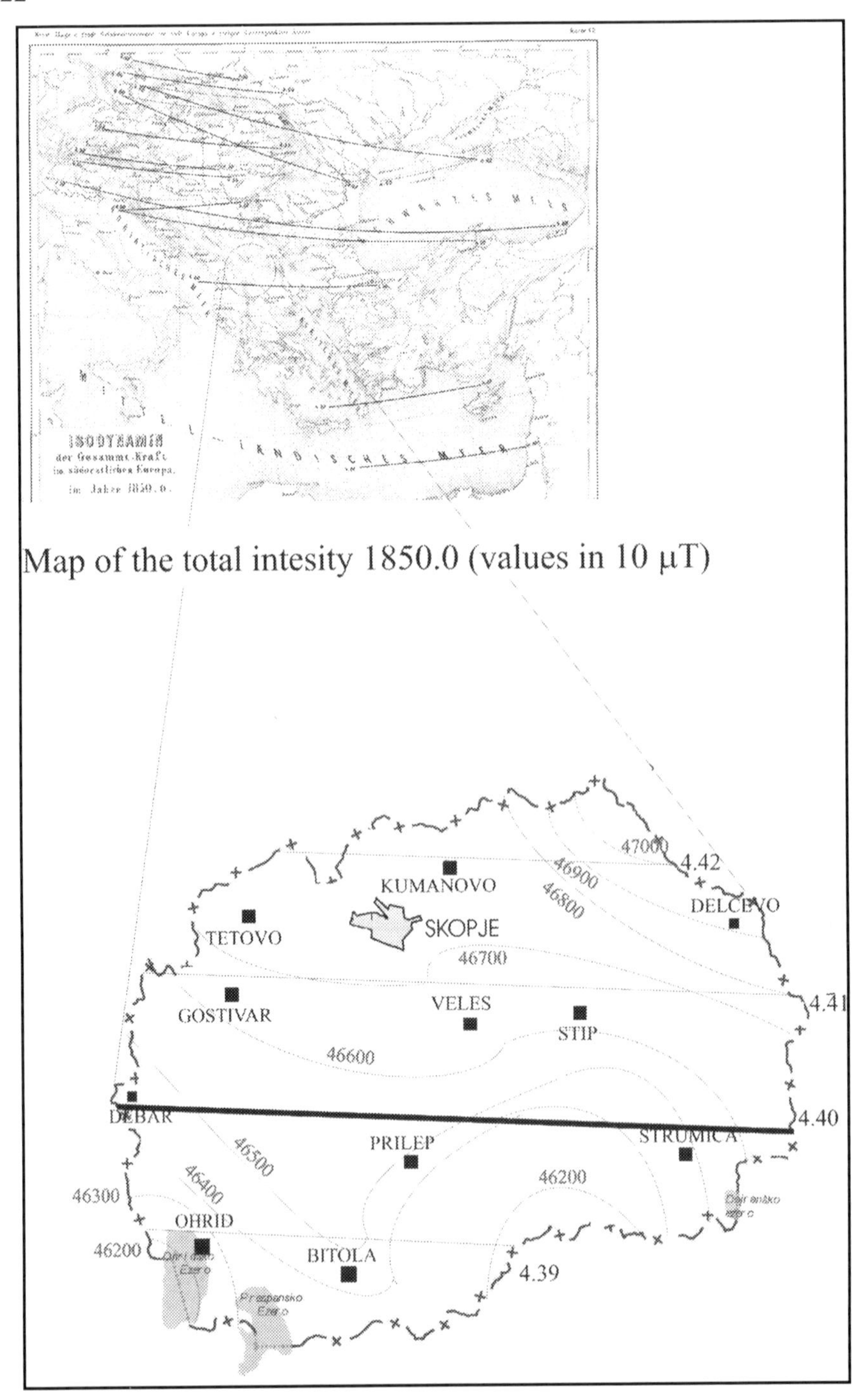


Figure 10. Map of total intensity in Macedonia 1850.0 and 2003.5.

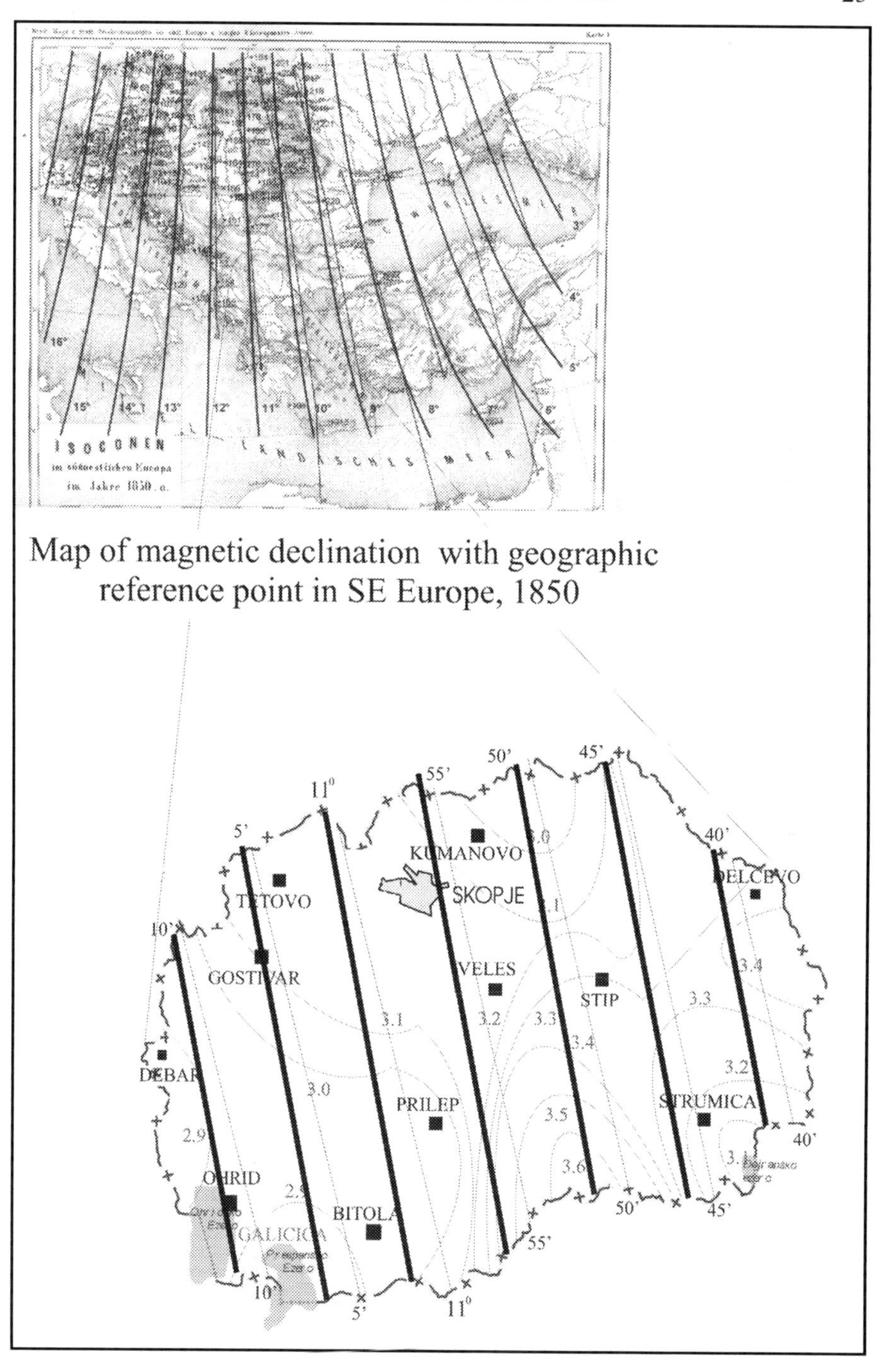


Figure 11. Presentation of lines of equal declination with geographic reference point in Macedonia, 1850.0 and 2003.5.

## 3.1. INVESTIGATION AND STUDY OF THE GEOMAGNETIC FIELD

Geomagnetic measurements carried out in the territory of Macedonia show that during the 19th century magnetic observations were made over a wide area of eastern Europe, although with a sparse grid. This paper presents the maps compiled for the 1850.0 epoch by Karl Kreil[8].

Black isolines on the maps indicate the result of Kreil's investigations and red isolines show results of the investigations carried out in 2002 and 2003 by Jean Rasson and Marjan Delipetrov[4] (see Figure 8 – Figure 11).

During the 1960's the territory of Yugoslavia was investigated with a relatively dense grid of field stations. A map of the vertical or Z component of the anomalous geomagnetic field of Yugoslavia with a scale of 1 : 500 000 was compiled[2] (Figure 12).

All investigations were centralized and marked as "top secret". Purchase of instruments for geomagnetic measurements was impossible.

During the 1970's a project was implemented for central and eastern Europe which included the observatories given in Table 2.

Table 2. Observatories in central and eastern Europe.

| IAGA Code | Name | Country | Operation | Latitude | Longitude |
|---|---|---|---|---|---|
| FUR | Furstenfeldbruck | Germany | 1939- | 48.17 | 11.28 |
| AQU | L' Aquila | Italy | 1960- | 42.38 | 13.32 |
| BDV | Budkov | Czech Republic | 1967- | 49.07 | 14.02 |
| WIK | Wien Kobenzl | Austria | 1851- | 48.27 | 16.32 |
| NCK | Nagycenk | Hungary | 1961- | 47.63 | 16.72 |
| THY | Tihany | Hungary | 1955- | 46.90 | 17.89 |
| HRB (OGY) | Hurbanovo | Slovakia | 1894- | 47.87 | 18.19 |
| GCK | Grocka | Serbia | 1958- | 44.63 | 20.77 |
| CST | Castel Tessino | Italy | 1965- | 46.05 | 11.65 |
| ROB | Roburent | Italy | 1964-1973 | 44.30 | 07.90 |
| PRU | Pruhonice | Czech Republic | 1946-1972 | 50.00 | 14.55 |
| REG | Regensberg | Germany | (1931-1956) | 47.50 | 08.45 |
| POL | Pola | Austrian Empire | 1847-1909 | 44.87 | 13.85 |
| PAG | Panagyurishte | Bulgaria | 1948- | 42.50 | 24.20 |
| LVV | Lviv | Ukraine | 1952- | 49.90 | 23.75 |
| PEG, PEG2 | Penteli | Greece | 1959- | 38.10 | 23.90 |
| SUA | Surlari | Romania | 1949- | 44.68 | 26.25 |
| ISK | Kandili | Turkey | 1947- | 41.06 | 29.06 |

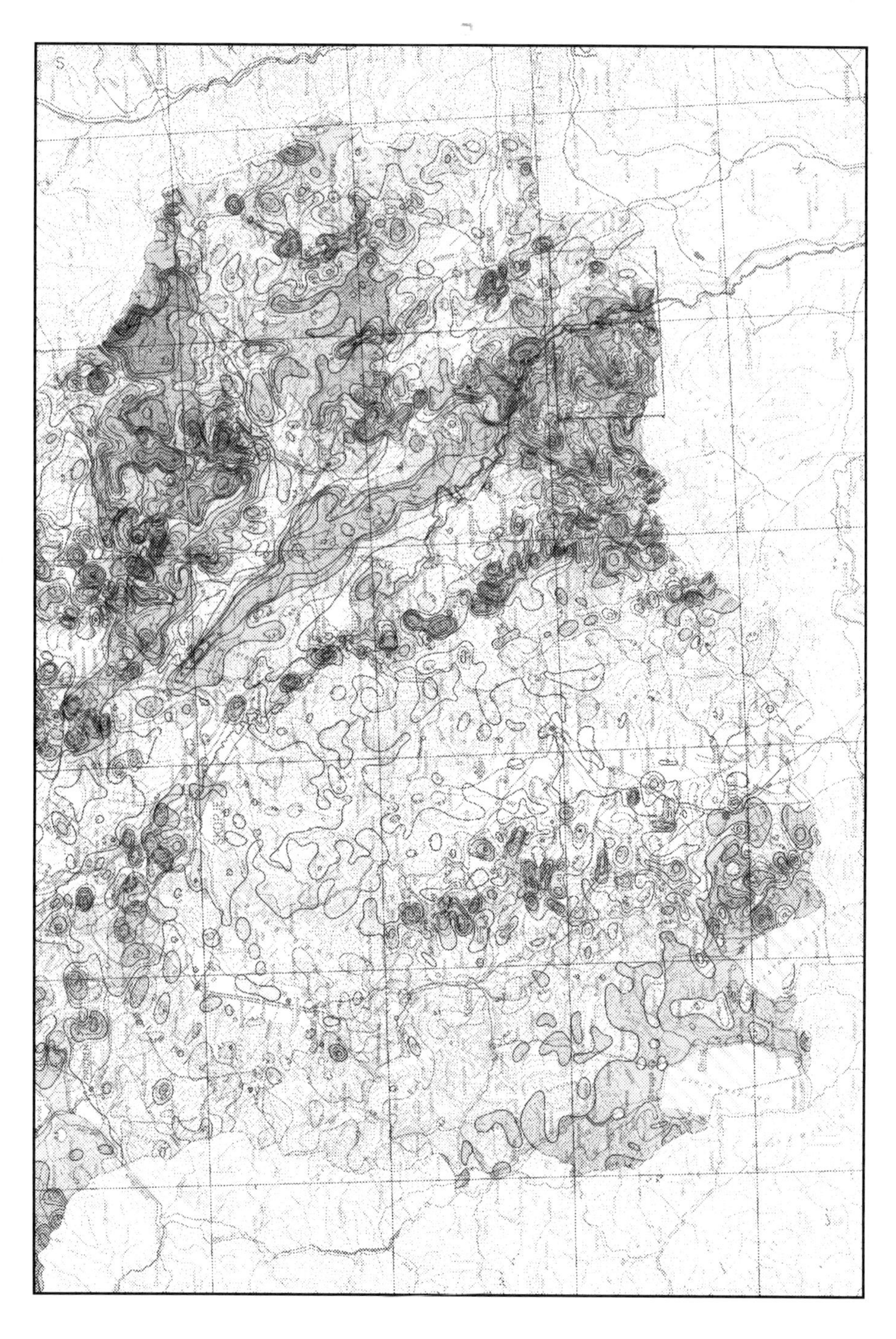

Figure 12. Map of the vertical or Z component of Macedonia.

The maps that follow (Figure 13 - Figure 16) show the location of observatories and components of the normal geomagnetic field for the 1970.5 epoch.

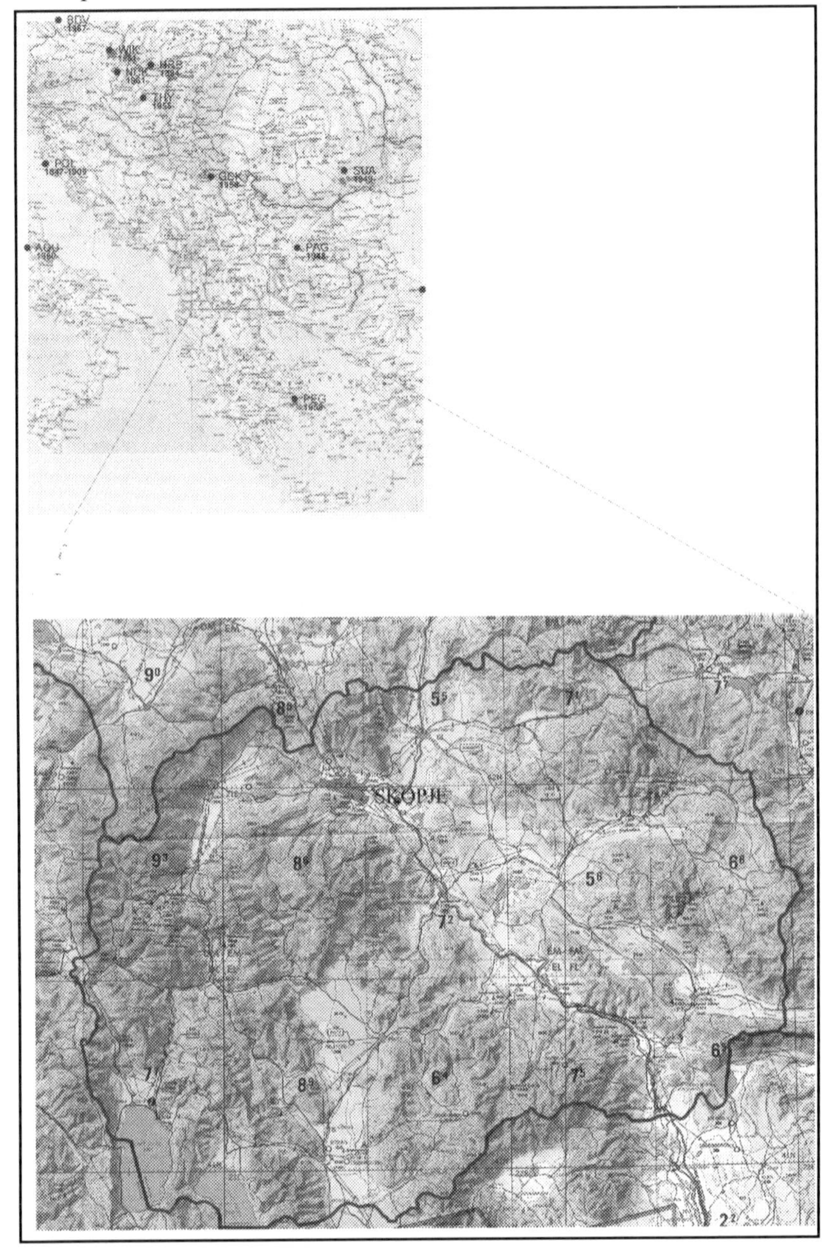


Figure 13. Map of geomagnetic observatories in SE Europe and their operation periods and map of Macedonia.

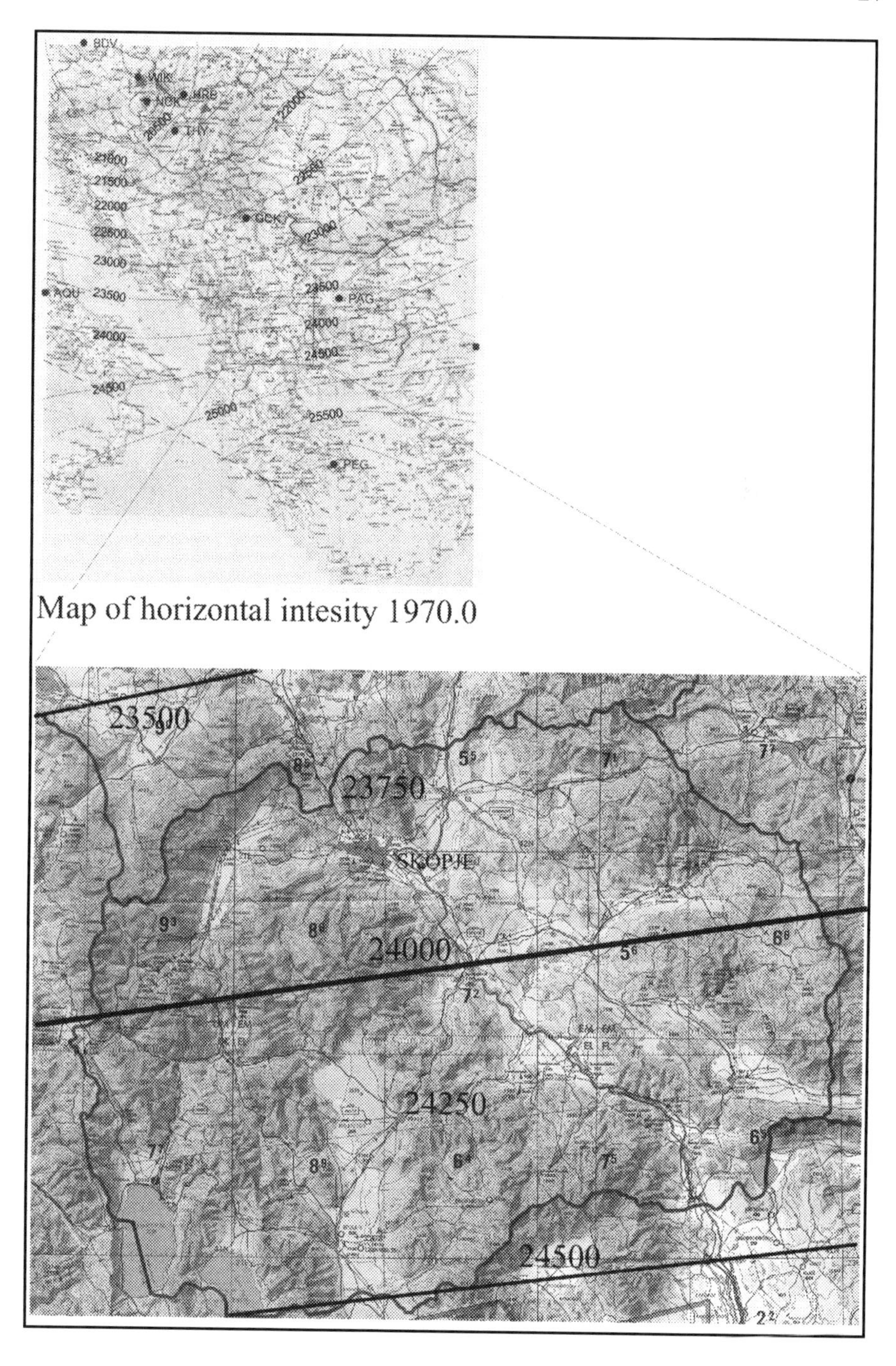


Figure 14. Map of horizontal intensity in Macedonia 1970.5.

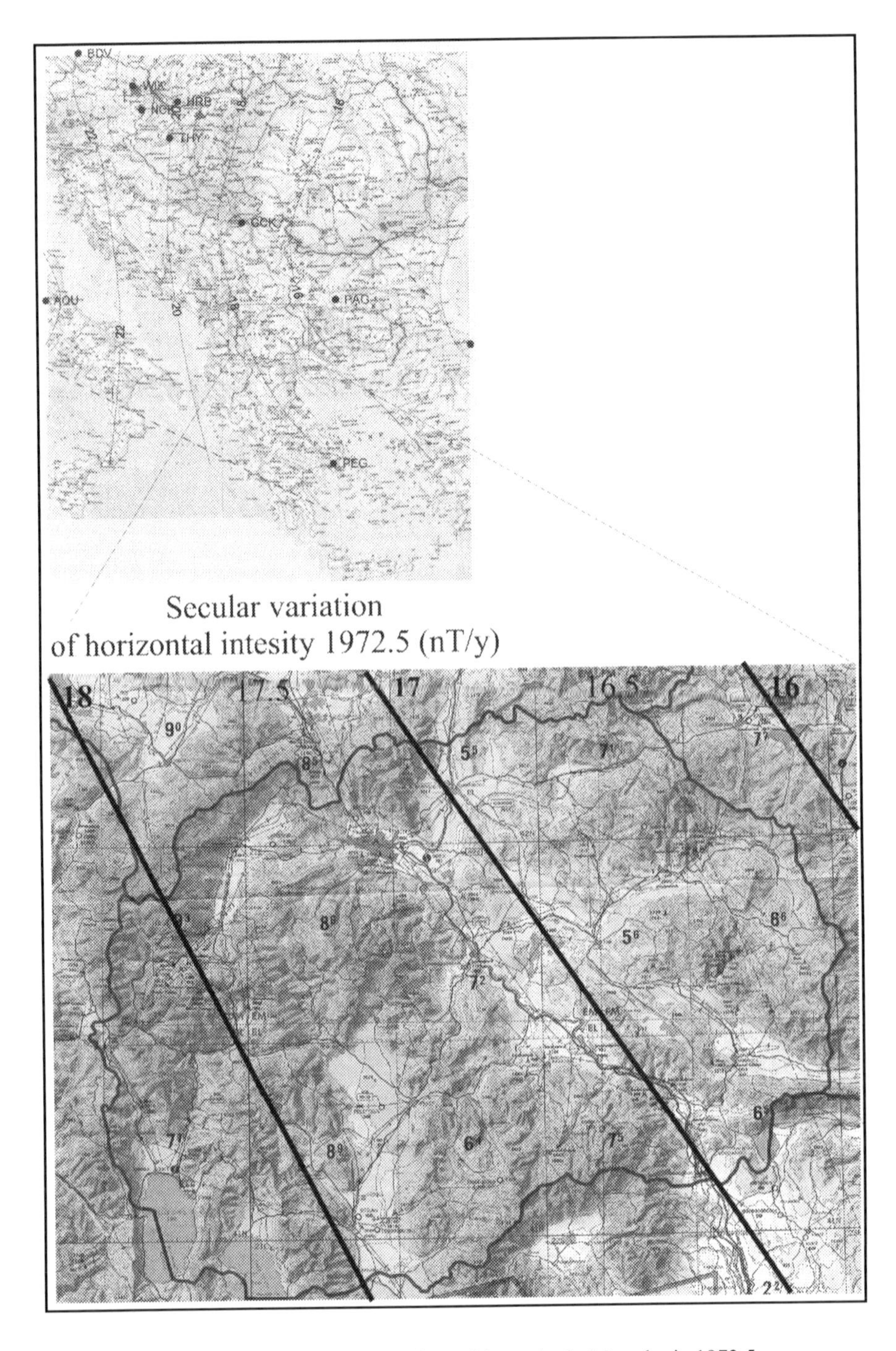


Figure 15. Map of secular variation of horizontal intensity in Macedonia 1972.5.

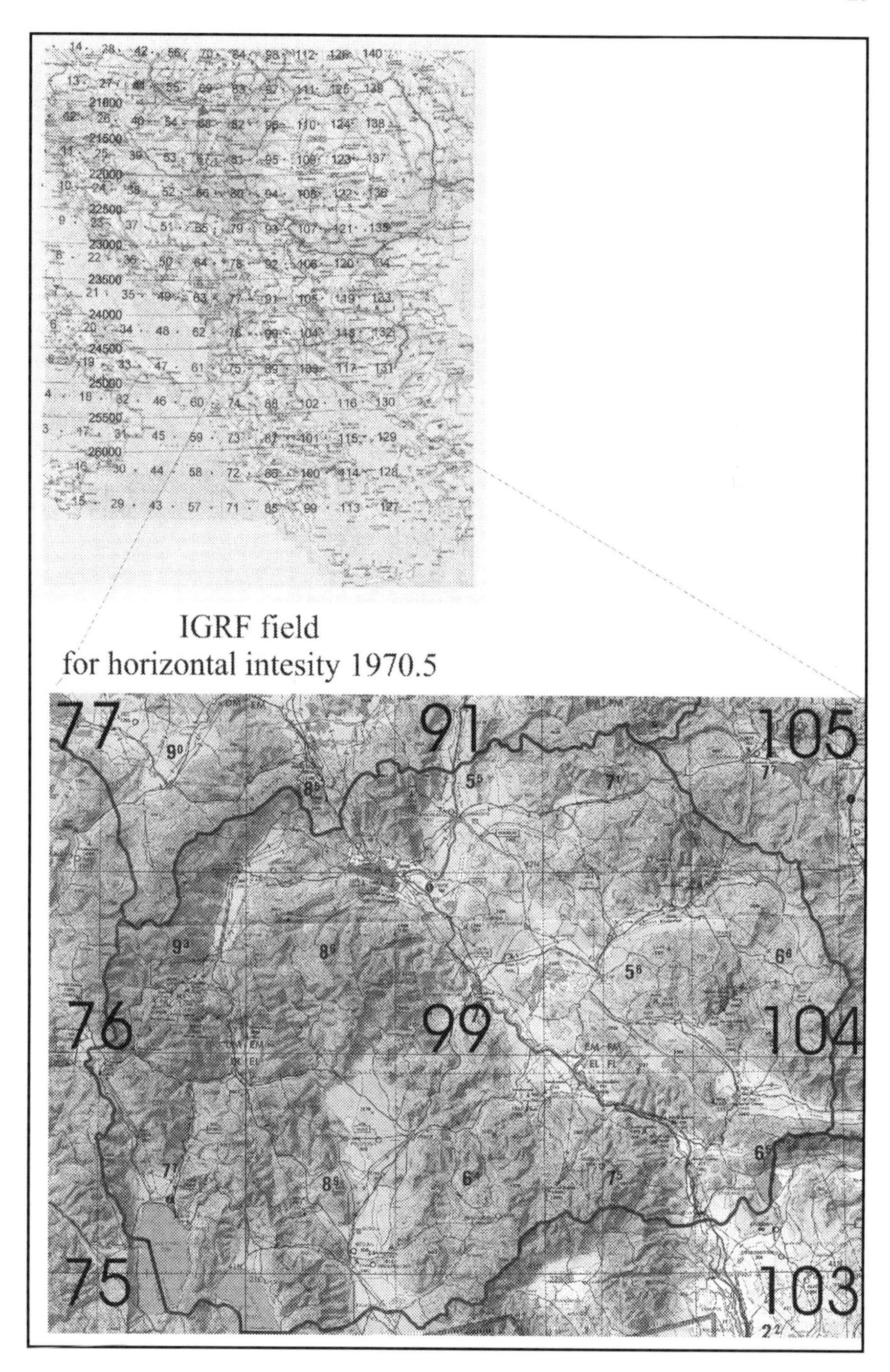


Figure 16. Map of IGRF field for horizontal intensity in Macedonia 1970.5.

In 1997 a project was implemented under the title "Eastern Europe Magnetic Project" (EEMP) in which Report 8[6] was carried out for Macedonia in cooperation with GETECH, Geophysical Exploration Technology. The collaborator from the Republic of Macedonia was Prof. Dr. Todor Delipetrov. The following maps offer a presentation of investigations carried out (Figure 17 – Figure 22).

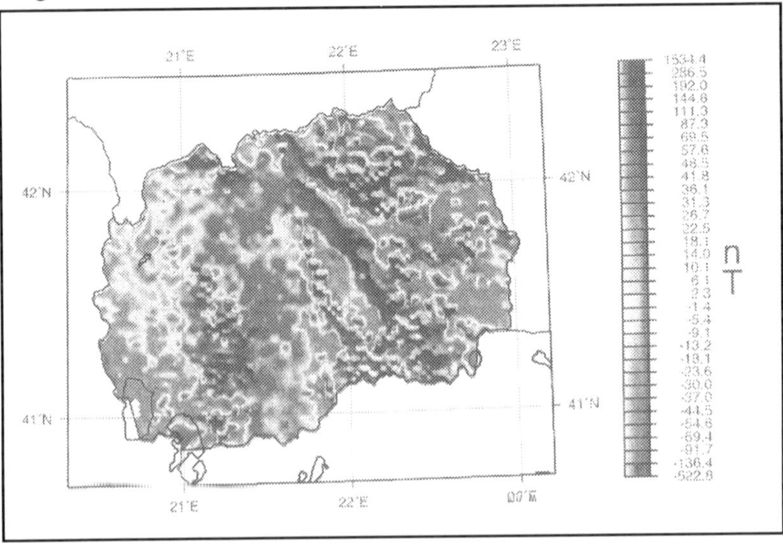


Figure 17. Vertical field (ΔZ) anomaly map of Macedonia based on 1 km grid at original survey elevation (~1 m above ground).

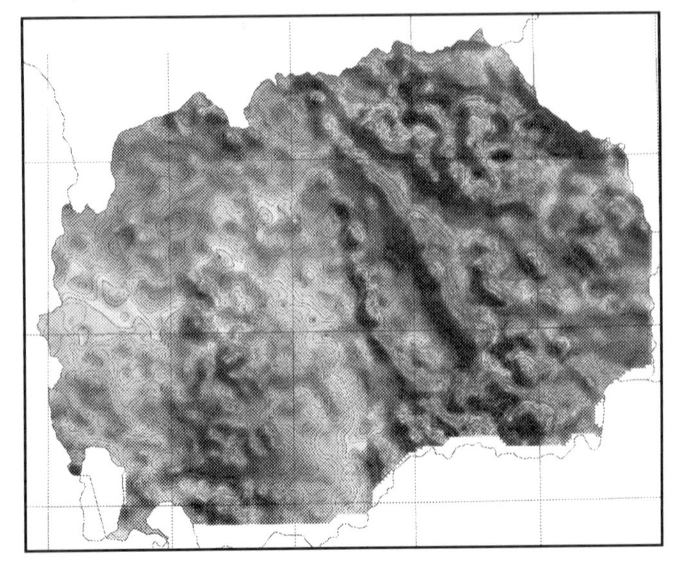

Figure 18. Topography map of Republic of Macedonia.

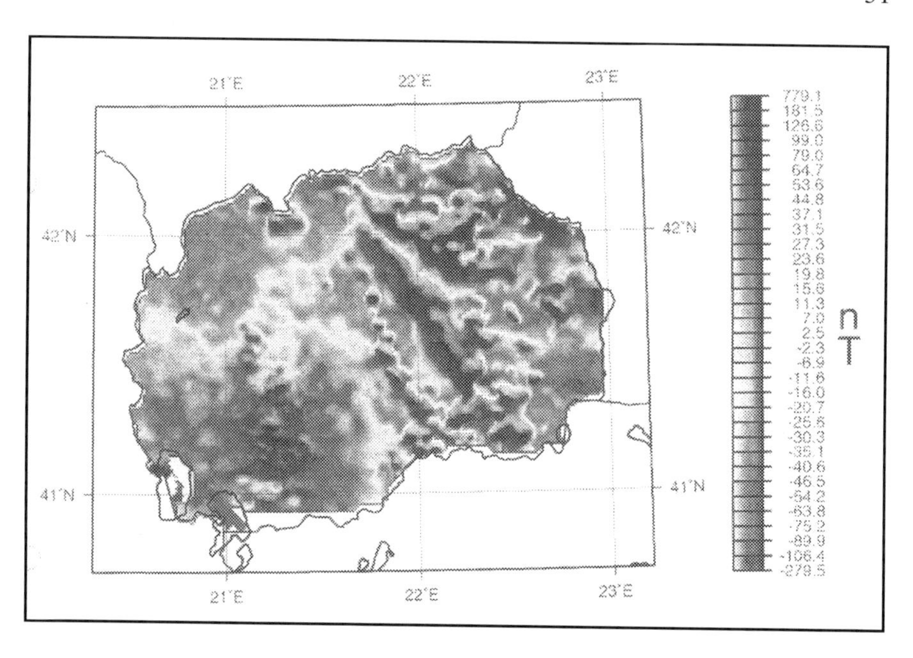


Figure 19. Derived total magnetic field anomaly map of Macedonia based on 1 km grid upward continued to 1 km above topography and linked to Albania, Serbia and Greece.

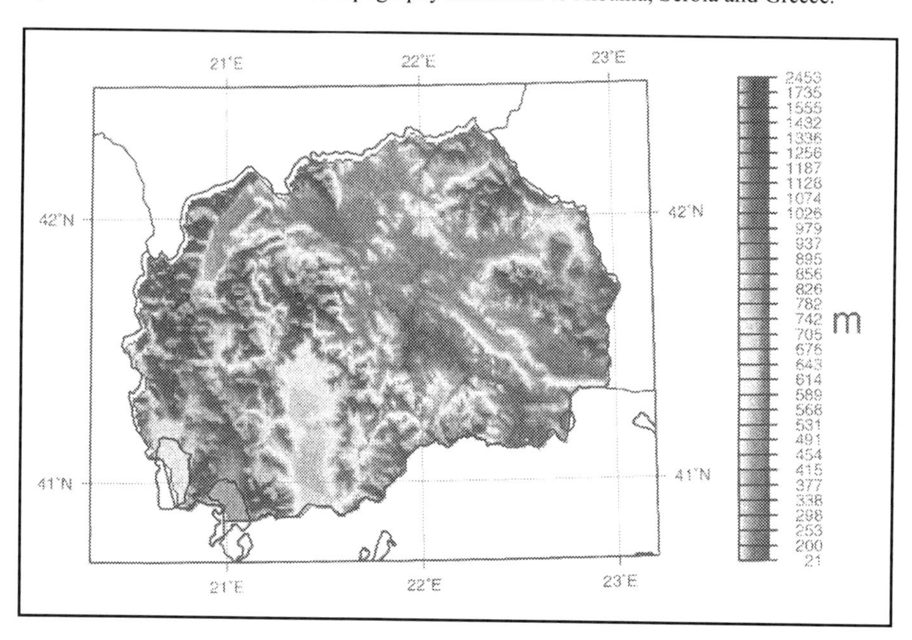


Figure 20. Digital terrain model of Macedonia at 1 km grid.

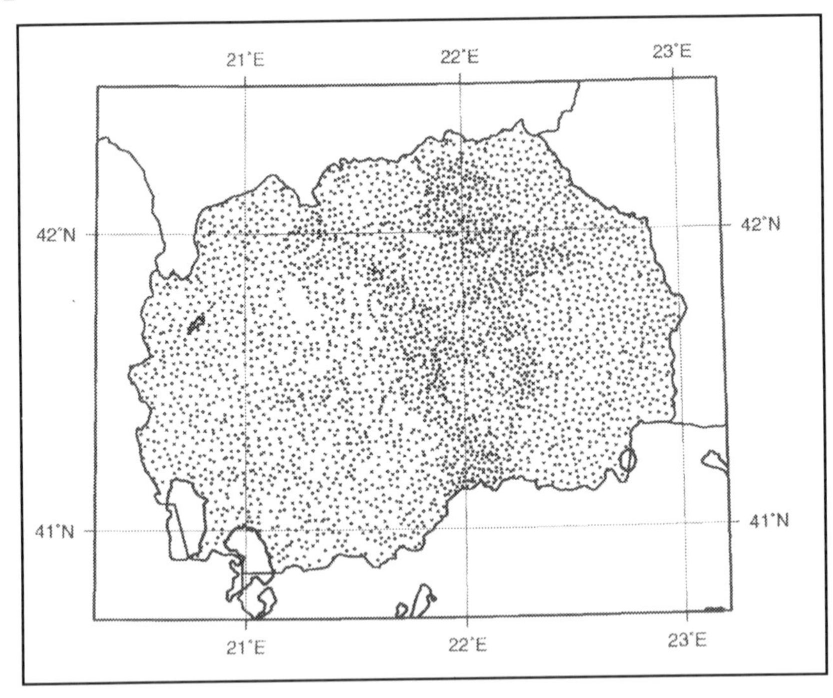


Figure 21. Gravity station distribution for Macedonia used to generate 8 km grid.

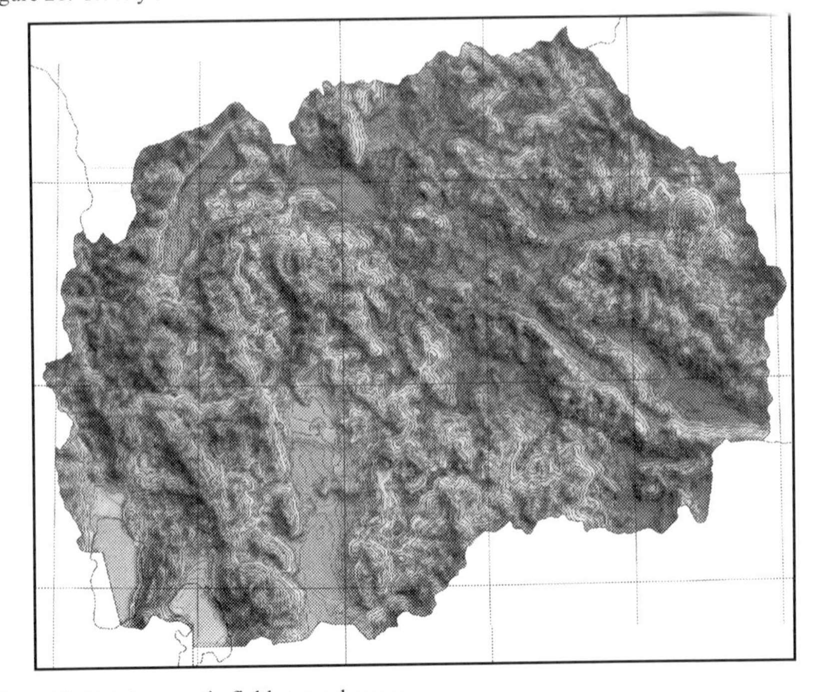

Figure 22. Total magnetic field anomaly map.

Vertical component magnetic field anomalies ($\Delta Z$ anomalies) were measured at approximately 8000 stations in Macedonia between 1954 and 1972. Analysis was carried out by Fourier transform technique. The field was upward and continued to 1 km above topography. The data were linked to data from Serbia, Albania and Greece.

$\Delta Z$ anomaly was first calculated with the formula

$$\Delta Z = Zm - Zn$$

Where Zm is the measured value of the vertical component and Zn is the normal value for the vertical component according to the point with latitude 44° and longitude 18.5° E. Zn was calculated with the formula

$$Zn = 39.964 + C1 * DF + C2 * DL + C3 * DFF + C4 * DFDL + C5 * DLL$$

where

C1 = 10.76986; C2 = 1.21625; C3 = 0.0023555; C4 = 0.005743; C5 = 0.0012558;

DF = [latitude (degrees) - 44 (degrees)] * 60 .... (in minutes)

DL = [longitude (degrees) - 18.5 (degrees)] * 60 .... (in minutes)

DFF = DF * DF

DFDL = DF * DL

DLL = DL * DL

For the purpose of this study, the original values of the vertical components of the field were cross checked (calculated again) and the values of IGRF were eliminated.

During the past several years magnetic measurements have been carried out in Macedonia to detail the geomagnetic field and to compile a new map of geomagnetic anomalies. The measurements were useful to determine locations for a grid of geomagnetic stations.

After 2000, cooperation with the Royal Meteorological Institute, Geomagnetic Observatory in Dourbes, Belgium and the Department for Geology and Geophysics at the Faculty of Mining and Geology in Stip was established. The study of the geomagnetic field of the Republic of Macedonia was undertaken. A basic grid of repeat stations was set up and a location for a new geomagnetic observatory was selected.

## 3.2. GEOMAGNETIC INVESTIGATIONS IN SOME LOCALITIES FOR THE DISCOVERY OF DEPOSITS OF MINERAL RAW MATERIALS[1, 5, 7]

Before World War II, geomagnetic investigations were undertaken to explore for mineral raw materials. These investigations intensified during 1950s.

In 1930, investigations were carried out with a magnetic balance instrument to determine vertical intensity in the Orasje and Ravniste mines near Skopje in order to determine whether this magnetic method could successfully be used in investigation of chromium ores.

During the 1950s investigations were carried out at several localities. In 1952, geomagnetic investigations were carried out in Tajmiste in order to determine the location of magnetic anomalies and the possible presence of schists with ore beneath limestones. In 1952 and 1953, geomagnetic investigations were carried out at Slopce in order to test the geomagnetic methods on known shamosite occurrences, and if they proved applicable, to continue investigations on new occurrences. In 1953 the area of investigation was widened to include Sopur and Damjan. Geomagnetic methods were employed to explore for iron ores and to delineate the areas of interest.

In 1954 geomagnetic investigations were carried out at Sloesnica, Valandovo to determine the extent of the shamosite zone. It was determined that the zone extended eastward of the original site.

In 1955 the localities of Valandovo and Tajmiste as well as those of Konce, Mitrasinci, Curkov Dol, and Galicnik were investigated. The goal of the investigations was to determine the possible application of geomagnetic methods to ore exploration.

In 1956 a number of investigations were carried out at several localities. At Damjan, investigations were made to determine the extension of the ore zone to the north. A study of the contact zone between flysch and andesite to the south and south-west of the area was also undertaken. Magnetic measurements were carried out at Demir Hisar to locate ore bodies at Sapotnicko, Sapotnicko Pole, and Seliste. Reconnaissance studies of the Majdaniste district were carried out to determine the extent of the ore zone like the occurrence of shamosite shists at Sv. Ilija. Geomagnetic investigations were carried out southeast of 'Rzanovo in order to distinguish nickeliferous oolites.

In 1959 similar geomagnetic measurements were carried out at Kozjavki Kamen near Alsar to determine the extent of the magnetic anomaly that was found during the geological prospecting.

In 1960 detailed geomagnetic investigations were carried out southeast of 'Rzanovo at the location of newly discovered outcrops of nickeliferous iron. Also, in 1960 the Smilevica, Algunja, Prespa, Kriveni and Pusta Reka areas were investigated. The objectives were to determine the size of known magnetite deposits, and to discover primary deposits of magnetite ore. The Figure 23 to Figure 26 show some anomalies in the geomagnetic fields of the Republic of Macedonia.

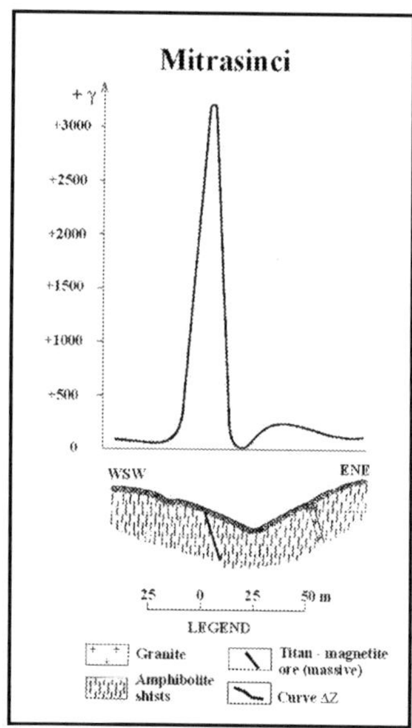


Figure 23. Geomagnetic anomaly.

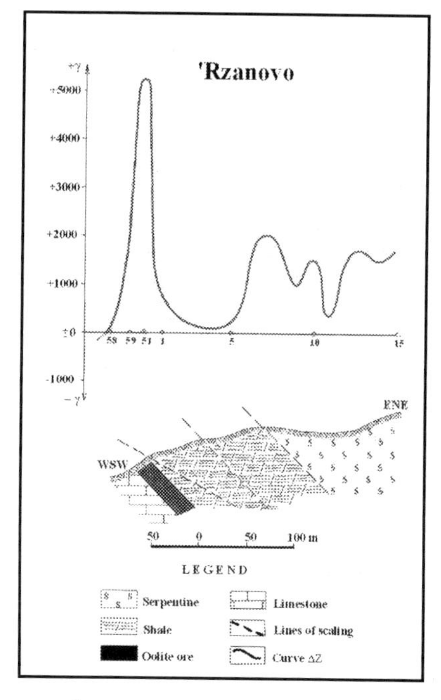


Figure 24. Geomagnetic anomaly.

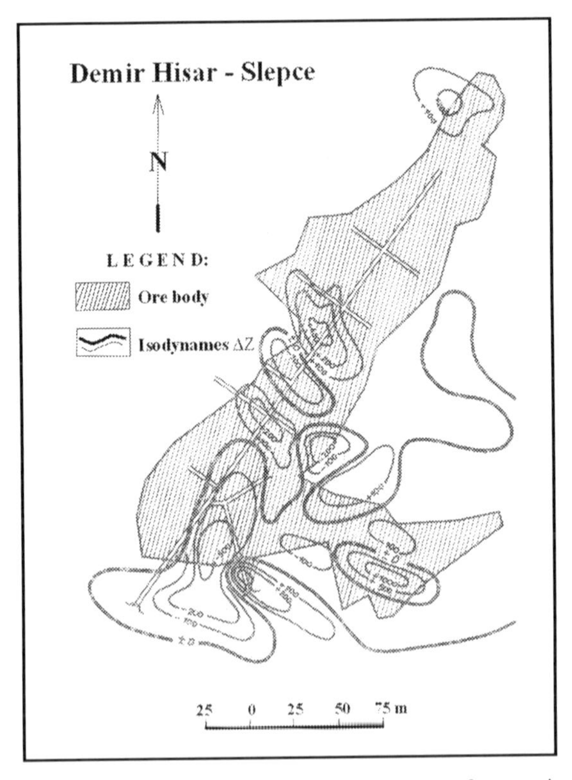


Figure 25. Isodynames ΔZ above ore body with uneven content of magnetite.

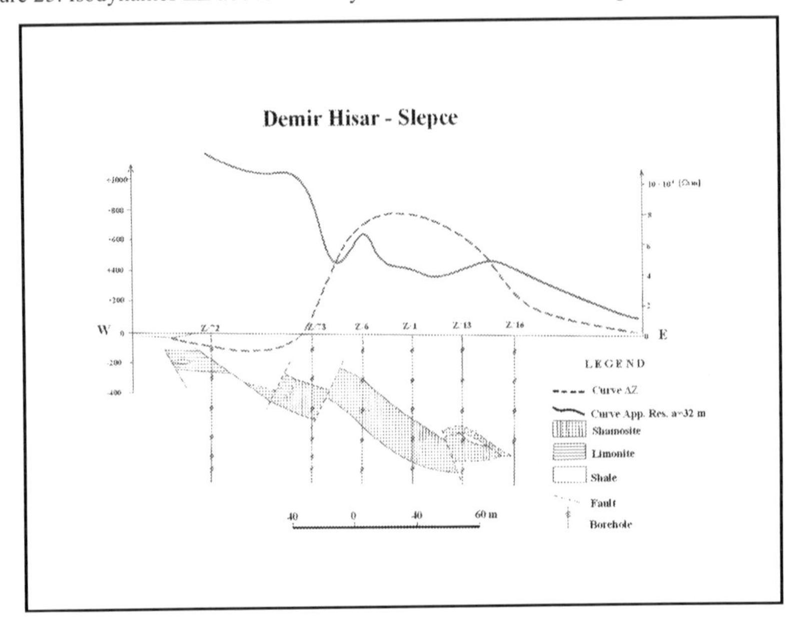


Figure 26. Geomagnetic anomaly and curve of apparent receptivity.

After independence in 1991, the Department of Geology and Geophysics at the Faculty of Mining and Geology in Stip purchased proton magnetometers in order to investigate magnetic anomalies caused by ore deposits. The instruments were also used to locate archeological sites and structural composition of certain areas. The technique was also used to locate archeological structures at Zajcev Rid and Skopje. They helped uncover and define the walls around the ancient city of Skupi. Measurements of stability of the terrain around the classical settlement at Bansko, Strumica were performed including archeological investigations using magnetic measurements at Vrsnik - Tarinci, Stip.

Magnetic measurements at Strumica, Berovo - Delcevo and the Kocani depressions were carried out to define their structural compositions.

## 4. Anomalous magnetic field of the territory of the Republic of Macedonia

Regional magnetic provinces have a multicomponent anomalous field. Many geologic structures at various depths of the earths crust create this effect. In other words, the magnetic field has an integral influence on the surface from many deeply seated geologic causes.

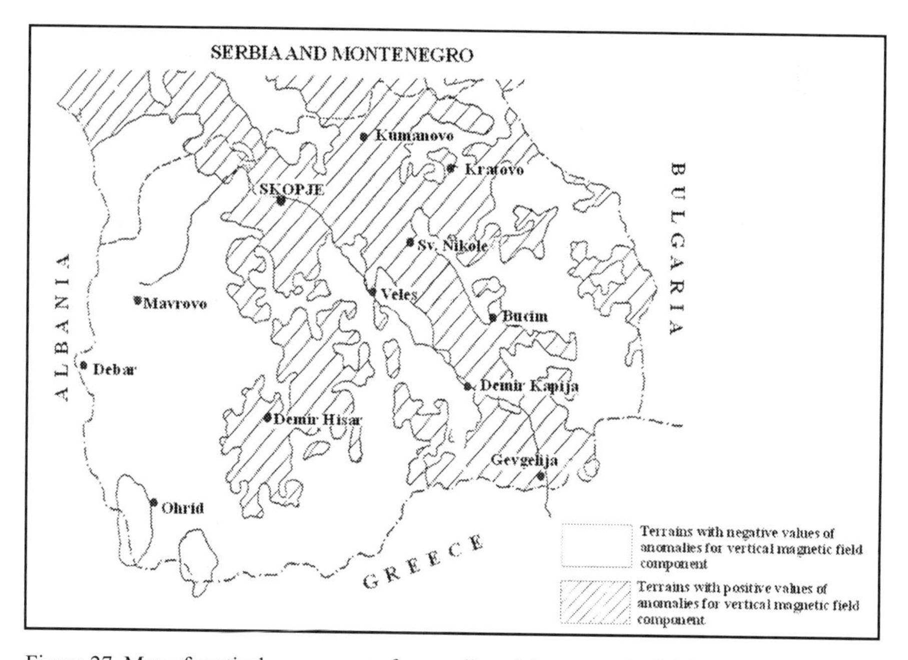


Figure 27. Map of vertical component of anomalies of the magnetic field of Macedonia.

## 4.1. THE DINARIDE - ALPINE MAGNETIC PROVINCE

This province dispalys a magnetic minimum across the Balkans. It can be traced from Rijeka to Pec, in Albania. Its influence at depth is manifested again in the Korab zone in the western part of Macedonia (Figure 27).

## 4.2. INNER MAGNETIC PROVINCE

The inner magnetic province occupies more than half of the territory of Serbia and Montenegro and stretches into Macedonia. The Macedonian part of this province has the following zones:

1. Western Macedonian zone,
   - Pelagonian zone
2. The Vardar zone,
3. Eastern - Macedonian zone
   - Strumska zone.

Essential magnetic features of the province are a clearly pronounced pattern of positive magnetic anomalies that alternate interspaces of negative sign. One pattern of magnetic anomalies starts in the region of Kozuf, continues towards Veles, and goes on to Gnjilane in Serbia. Another pattern of magnetic anomalies enters Macedonia from neighboring Greece at Gevgelija and goes further to the northwest via Sveti Nikole. It then goes south of Kumanovo where it diverges and translaterally goes further for about 20 km to the west. Then from Skopje it continues over the eastern slopes of Mt. Sara, passing further into the territory of Serbia and joins the known gravimetric and magnetic anomaly of the Metohija ultramafites that enter our region from Albania. The interception and translateral movement of this pattern to the west, some 20 km from the Kumanovo - Skopje region could be related to the regional fault of transverse strike to the Vardar strike (Debar - Mavrovo - Skopje - Kumanovo - Kustendil), or with magmatites deposited along that area.

Above were described two marked patterns of geomagnetic anomalies that, in part, traverse the territory of Macedonia. However, their extension exceeds the area under consideration. South of Kumanovo, a geomagnetic anomaly with the Vardar extension is distinguished. The territory that it occupies coincides with that of the Vardar zone. The reason for the occurrence of this anomaly may be in some deep magnetic formations. It may also be from the continuing presence of basites below Tertiary sediments (Vukasinovic, 1965, Jancic, 1970) or from the complex influence of the deep causes of magnetic formations at the base of the basic and ultrabasic members of the Jurassic complex and Tertiary eruptions (Ciric, 1970). The Tertiary sediments, whose thickness to the magnetic base amounts to 0.9 - 2 km (Jancic, 1970), probably overlie young dacite - andesites. These formations, where exposed, have high magnetic anomalies. The values of the vertical components of the geomagnetic field of the anomalous field in the zone depend on the composition of the deep parts of the Earth's crust, bearing in mind that the Vardar zone, as a rift zone, is crosscut by deep faults along which penetrate material of the deep parts of the earth crust and the upper mantle. The normal continuous extension of the Vardar anomalous zone was distinguished by the method of profiling when some young eruptive occurrences were neglected.

The Pelagonian magnetic zone, mostly overlaps the Pelagonian massif. It is characterized by a quiet magnetic field and Z anomalies of small intensity and irregular form. A magnetic field with such characteristics is an indicator of homogenous geology. Exceptions are the anomalies of some dacite - andesite intrusions and tuffs.

The characteristics of the magnetic field of the Pelagoian zone, which is built up of Precambrian metamorphic rocks and granitoides and Neocene - Quaternary mollases in young depressions, can be interpreted by purely expressed magmatism in the formations mentioned. Additionally, the Pelagonian massif is a homogenous block of the Earth's crust deeply embedded in the upper envelope at significant depth. The faults in the block are shallow.

The Porecka - Demir Hisar zone extends to the west of the Pelagonian zone with high anomalies (Ciric, Stojkovic, Veljkovic, Jancic, 1971).

The composition of the zone is variable and crosscut with faults. However, the magnetic field is rather homogenous and its intensity does not correlate to the discovered geological structures.

Most probably it is a reworked block of the Earth's crust from the Tertiary, and separated by deep faults in the Pelagonian and Western Macedonian zones into which the granitoides are forced with apical parts with common bases. The magnetic anomalies of high intensity correlate to the apical parts of the granitoides. The extension of the anomalies of the vertical component of the geomagnetic field may lead us to conclude that the major fault at the western edge of the Pelagonian depression is of meridian strike.

The Western Macedonian magnetic zone occupies the area of the Porec - Demir Hisar zone to the Albanian border. It is characterized by a uniform magnetic field with anomalies of vertical component with a negative sign. The geology of the zone consists of several formations of various composition and magneticity. The recorded anomalies are almost the same values for all the different formations. This opens the possibility of many variations for their interpretation. However, the occurrence of magnetic anomalies of the same values of non-magnetic limestones at Galicica and the neighboring diabase-chert formation and its basic and ultrabasic magmatites could be interpreted as the existence of an overthrust of the limestones over the diabase - chert formation at Jablanica. This possibility needs to be proven.

The Serbo - Macedonian magnetic zone, which, to the west is in contact with the Vardar zone and to the east extends to the border with Greece can also be distinguished. The zone coincides with the Serbo-Macedonian massif and is distinguished as an individual geotectonic unit. The Serbo - Macedonian massif is composed of high crystalline schists, amphiboles, gneisses, michashists, greenshists, marbles, and quartzites. The schists are occasionally intruded by granites so there is a contact metamorphosis of the neighboring granite - gneiss rocks. Occasionally the mass is intruded by basic and ultrabasic rocks, young Tertiary granites and granitoides, and dacite - andesites. With neotectonic movements the Serbo - Macedonian massif was broken up and turned into a series of trench and horst structures. The grabens are filled with Neocene layers up to 3 km depth.

In the area made up of various formations, the anomalies of the vertical component of the geomagnetic field correlate with the geological structures. Crystalline schists that are homogenous have a relatively quiet and balanced field Intercalations of amphibolites cause significant increases in the intensity of Z anomalies.

Other causes of high Z anomalies that do not fit the characteristic magnetic pattern of the crystalline basement and the earlier granite - gneiss cores, are always linked with young magmatic phenomena, visible on the surface or inside the Earth. Characteristics of the anomalies of the vertical component connected to the occurrence of younger magmatic rocks are "swarms" with alternating sign and high intensity, most probably connected to the amounts of magnetic minerals in the rocks. With Neocene depressions in the Serbo - Macedonian massif, anomalies express magneticity of the basement and their intensity is reciprocal to the depth of the depression.

The described magnetic zones have a characteristic elongated shape. In the explanation that follows, description will be given on some local anomalous areas that deviate from the general anomalous model. This anomalous group is characterized by rapid change of the sign and intensity of magnetic field. In the main anomalous zones, a ring-like magnetic structure is distinguished in the Vardar zone and the Serbo - Macedonian zone. They are located in the Kozuf, Demir Kapija, Bucim and Kratovo areas.

It can be concluded that the system of the anomalous magnetic model described covers the area of Macedonia that was affected by tectonic - magmatic atomization during the long period of the Hercynian volcanic phase (Permian - Carbon) through the youngest volcanism.

## 5. Conclusion

For further scientific and applied studies, we must address the problem of establishing a geomagnetic observatory. Therefore, our efforts and activities will be directed to the construction of a geomagnetic observatory.

## References


1. College of Geological, Hydrogeological, Geophysical and Geotechnical investigation - Belgrade, Geophysical Institute, *"Synthesis of geophysical investigation on the territory of the Republic of Macedonia"*, Belgrade, Serbia and Montenegro, 1976
2. Delipetrov, T., *"Basics of Geophysics"*, Faculty of Mining and Geology, Stip, R. Macedonia, 2003
3. Delipetrov, T., *"Correlation between crusts and subcrusts structures on the territory of Macedonia and seismicity"*, Doctor thesis, Stip, R. Macedonia, 1991
4. Delipetrov, T., *Report: "Establishing geomagnetic observatory in the Republic of Macedonia according to INTERMAGNET standards"*, Stip, R. Macedonia, 2004
5. Federal geological college, *"Informative bulletin - Geophysics, Magnetometry"*, Belgrade, Serbia and Montenegro, 1985
6. GETECH (Geophysical Exploration Technology), *Eastern Europe magnetic project (EEMP) Report 8,* Leeds, United Kingdom, December, 1997
7. Jancic, T., *"Geophysical investigations of iron ore deposites in Yugoslavia"* - Workshop of applied geophysics, Belgrade, Serbia and Montenegro, 1961
8. Duma, G., *"Historical Geomagnetic data and maps of Austria and southeast Europe"*, ZAMG, Viena, Austria


**DISCUSSION**

Question (Jean Rasson): There is a repeat station near Kavadarci where we have anomalous results. Can you give me the geological causes of this anomaly?

Answer (Todor Delipetrov): The anomaly occurring at the Gradot Island repeat station is due to the central deep fault of the Vardar Zone (The Vardar Zone, on the west is separated by a fault form the Pelagon, and in the east from the Serbian Macedonian massif. The zone is divided into three subzones, the left deep fault being close to the repeat station). Along the fault and on the surface on the terrain tectonitic serpentinites have been mapped. From the geological point of view they are the cause for the anomaly.

Questions (Angelo De Santis):

1. As is known, the close presence of a railway can heavily affect the recording of any magnetic observatory. Are there railways in your country?
2. I saw from your presentation that in your country gravity and magnetic measurements were made. From these measurements you propose also

some interpretations in terms of structure models of density and/or magnetization. Were there also magnetovariational or magnetotellurics measurements made in order to get some information about the possible conductivity structure underneath the Republic of Macedonia?

3. In one of the lat slides you present a simple field map of the magnetic anomalies in your country. Do you know which reference field was used for the reduction of data?

Answers (Todor Delipetrov):

1. There is a railway, but it is not close to the terrain considered for the construction of the Observatory at Mt. Plackovica.

2. Magnetovariational and magnetotellurics measurements have not been carried out in Macedonia.

3. It is a scheme of magnetic measurements carried out in 1965. The measurements help to compile the map for geomagnetic anomaly of the Republic of Macedonia.

# GEOMAGNETIC FIELD MEASUREMENTS AT MAGNETIC REPEAT STATIONS IN FORMER YUGOSLAVIA

SPOMENKO J. MIHAJLOVIC[4]; DRAGAN POPESKOV; CASLAV LAZOVIC; NENAD SMILJANIC
*Geomagnetic Institute Grocka*

**Abstract.** A brief historical review of the geomagnetic field measurements in the territory of former Yugoslavia follows. In 1957 continuous magnetic recordings began at Grocka Geomagnetic Observatory (GCK). Field measurements at repeat stations were made at the same time. Grocka Observatory is still the only institution of its kind in the territory of former Yugoslavia. The frequency of the repeat station measurements and the instruments used will be discussed. A new network of magnetic repeat stations was designed for the territory of the Federal Republic of Yugoslavia (FR Yugoslavia). Between 1994 and 1998, geomagnetic surveys and first order surveys have been done at repeat stations (secular stations). This paper reports on the variometer methodology used in magnetic repeat station surveys. The diurnal variations of the geomagnetic field are analyzed at the secular stations in the northern part of Yugoslavia. Also, the spectrum of the diurnal variations, recorded at secular stations, is shown.

**Keywords:** geomagnetic survey, repeat stations, secular variations, diurnal variations, spectral analysis

## 1. Introduction

Since 1958, contemporary geomagnetic field observations have been carried out at the Grocka (GCK) geomagnetic observatory. The observatory is part of the Geomagnetic Institute situated in the village of Brestovik near Grocka, about 36 km east of Belgrade. The Observatory is, the only

[4] To whom correspondence should be addressed at: Geomagnetic Institute, 11306 Grocka, Belgrade, Serbia and Montenegro, e-mail: mihas@sezampro.yu



*J.L. Rasson and T. Delipetrov (eds.), Geomagnetics for Aeronautical Safety,* 43–60.

institution of its kind in the territory of former Yugoslavia. Supported by the federal budget, it has been operated continuously since 1958, in spite of many difficulties, particularly in the last decade. Its present status in the new state of Serbia and Montenegro (SCG) has not been defined.

The oldest geomagnetic measurements in this territory were carried out by L. Steiner between 1902 and 1904. He measured declination and horizontal intensity in Srem, a part of Serbia between the Sava and the Danube rivers.

In 1938, the Military Geographical Institute of Belgrade carried out declination measurements at some 20 points but did not reduce the data.

The measurements made before World War II are mostly of historical interest. In 1950, the Geophysical Institute of Zagreb carried out declination measurements on the Adriatic coast.

The first contemporary three-component geomagnetic survey was performed between 1958 and 1960 at about 480 stations, in a 260.000 $km^2$ area of former Yugoslavia. A second survey of the same type was made between 1964 and 2001 at 180 stations in the territory of FR Yugoslavia, now known as Serbia and Montenegro (SCG).

This survey was not completed. About 100.000 $km^2$ or approximately 40% of the territory of former Yugoslavia was measured.

Measurements from a network of 16 repeat (secular) stations, evenly distributed over the territory of former Yugoslavia, were carried out on a 3-5 year interval between 1960 and 1989. Due to economic crises, geomagnetic surveys in the territory of FR Yugoslavia (now SCG) were not performed again until 1994.

### 1.1. GEOMAGNETIC FIELD SURVEY AT SECULAR REPEAT STATIONS

Geomagnetic field surveys in the territory of the former Yugoslavia / FR Yugoslavia/ SCG can be categorized as follows:

- GCK Observatory surveys ;
- Repeat station surveys with the variometer method;
- First-order surveys with the variometer method;
- Second-order surveys;
- Ground surveys.

Permanent magnetic observatories are the most accurate sources of secular variation data. Repeat station measurements also give good secular variation data. Ground survey measurements (vector surveys) are only used for spatial mapping of the geomagnetic field.

According to the territory they cover, these surveys can be divided into:

- surveys in the territory of former Yugoslavia carried out before 1990, and
- surveys in the territory of FR Yugoslavia (now SCG) carried out after 1990.

Repeat station surveys in the territory of former Yugoslavia (SFRJ/SRJ/SCG) are either Class-B, (stations occupied every 5 years) or Class-C (stations occupied every 10 years).

IAGA has recommended that surface measurements of the magnetic elements for main-field mapping be made at a spacing of about 200 km (Vestine, 1961; Newitt L.R., et al.; 1996). The geomagnetic surveys of SFRJ/SRJ/SCG used an average station spacing of 100 to 150 km.

16 repeat stations were surveyed prior to 1990 in the territory of Former Yugoslavia (SFRJ). The average station spacing is about 150 km. 19 repeat stations were surveyed after 1990 in the territory of FR Yugoslavia. The average spacing of these stations is 100 km. First–order surveys, in the territory former Yugoslavia, from 1958 to 1960, were done at 480 points. The average point spacing was about 30 km); First–order surveys in the territory of Yugoslavia (SCG) from 1995 to 2001 were done at 180 points. The average point spacing for these surveys was also 30 km.

Previously, from 1960-1989, surveys were carried out over a 260 000 $km^2$ area. There were 16 repeat stations and the GCK observatory, evenly distributed at an average distance of 150 km.

Repeat station locations (secular stations) were selected according to a special procedure. The locations are on state territory, and are reoccupied every 3-5 years. Each station consists of two measuring sites less than 5 km apart where absolute measurements are done. These sites were carefully chosen in order to be representative of the geomagnetic field. Horizontal gradients, measured at 50 m in N-S and E-W directions had to be less than 1-3 nT/m. Vertical gradients had to have the same constraints.

For three-component, continuous recordings of the geomagnetic field variations, each secular station was occupied for at least three whole days. Absolute measurements were carried out twice a day at the secondary stations and near the position of the variometer.

Figure 28 shows the network of repeat stations established before 1990. Final data reduction was done for these nine epochs: 1960.0, 1965.0, 1971.0, 1974.0, 1977.5, 1980.0, 1983.5, 1986.5 and 1989.5.

After the disintegration of Yugoslavia and formation of the new state, the FR Yugoslavia, it was necessary to plan a new network of repeat (secular) stations. Along with applying well known rules for choosing the most suitable sites for repeat stations, we tried to use existing sites from the old network. We assumed these stations to be the most important in the new network. Finally, there were 19 repeat stations and the GCK observatory, evenly distributed with an average distance of 100 km covering about 100 000 km$^2$. The network of repeat stations established after 1994 is shown on Figure 29.

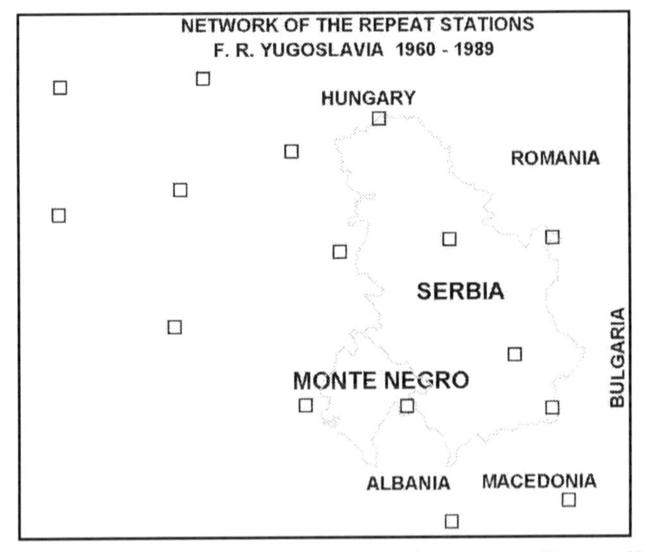


Figure 28. Distribution of the magnetic repeat stations in the territory of Former Yugoslavia. Open squares indicate the position of the repeat station.

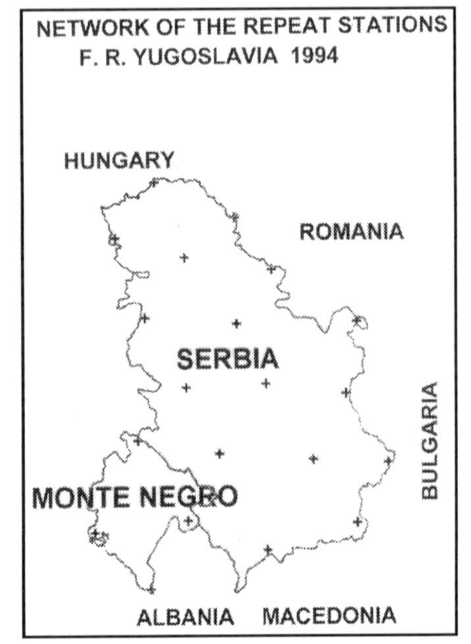


Figure 29. Distribution of the magnetic repeat stations in the territory of the Federal Republic of Yugoslavia. Symbol + indicates the position of the repeat station.

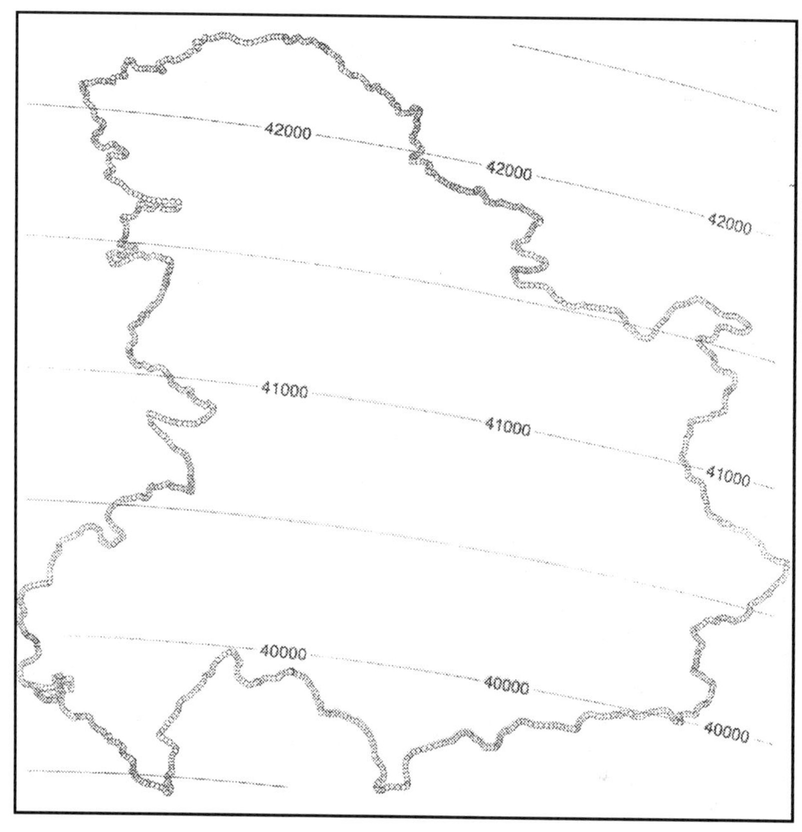


Figure 30. Distribution of the normal geomagnetic field (Z: vertical component) measured at the Magnetic Repeat Stations in the territory of Yugoslavia. Isolines indicate the values of vertical component of the geomagnetic field.

### 1.1.1. *Instruments*

As mentioned above, at the time of the 1994 survey we were using the three-component variometer EDA FM 100-C, but the data acquisition system was modified so that a PC motherboard was used instead of magnetic cassettes. Recordings were saved on standard 3.5 " diskettes.

A site for continuous measurements is precisely located so that it can be relocated easily and accurately for future surveys. The variometer sensor is placed in a plastic cylinder, leveled, oriented, and buried in the ground in a 1 meter deep hole to protect the sensor from temperature variations and rain. Approximately 6-12 hours are necessary for the sensor to achieve physical and electronic stability. After this time, X, Y, and Z values are continuously recorded for three days.

Variations of F are measured by a proton magnetometer, an Overhauser effect type GSM-10, produced by GEM SYSTEMS. Its resolution is

±0.1 nT and the specified absolute accuracy is ± 0.2 nT. The electronic part of the magnetometer is located about 30 m from the sensor.

For absolute measurements a standard instrument, commonly called the DIM magnetometer (Declination-Inclination Magnetometer) is used. A single-axis fluxgate sensor is mounted parallel to the axis of the telescope on a Zeiss 010B non-magnetic theodolite. The sensor and its electronics are produced by ELSEC. F is measured with the GSM-10 magnetometer. Absolute measurements were carried out twice a day at the variometer site and at the secondary stations.

#### 1.1.2. *Data reduction*

Absolute observations made at the repeat stations must be combined with the continuous variation data from the reference observatory in order to obtain an estimate of either the normal field, or an equivalent annual mean value at the repeat station. Sufficient data reduction should be done on-site to check the validity of the observations, paying special attention to the consistency between results from successive sets of absolute measurements.

Final data reduction assumes comparison between hourly mean values from the secular station and the observatory. First, hourly mean values are determined for each day of field recordings and then the mean value is found for all three days. At the end we find the value of all geomagnetic components observed or calculated, that determine values of the geomagnetic field for an exact determined moment, i.e. the value is reduced to a particular epoch. The obtained data enable normal field calculation, plotting and modeling of the geomagnetic field.

Absolute observations made at the repeat station must be combined with continuous variation data from an on-site variometer in order to obtain an estimate of either the normal geomagnetic field, or an equivalent annual mean value at the repeat station (secular station).

The repeat station surveys in FR Yugoslavia (interval 1994-1998) were performed using:

- An on-site variometer with digital output;
- Analog observatory's records;
- X, Y, and Z elements were recorded by the variometer and the D, H, Z, and F elements were recorded by the observatory's variometers. (The X, Y, Z and F elements were being observed for absolutes);

The reduction method used for the repeat station survey data is based on the assumption that transient (including diurnal) variations of the geomagnetic field are identical at both the repeat station and reference observatory. This method is usually applied to calculate an effective annual mean value at the repeat station. The size of the associated error in data reduction depends on the distance between the station and the observatory. The data reduction procedure uses the relationships (Newitt L.R., et al. 1996):

$$E(t) : E = E0(t) - E0 \tag{1}$$

$$E = E0 + E(t) - E0(t) \tag{2}$$

$$E = E0 + (E0(t) - E0) + (SV - SV0)\Delta T \tag{3}$$

where::

- E : annual mean values at repeat station;
- $E_0$ : annual mean values at observatory
- $E_{(t)}$ : instantaneous value at repeat station
- $E_{0(t)}$ : instantaneous value at observatory
- SV; $SV_0$ : secular variation at repeat station and at observatory;
- $\Delta T$ : the time difference

The most likely case in a practical geomagnetic survey of repeat stations is Case C: The secular change is non-uniform and different at both the repeat station and the observatory (Newitt L.R., et al. 1996):

$$E_{(t)} - E_{0(t)} = E - E_0 - (a\text{-}b) \tag{4}$$

$$(a\text{-}b) = (dE/dt - dE_0/dt)\ \Delta T \tag{5}$$

In this practical case, it is important to know the form of the secular variation functions. The error (a-b) will depend, in relationships (4) and (5), on factor $\Delta T$, and also on the difference in secular variation throughout the entire year in question.

On the basis of the reduced values of the geomagnetic field from the observatory and repeat stations, we can compute the values of the normal geomagnetic field, for the territory FR Yugoslavia, and for the epoch 1994.5. Normal geomagnetic field values can be expressed by spherical harmonic functions or by polynomial functions.

Prospectors use the Gauss method of the spherical harmonic analysis, in the processing and analysis of values of the normal geomagnetic field. Normal geomagnetic field values for the territory of FR Yugoslavia and epoch 1994.5 were computed using the polynomial functions of the second order with six coefficients:

$$F(\Delta\phi, \Delta\lambda) = a1 + a2\Delta\phi + a3\Delta\lambda + a4\Delta\phi^2 + a5\Delta\lambda^2 + a6\Delta\phi\Delta\lambda$$

$$\Delta\phi = \phi 1 - \phi 0\,;$$

$$\Delta\lambda = \lambda 1 - \lambda 0\,;$$

$F(\Delta\phi, \Delta\lambda)$: component of the geomagnetic field;

$a_1$ to $a_6$ :polynomial coefficients

$\phi 1, \lambda 1$ : longitude and latitude of the magnetic repeat station;

$\phi 0, \lambda 0$: longitude and latitude of the reference geomagnetic observatory

Latest data reduction was done only for the epoch 1994.5. On Figure 30 the distribution of the normal geomagnetic field (Z- vertical component) determined on the magnetic repeat stations survey of the territory of Yugoslavia is shown.

## 2. Survey Methodology

A magnetic repeat station (secular station) is chosen using the same criteria used for establishing a magnetic observatory. Several criteria are listed in the "Guide for magnetic repeat station surveys" (Newitt L.R., et al. 1996):

- The values of the magnetic elements should be representative of the region;
- The magnetic field at the site should not be influenced by magnetic anomalies caused by geological structures;
- The subsurface in the surrounding region should be electrically homogenous;
- The magnetic field should be uniform in the vicinity of the station marker;
- The site should be free from sources of artificial disturbances (electric railways, generating stations, power lines, transmitters, etc.).

A geomagnetic survey must determine the total-field gradient at the location of the repeat station. If the geomagnetic field varies less than 50 nT within a radius of 10 m, the site of the secular station is acceptable. At the location of the repeat station a more detailed total-field gradient survey should be carried out. The point at which the secular station will be installed should be marked. An initial F reading at the mark should be taken, then; proton magnetometer readings should be made in the four cardinal directions at 0.5 m, 1.0 m and every meter thereafter out to 10 m. Proton magnetometer observations should also be taken vertically above the station marker at 20 cm intervals.

If the total-field gradient changes are a few nanoteslas per meter (1-3 nT), the site is satisfactory.

A repeat station (secular station) must be permanently marked. It is recommended that a secondary station be installed some distance from the primary station. The stations should be at least 200 m apart. Parallel observations should be carried out at both stations during each occupation, to check the station differences, and to test that no magnetic contamination exists at either station.

The repeat stations at which continual vector recordings and measurements of the absolute values of the geomagnetic field elements are projected, are designated as secular stations. The amount of change of the geomagnetic field at secular stations is called the secular variation. The measurements of the geomagnetic field elements at secular stations are repeated every three to five years. At secular stations, continuous 3-5 day recordings of variations of the geomagnetic field elements (D, H, Z, or X, Y, Z geomagnetic field components) and measurements of the absolute values of the geomagnetic field elements are done.

Absolute values of the geomagnetic field elements are measured at two points near the secular station, twice per day (in the morning and in the evening), during the three to five days when the stations are occupied.

The secular variations of the geomagnetic field elements at the repeat station locations are the differences of normal or yearly mean values. These variations are "visible" at geomagnetic observatories after repeat station measurements are made through two, three, or more Solar cycles. Secular variations can be determined from results of measurements at observatories and at secular stations, by analysis of repeat station surveys, and by "correlation" of magnetic maps from different epochs.

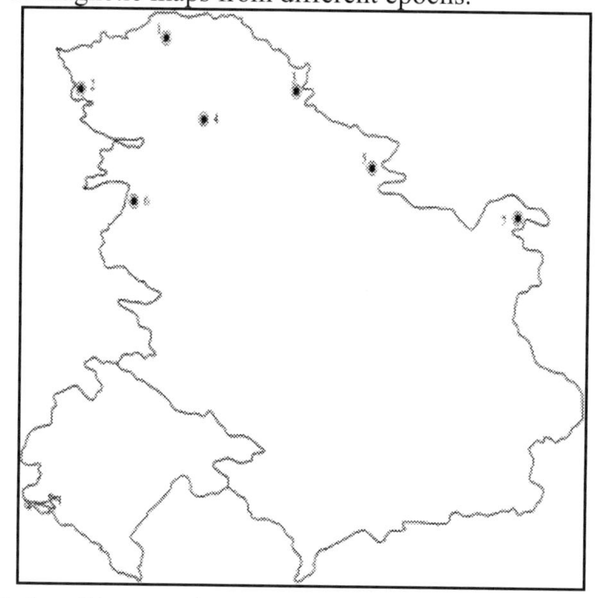

Figure 31. Distribution of the magnetic repeat stations, in the Northern part of Serbia, 1994.

Repeat stations surveys were done in 1994 in the FR Yugoslavia. Five secular stations in Vojvodina were occupied and surveys at secular stations in north-west and north-east Serbia were done. Figure 31 shows, the distribution of magnetic repeat stations in northern Serbia (Vojvodina).

## 2.1. DAILY VARIATIONS AT SECULAR STATIONS

The periodic change of the geomagnetic field elements with a period of twenty-four hours is called the regular diurnal variation $S_R$ (sometimes this variation is called the Sun's variation).

The morphology of the daily variation $S_R$ is different at different latitudes. The regular daily variation at middle latitudes has a maximum at about 12 noon (local time) on the X component of the geomagnetic field. The vertical component has a maximum in the afternoon. Amplitudes of regular daily variation $S_R$ have a maximum during the summer and a minimum in the winter. This quality is due to the seasonal character of regular daily variation and is dependent on the Sun's activity level.

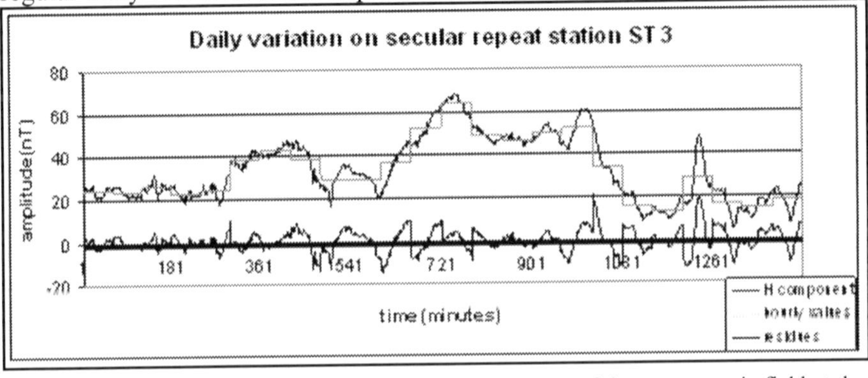


Figure 32. The diurnal variation of the horizontal component of the geomagnetic field at the magnetic repeat station (secular station) ST(3)

Diagram (1): the diurnal variation H component of the geomagnetic field (minute values)

Diagram (2): the diurnal variation H component of the geomagnetic field (average hourly values)

Diagram (3): the differences between values (1) and (2).

The changes of the minute values of the horizontal and vertical component of the geomagnetic field were observed over three to five day periods at each of the magnetic repeat stations, from June-November 1994. The minute values of the diurnal variations of the horizontal and vertical components of the geomagnetic field at all secular stations are compared with variations at the GCK Observatory.

The hourly mean values of the diurnal variations were calculated on the basis of the minute values. The differences between the hourly mean values and the minute values (the residuals) were computed, for every day the magnetic repeat stations were occupied, illustrating the variable magnetic conditions. The diurnal variation of the horizontal component at secular station ST (3) is shown on Figure 32. On the diagram (3) the composite signal of the transient contributions of the geomagnetic field is illustrated.

### 2.1.1. *Spectral analysis of the diurnal variations at secular stations*

In spectral analysis the residuals (the differences between the hourly mean values and the minute values) of the observed field are used. The signal includes the mean values of differences between hourly and minute values of the geomagnetic field components, for every day the magnetic repeat stations are occupied. The aim of the applied spectral analysis of the diurnal

Table 3. The groups of periodic changes of short-period, mean-period and long-period part of the spectrum of the horizontal and vertical components of the geomagnetic field.

| Variation of the GMF | GMF | Station S1 Period T( minute) | Station S2 Period T( minute) | Station S3 Period T( minute) | Station S5 Period T (minute) | Station S6 Period T( minute) |
|---|---|---|---|---|---|---|
| Short-periodical spectrum | X component | T1=80-90<br>T2=120<br>T3 =150<br>T4 =170 | T1 =40<br>T2 =50<br>T3 =100<br>T4 =140<br>T5 =160 | T1=45-48<br>T2=85-90<br>T3=100<br>T4=120<br>T5=150 | T1=130<br>T2=140<br>T3 =170<br>T4 =85<br>T5=140 | T1=50-58<br>T2=70-80<br>T3=90-95<br>T4 =105<br>T5 =180 |
| | Z component | T1=65<br>T2=140-180 | T1 max=110<br>T2 max=130<br>T3=70-80 | | T1=62-72<br>T2=85-93<br>T3 =140<br>T4 =150 | T1=48-58<br>T2=65-80<br>T3=98-130 |
| Mean-periodical spectrum | X component | T1 max=380<br>T2 max=410<br>T3 max=500<br>T4 max=550 | T1=210-240<br>T2 max=330<br>T3 max=400<br>T4 max=500 | T1=360-390<br>T2max=210<br>T3 =390<br>T4 =470 | T1 =200<br>T2 =290<br>T3 =320<br>T4 =410<br>T5 =500 | T1 max=320<br>T2 max=640 |
| | Z component | T1-250-260<br>T2=390-410<br>T3 max=490<br>T4 max=505 | T1=230-240<br>T2 max=330<br>T3 max=450 | | | T1 max=210<br>T2 max=400<br>T3 max=490<br>T4 max=530 |
| Long-periodical spectrum | X component | T1 =720<br>T2 =850<br>T3 =1020 | T1=970 | T1 =710<br>T2 =900<br>T3 =1040 | T1 =700<br>T2 =910<br>T3 =1010<br>T4=1120 | T1 max=890<br>T2 =1020 |
| | Z component | T1=880<br>T2 =950<br>T3 =1020 | T1 =840<br>T2 =1020 | | | T1=800-970<br>T2 max=800<br>T3 max=900<br>T4 max=970 |

variations of the geomagnetic field is to obtain the signal of the short-periodic part of the spectrum with cycles in the 3 < T < 200 minute band, the mean-periodic part with cycles in the 200 < T < 600 minute band and the long-period part of the spectrum with recurrent changes in the 600 < T < 1200 minute band (Mihajlovic J. S. et al.; 1998,2003). On Figure 33 - Figure 35 the structure of the three parts of the spectrum of diurnal variations of the horizontal component of the geomagnetic field, at the secular stations in Vojvodina is shown. Table 3 shows the groups of periodic changes of short-period, mean-period, and long-period part of the spectrum of the horizontal and vertical components of the geomagnetic field.

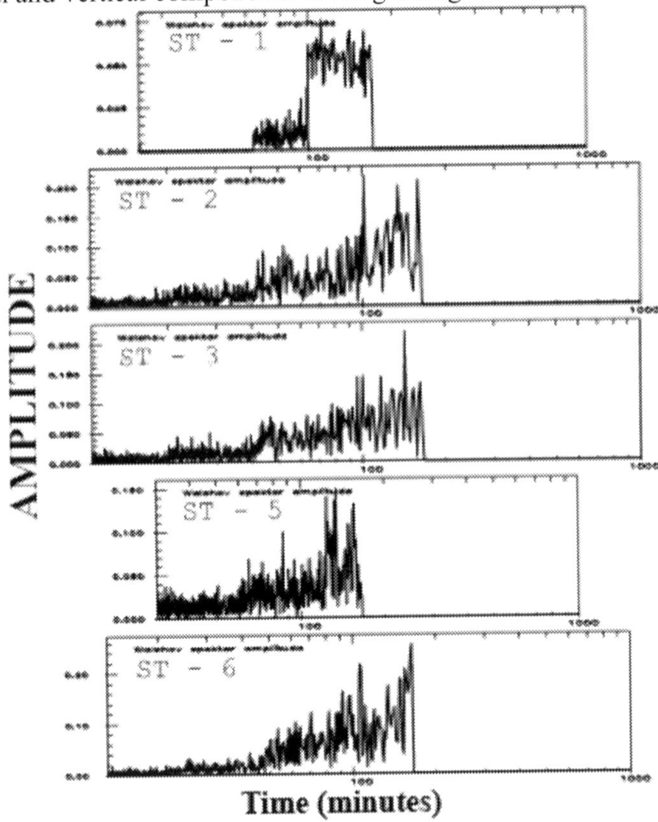


Figure 33. Short- periodic part of the spectrum of the geomagnetic field, horizontal component.

In the short-period part of the spectrum of the horizontal component of the geomagnetic field periodic changes in bands T = 40-60, T = 70-90, T = 100-120, T = 130-170 and T = 180 minutes occur (Figure 33; Table 3). In the short-period part of the spectrum, the vertical component shows cyclic variations in bands T = 50-70, T = 100-130 and T = 140-180 minute (Table 3).

The mean-periodic part of the spectrum of the horizontal component of the geomagnetic field has periodic changes in bands T = 210-290 and T = 320-390,T = 410-500 and T = 550 minute (Figure 34; Table 3). In this part of the spectrum of the diurnal variation of the vertical component, cyclic variations in bands T = 210-260, T = 330-410, T = 450-490 and T = 505 minute are found (Table 3).

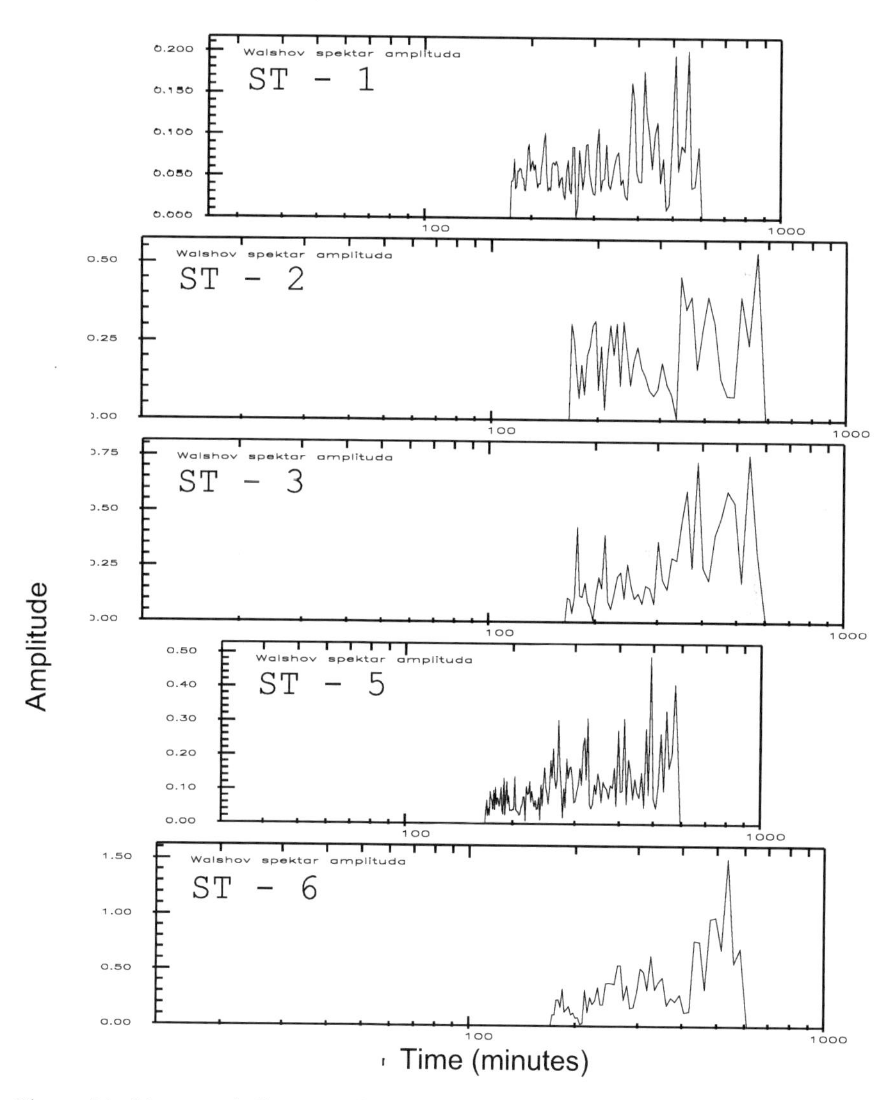


Figure 34. Mean- periodic part of the spectrum of the geomagnetic field, horizontal component.

The long-period part of the spectrum of the horizontal component has groups of periodic changes in bands T = 700-850, T = 890-970, and T = 1010-1120 minute (Figure 35; Table 3). The same part of the spectrum of the vertical component contains the cyclic variations in bands T = 800-880, T = 950-970, and T =1120 minute (Table 3).

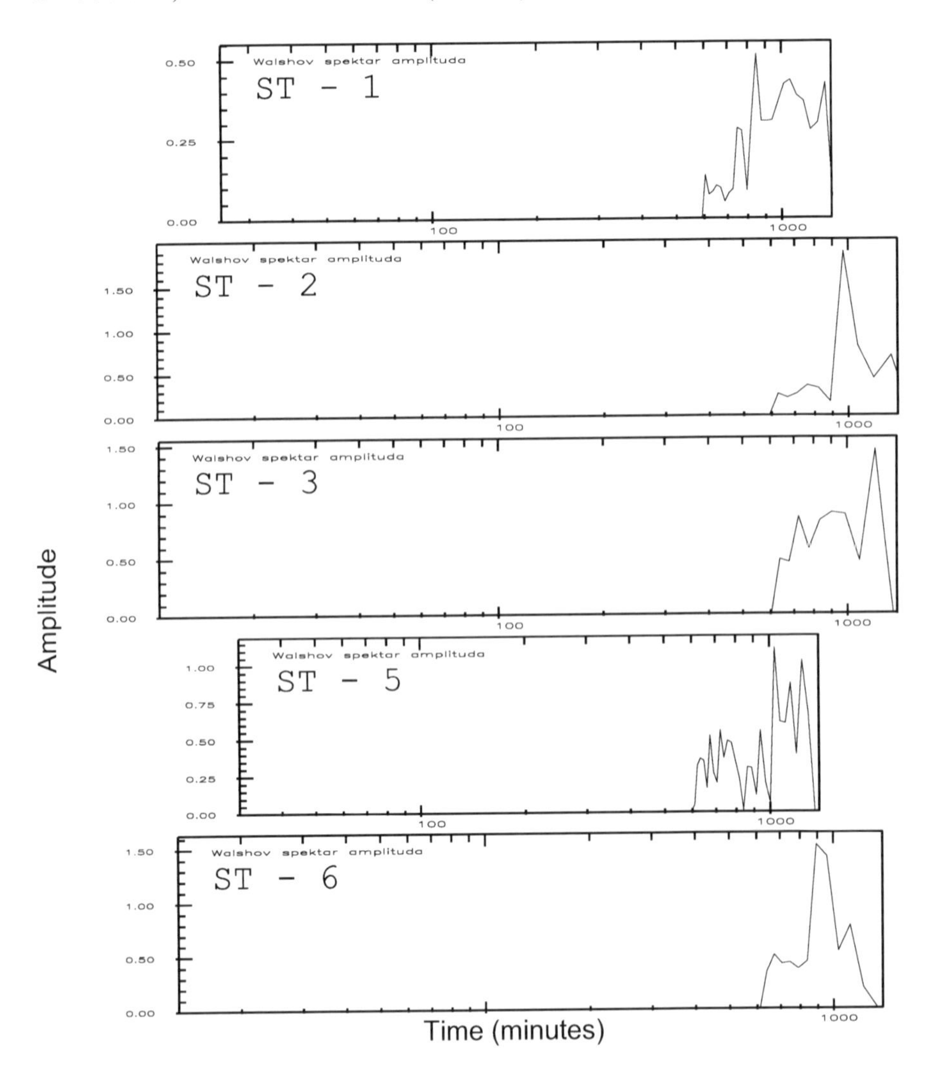


Figure 35. Long - periodic part of the spectrum of the geomagnetic field, horizontal component.

## 3. Conclusion

The results of the analysis of the short-period, mid-period, and long- period part of the spectrum of diurnal variations show the typical variability of the field over the region covered by the magnetic repeat stations. We observed the distribution of geomagnetic activity into space and time (several days). In particular parts of the spectrum of the diurnal variations of the horizontal and vertical components of the geomagnetic field, two groups of recurrent changes in the geomagnetic field are seen:

- The groups of cycles which arise from the Sun's activity (external source), and
- The groups of cycles which arise from the different conductivity zones in the area surrounding the repeat stations.

The analysis of the daily variations spectrum at the repeat stations aided in:

- Determining the measuring interval between sets of absolute measurements at the points near the magnetic repeat stations;
- Determining the geomagnetic conditions in the area of the magnetic repeat stations;
- Determining the transient contributions to the observed field on the magnetic repeat stations data.

## References

Bartels J., 1963: Discussion of time-variations of geomagnetic activity indices $K_p$ and $A_p$. *1932-1961, Annales Geophysics 19, pp.1.*

Chapman S. and Bartels J., 1951 : Geomagnetism. *(Vol.I Geomagnetic and related phenomena), Oxford.*

Janovskij B.M., 1972 : Zemnoi magnetizm I. *(glava IX) Izdatelstvo Leningradskogo Univerziteta.*

Mihajlovic J.S., 1993 : Spektralna anliza sekularnih varijacija i magnetskih bura na Geomagnetskoj opservatoriji Grocka. *magistarski rad, Beograd.*

Mihajlovic J. S., Djordjevic A., Jankovic J., Obradovic M., 1998 : Spektar dnevnih varijacija geomagnetskog polja registrovanih na sekularnim stanicama u Jugoslaviji. Prvi kongres geofizicara Jugoslavije - Zbornik; Beograd; E - 10, 559-564.

Mihajlović, J.S., Rasić, M., & Smiljanić, N., 2000, 'Spectral analysis and determination transferal functions of electromagnetic field of the Earth': Contribution of YU Association of Physicists, Book 2, pp. 921-924, Vrnjačka Banja-Yugoslavia.

Mihajlović, J.S. et al. 2000, 'Solar Geophysical Processes and Geomagnetic Disturbances', Contribution of YU Association of Physicists, Book 2, pp. 913-920, Vrnjačka Banja-Yugoslavia.

Mihajlovic J.S.; Obradovic M., 2003; *Metodologija magnetovarijacionih merenja*; Zbornik radova LVII Konferencije ETRAN-a; Sv.II; pp.; Beograd, 2003.

Mihajlovich, J.S.; Cholakov, I., 2001; SECULAR VARIATIONS DURING THE 19-21 SOLAR CYCLES; Contributoins to Geophysics and Geodesy 2001; Vol. 31/1; pp. 261-263; Publications of SAS, Bratislava; Slovakia.

Newitt L.R. et al. 1996; Guide for magnetic repeat station surveys, Copyright by IAGA, NOAA Space Environment Center, Boulder, CO, USA.

Rikitake T., 1966 : Electromagnetism and the Earths interior. (Chapter 19. Conductivity anomaly in the crust and mantle), Elsevier Publishing Company, Amsterdam-London-New York.

## DISCUSSION

Question (Jean Rasson): In the previous talk by T. Delipetrov a very detailed Z map of Macedonia was mentioned. Do you know if this data is available? If yes, how can we access this data?

Answer (Spomenko J. Mihajlovic): The magnetic map of the Z component of geomagnetic field anomalies of Republic Macedonia is made on the program of geomagnetic surveys on the territory of the former Yugoslavia, and has been executed by the Geomagnetic Institute Grocka. Geomagnetic surveys of the territory of Republic Macedonia involved two secular stations (repeat stations), 45 points of first-order survey and 150 points of second-order survey.

The detailed magnetic map of the Z component of Macedonia is done on the basis of a three category geomagnetic survey and on the basis of the reduction and the data processing of measurements for epoch 1960.0 and epoch 1980.0.

The use of the data which are involved in that magnetic map is possible only with the necessary permission of the state institutions of the Republic of Serbia.

Note: Request to Geomagnetic Institute Grocka for the use of the results of geomagnetic surveys of the former Yugoslavia have been made only by the Republic of Slovenia. Prof. dr Todor Delipetrov and dr. Jean Rasson have made the request to Geomagnetic Institute for using the daily/diurnal variations of the geomagnetic field of the Geomagnetic observatory Grocka (GCK).

Question (Marjan Delipetrov): Is it possible to make one map for all Balkan? Do you agree with this idea? What kind of instruments do you use in Grocka observatory and for field work?

Answer (Spomenko J. Mihajlovic): The production of the magnetic map for Balkan region is a very important project. I think it is a complex and challenging problem for all researchers from all countries in the Balkans. In that project in future should be involved as many researchers as possible from the Balkan countries and that activity should be realized as an international project. The Geomagnetic Institute and Geomagnetic observatory Grocka (Serbia) will accept all activities in this project about production the magnetic map of Balkan.

At the Geomagnetic Observatory Grocka following instruments are used for the absolute measurements of the geomagnetic field elements:

- Declination-inclination magnetometer DIM-810 (measurement of declination, inclination);
- Proton magnetometer GSM-10 / GSM- 19 (total intensity of the geomagnetic field).

The system of variometers "Askania" have been used for continuous recording of the geomagnetic field variations. At the Geomagnetic Observatory Grocka are used since 2004, FGE - Fluxgate three axial magnetometer normal version (produced by DMI- Danish Meteorological Institute) with DIMARK acquisition system (produced by ELGI institute-Hungary).

We use the following set of instruments for geomagnetic surveys:

- Three-component variometer EDA FM 100-C, with corresponding data acquisition system (on-site variometer or local variometer);
- The variations of the F component are registered with proton magnetometer type GSM-10 or GSM-19 (produced by GEM SYSTEMS-Canada);
- For absolute measurements we use standard instrument DIM–810 (Declination-Inclination Magnetometer).

Question (Angelo de Santis): According to the Bullard's rule (1967) that states that the number of coefficients of a polynomial reference field must be equal to the number of coefficients of a spherical harmonic model that contains the same "information" you want to get in your region, you could consider to use a first order polynomial instead of a second order one. Have you tried it in your analysis?

Some authors, for instance Chapman and Bartels (1940), suggested considering polynomial reference fields for the Cartesian magnetic components with coupled coefficients satisfying this condition of vertical current-free region. This implies to impose that

$$\partial X / \partial \lambda = \partial (Y \cos \phi) / \partial \phi$$

with X, Y horizontal magnetic components and $\lambda, \phi$ longitude and latitude. Have you considered applying this condition to your polynomial reference fields?

Answer (Spomenko J. Mihajlovic): Absolute observations made at the repeat station must be combined with continuous variation data from an on-site variometer in order to obtain an estimate of either the normal geomagnetic field, or an equivalent annual mean value at the repeat station (secular station).

The method of reduction of the repeat station survey data is based on the assumption that transient (including diurnal) variations of the geomagnetic field are identical at both the repeat station and reference observatory and is usually applied to calculate an effective annual mean value at the repeat station.

On the basis of the reduced values of the geomagnetic field on geomagnetic observatory and on the magnetic repeat stations, we can compute the values of the normal geomagnetic field for the territory FR Yugoslavia and for the epoch 1994.5. Normal geomagnetic field values could be represented by the polynomial functions.

The prospectors have used Gauss method of analysis in the processing of the values of the normal geomagnetic field. Normal geomagnetic field values for the territory of FR Yugoslavia and epoch 1994.5 are computed using the polynomial functions of the second order with sixth coefficients:

$F(\Delta\varphi, \Delta\lambda) = a1 + a2\,\Delta\varphi + a3\,\Delta\lambda + a4\,\Delta\varphi^2 + a5\,\Delta\lambda^2 + a6\,\Delta\varphi \bullet \Delta\lambda$

where

$\Delta\varphi = \varphi 1 - \varphi 0$;

$\Delta\lambda = \lambda 1 - \lambda 0$;

$F(\Delta\varphi, \Delta\lambda)$ is the component of the geomagnetic field;

a1 to a6 are the coefficients;

$\varphi 1$, $\lambda 1$ is the longitude and latitude of the magnetic repeat station and

$\varphi 0$, $\lambda 0$ is the longitude and latitude of the reference geomagnetic observatory.

# FIELD AND OBSERVATORY GEOMAGNETIC MEASUREMENTS IN BULGARIA

IVAN A. BUTCHVAROV[5]
*Geophysical Institute*
ILIYA V. CHOLAKOV
*Geomagnetic Observatory of the Geophysical Institute*

**Abstract.** Historical and recent geomagnetic measurements and geomagnetic maps of Bulgaria are given. A brief description of the Geomagnetic Observatory – Panagjuriste is included. The observatory measurements, records, calculations of the geomagnetic elements, and the geomagnetic data organization in the observatory are described.

**Keywords:** geomagnetism; geomagnetic field; geomagnetic measurement; geomagnetic observatory; Panagjuriste.

## 1. Historical geomagnetic measurements in Bulgaria

The first geomagnetic measurements in the territory of Bulgaria were performed during the Russian-Turkish wars at the end of the 18$^{th}$ and the beginning of the 19$^{th}$ centuries: 1787 – 1791 and 1828 – 1832. Only the declination, D, was measured. In 1858 Dr. K. Kreil, director of the Central Meteorological and Magnetic Survey in Vienna, made measurements of the declination, D, the horizontal intensity, H, and the inclination, I, and in 1859 the Russian military officer Dikov measured D. A certain number of geomagnetic measurements from the Black Sea were made too.

During the 1890's, P. Bahmetiev, professor of the Sofia University, measured the diurnal variations of the declination, D, near Sofia (❖, Figure 36), Petrohan and Berkovitza (❶, Figure 36).

[5] Address for correspondence: Geophysical Institute, Acad. G. Bonchev St. Bl. 3, Sofia 1113, Bulgaria. Email: buch@geophys.bas.bg



*J.L. Rasson and T. Delipetrov (eds.), Geomagnetics for Aeronautical Safety,* 61–82.

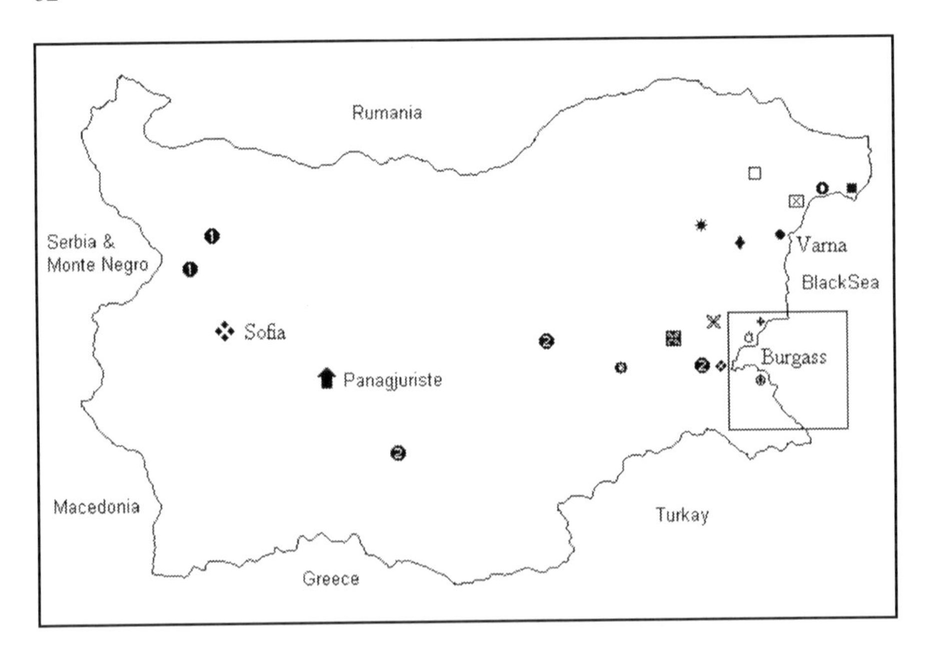


Figure 36. Position of the measured geomagnetic stations.

In 1911, specialists from the Carnegie Institution made measurements of D, H, and I near Sofia (❖, Figure 36), Burgas, Nova Zagora and Plovdiv (❷, Figure 36).

Some of the measured values are presented in Table 4:

Table 4. Measurements of the declination D in the end of the 18th and first half of the 19th century.

| Station | | Month | Year | D |
|---|---|---|---|---|
| Port Varna | ● | | 1787 | -15°00' |
| Varna Bay | | | 1829 | -11°00' |
| On sea | | | 1834 | -11°00' |
| Varna city | | IX | 1829 | -9°50' |
| Varna city | | | 1859 | -7°00' |
| Baltchik | ☒ | | 1859 | -6°43' |
| Kavarna | ○ | VI | 1830 | -10°12' |
| C. Kaliakra | ■ | X | 1858 | -6°42' |
| Novi Pazar | ✸ | V | 1830 | -11°06' |
| Provadia | ♦ | VI | 1830 | -14°41' |
| Dobritch | □ | V | 1830 | -10°41' |

| Station | | Month | Year | D |
|---|---|---|---|---|
| Karnobat | ▩ | IX | 1829 | -11°20' |
| Aitos | ✕ | IX | 1829 | -11°32' |
| Burgas | ❖ | IX | 1829 | -11°25' |
| Burgas | | IX | 1858 | -6°59 |
| Burgas | | | 1859 | -6°36' |
| Burgas Bay | | | 1829 | -9°30' |
| Pomorie | ✡ | IX | 1829 | -11°19' |
| Sozopol | ❄ | | 1859 | -6°28' |
| Nesebar | ✠ | IX | 1829 | -10°48' |
| On sea | | | 1834 | -9°30' |
| Yambol | ✹ | IX | 1829 | -11°35' |

The positions of the measuring stations are shown in Figure 36.

All data are authentic, taken from Russian archives. A detailed description of the literary sources and the data are available in the monograph of Kostov and Nozharov (1987).

## 2. Geomagnetic measurements in Bulgaria during the first half of 20th century

Acad. K. Popov, professor of the Sofia University, carried out the first systematic geomagnetic measurements in Bulgaria between 1917 and 1920. The geomagnetic elements D, H, and I were measured at 76 stations. The reduction from the diurnal variations was made using the data from the Pula Geomagnetic Observatory. On the basis of these measurements, the first geomagnetic map (of declination, D) in the territory of Bulgaria for epoch 1921.0 was drawn (Figure 37).

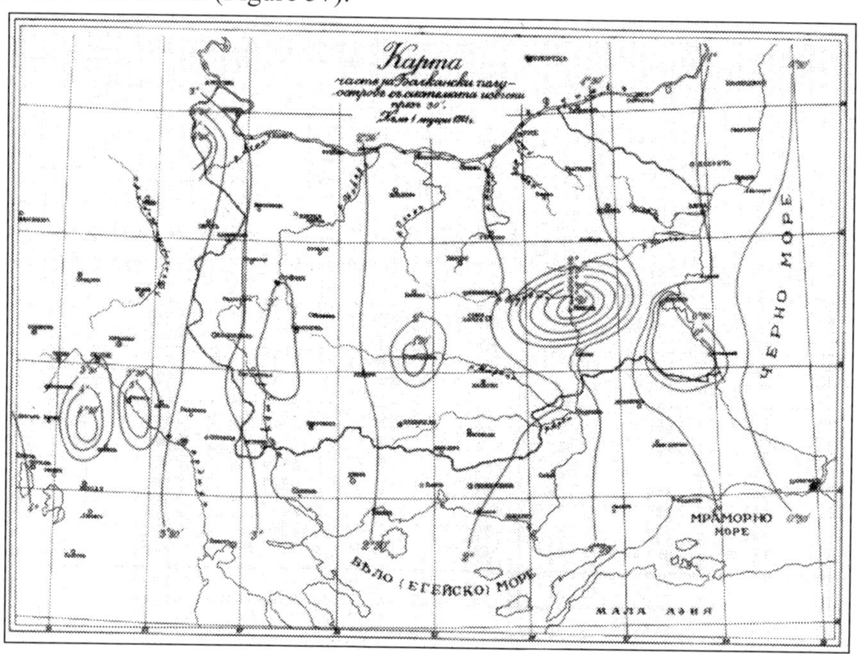


Figure 37. The declination, D, in Bulgaria, epoch 1921.0 (the values are negative).

In the 1930's and in the beginning of the 1940's, many geomagnetic measurements were performed in the territory of Bulgaria and on the Black Sea shelf. A map of the declination on the Black Sea shelf was drawn for epoch 1938.5 (Figure 38). The investigated area is shown in Figure 36 (the rectangle).

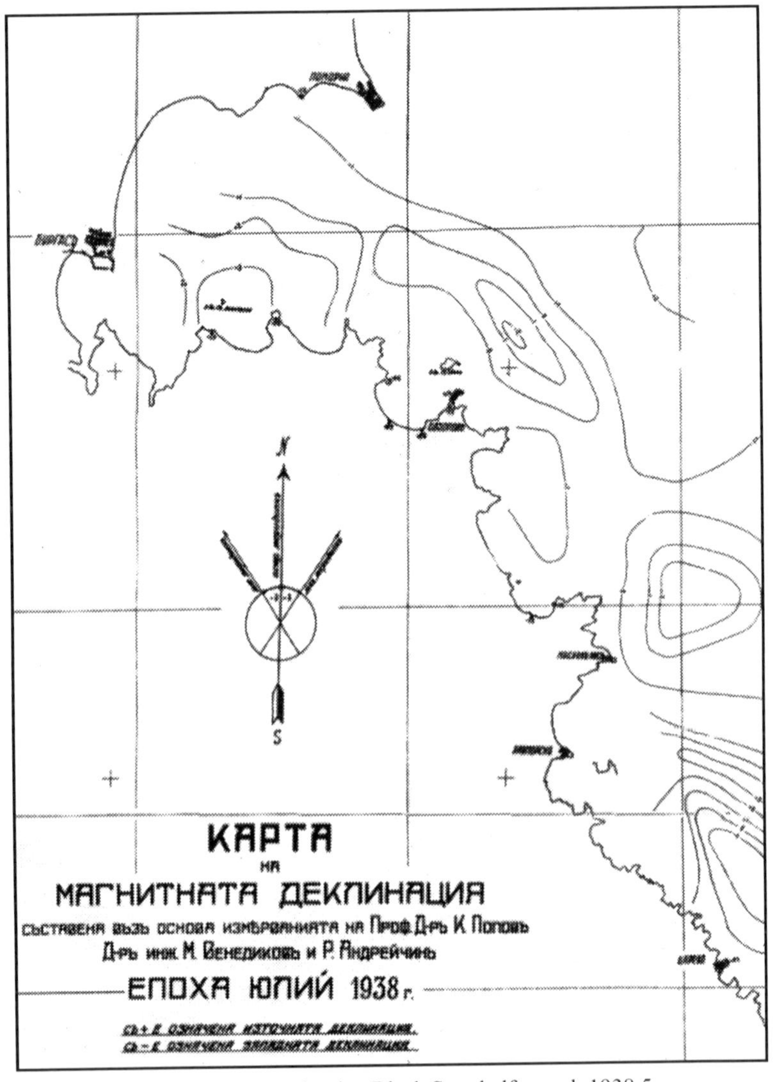


Figure 38. The declination, D, on Bulgarian Black Sea shelf, epoch 1938.5.

## 3. Building of the Panagjuriste Geomagnetic Observatory

The measurements described above were used to select a place for building a geomagnetic observatory. Thirteen sites were proposed and the town of Panagjuriste was preferred. At that time, the Bulgarian Geomagnetic Survey was part of the Military Geographical Institute. The Geophysical Institute of the Bulgarian Academy of Sciences took over the administration of the Observatory in 1961.

The Observatory was built in 1937 with the assistance of Dr. Fanselau from the Geomagnetic Institute, in Potsdam, Germany. There are 4 main buildings comprising the Observatory: the office building, two absolute houses and a relative house. All buildings were constructed very carefully. All building materials were tested for magnetic properties. The relative house is dug into the ground approximately 8 feet (about 1.80 m). The relative house has a constant temperature. The annual temperature variation is less than 2°C.

A detailed description of the measurements and of the construction of Panagjuriste Geomagnetic Observatory is given in the monograph of Kostov and Nozharov (1987).

## 4. Equipment used in Panagjuriste Geomagnetic Observatory

### 4.1. EQUIPMENT USED IN THE PAST

In the beginning, the absolute values of the geomagnetic elements D, H, and Z were measured by a "Shultze-545" geomagnetic theodolite. The recording of variations was made by a single series of "Askania-Werke-AG" variometers, and an "Edelton" recording system using photo paper. An "Askania-Werke-AG" observatory earth inductor with an optical galvanometer was supplied in 1945 and until 1965 the inclination I was measured with it.

In 1956 an additional series of "Mating & Wiesenberg" variometers were purchased for recording the geomagnetic field variations.

### 4.2. EQUIPMENT AND TECHNOLOGY USED NOW

In October 1961 the Geophysical Institute of the Bulgarian Academy of Sciences took over the administration of the Observatory. The following scheme of absolute geomagnetic measurements and recording of geomagnetic field variations was accepted:

#### 4.2.1. *Absolute measurements*

- The horizontal intensity, H, is measured by an absolute "Mating & Wiesenberg" geomagnetic theodolite mounted on pillar № 1 according to the Gauss-Lamont method. The measurements are performed with three deviating magnets. The semi-period of oscillation of the deviating magnet is determined on pillar № 2 by using an electronic periodometer. Three QHM's are also used. The accuracy is ~ 1 nT.

- The declination, D, is measured by the same "Mating & Wiesenberg" geomagnetic theodolite. The accuracy is ~ 0.15'.
- The inclination, I, is measured by a "Mating & Wiesenberg" observatory earth inductor mounted on pillar № 6.
- The total intensity, F, is measured with a Polish proton magnetometer, model PMP-2P. The sensor is placed on pillar № 6 above the earth inductor only during the measurements. The accuracy is ~ 1 nT.

### 4.2.2. *Recording of geomagnetic field variations*

There are two series of variometers in the variation house. The main series is the western one. There are 4 quartz type "BOBROV" variometers in each of the series to record D, H, Z and F. The variations are recorded on photo paper: standard 48 x 20 cm magnetograms, 20 mm/h. The recording instrument in the eastern series has two drums. The first one records at a speed of 20 mm/h, while the second one records at 20 mm/h, 60 mm/h, and 240 mm/h. The fast recording is used only when making absolute measurements.

## 5. Comparative geomagnetic measurements

The magnetic level of the Panagjuriste Observatory has been practically connected to the level of the "Adolf Schmidt" Niemegk Observatory since the foundation of the Panagjuriste Observatory to the present. From 1934 to 1943 five comparative measurements were made in Niemegk with "Schulze-545" geomagnetic theodolite and the earth inductor. Many measurements were done after WW II in Panagjuriste and in other observatories (GCK, THY, MOS, KPR, SUA, CLF etc.). Details are available in Cholakov and Butchvarov (1995). Table 5 shows the comparative geomagnetic measurements in Panagjuriste and Niemegk.

Table 5. Comparative geomagnetic measurements PAG – NGK.

| Year of measuring | Obs. host | Operators | ΔD ' | ΔH nT | ΔZ nT | ΔF nT |
|---|---|---|---|---|---|---|
| 1963 | NGK | K.Kostov | +0.42 | 0.0 | 0.0 | - |
| 1964 | PAG | A.Grafe, W.Zander | -1.30 | +3.2 | -1.0 | - |
| 1966 | NGK | K.Kostov | -0.43 | +2.2 | - | - |
| 1967 | PAG | A.Grafe, W.Zander | -0.02 | +0.3 | -5.6 | - |
| 1969 | NGK | K.Kostov | +0.7 | -0.8 | -1.5 | -3.0 |

| Year of measuring | Obs. host | Operators | ΔD ' | ΔH nT | ΔZ nT | ΔF nT |
|---|---|---|---|---|---|---|
| 1971 | PAG | K.Lendning, W.Zander | -0.80 | +0.5 | -2.0 | -1.8 |
| 1974 | NGK | K.Kostov | 0.0 | -0.4 | +0.8 | -0.3 |
| 1975 | PAG | K.Lendning, W.Zander | -0.48 | +1.3 | +0.3 | +0.4 |
| 1976 | NGK | K.Kostov | -0.04 | -1.2 | - | -0.1 |
| 1978 | NGK | K.Kostov | +0.06 | -1.1 | 0.0 | -0.3 |
| 1980 | NGK | K.Kostov, I.Butchvarov | -0.02 | +0.2 | - | -0.4 |
| 1984 | PAG | E.Ritter, W.Zander | +0.11 | +1.2 | - | +0.7 |
| 1986 | NGK | I.Cholakov, Ch.Georgiev | +0.18 | +0.1 | - | +0.3 |
| 1987 | PAG | E.Ritter | -0.86 | -2.2 | - | - |
| 2003 | NGK | I.Cholakov, B.Srebrov | + 0.23' | -2.6 | + 1.0 | + 0.16 |

## 6. Recent geomagnetic field measurements in Bulgaria

From 1934 to 1947, H. Kalfin, physicist of the Military Geographical Institute, measured the declination, D, at 750 points by using the "Shulze" geomagnetic theodolite, horizontal intensity, H, and inclination, I, were measured at 350 points. The reductions of the observations, in the period 1934 – 1937, were made using records of the observatory in Vienna. The rest of the observations were reduced using the records of the Panagjuriste Observatory. Maps of D for epochs 1940.0 and 1950.5 were drawn (Kalfin, 1939).

K. Kostov, physicist of the Panagjuriste Observatory, performed a new geomagnetic survey of Bulgaria during the years 1958 – 1961. The declination, D, was measured at 2000 points, and H and I were measured at 342 of the points. The "Schulze-545" geomagnetic theodolite was used as well as three QHM's. The inclination, I, was measured with the "Shulze" earth inductor. The observations were reduced to epoch 1960.0 and maps of the geomagnetic field elements, and its normal and anomalous fields were drawn (Kostov, 1971; Kostov and Butchvarov, 1969; Butchvarov and Kostov, 1981).

The authors of the present paper performed the latest absolute geomagnetic survey during the period 1978 – 1980. The geomagnetic elements D, H, and F were measured at 473 points. The "Shulze-545" geomagnetic theodolite, three QHM's, and two PMP-2A proton magnetometers were used. The geographic azimuth to the mark in 90 % of the points was determined by a gyrotheodolite with two gyroblocks. The geographic azimuth for the remaining points was determined by geodetic methods. The geographic azimuth accuracy determination (the standard deviation) was ~ 0.2' (Mintchev, 1979; Kostov et al., 1991).

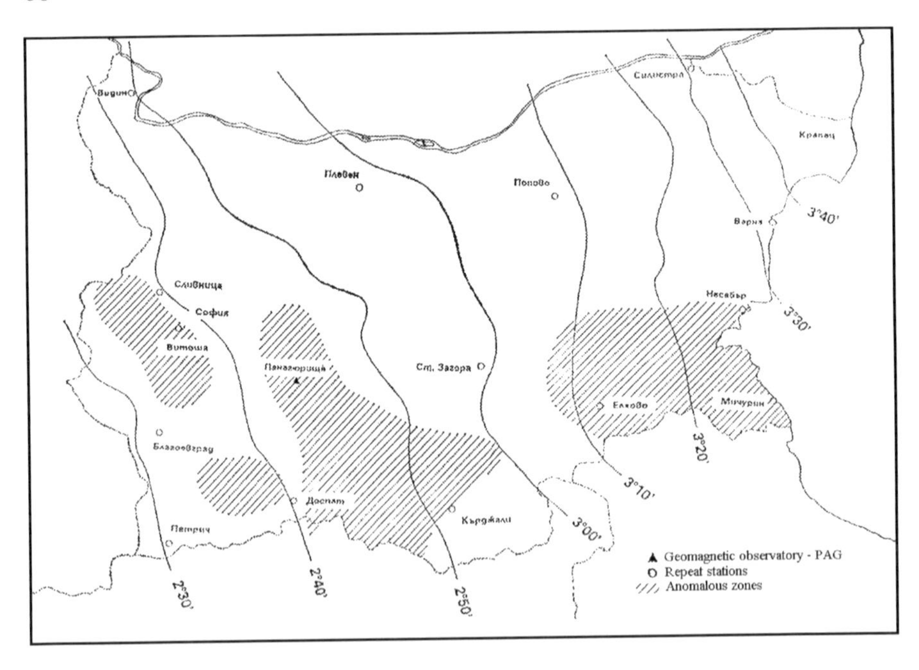


Figure 39. The declination, D, in Bulgaria, epoch 1990.0.

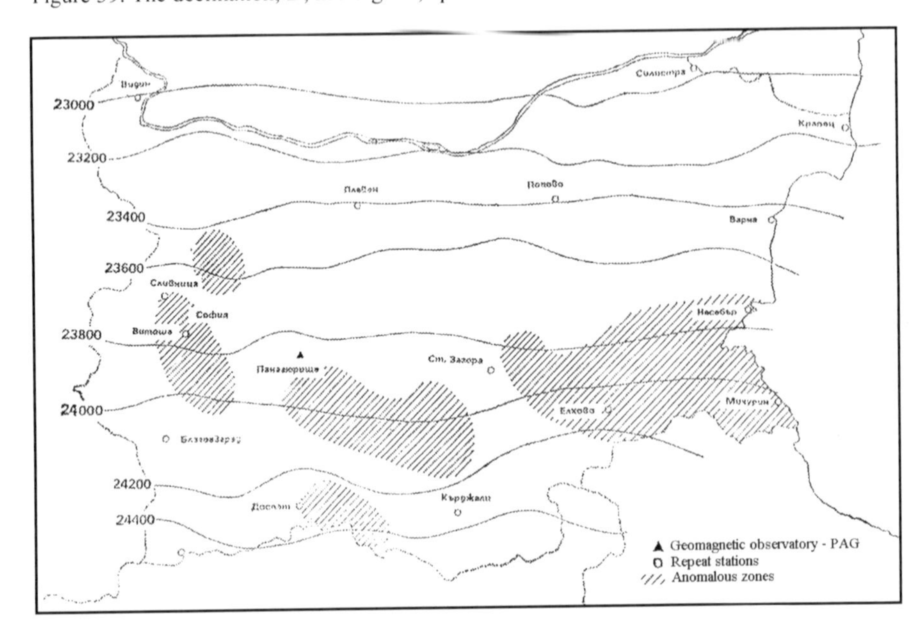


Figure 40. The horizontal intensity, H, in Bulgaria, epoch 1990.0.

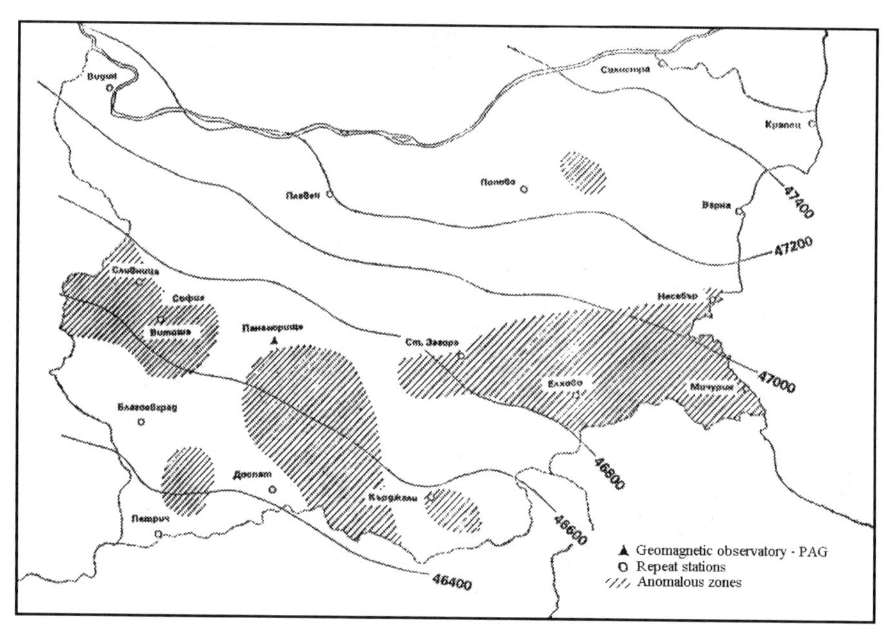


Figure 41. The total intensity, F, in Bulgaria, epoch 1990.0.

The reduction to epoch 1980.0 of the observations was made according to the Panagjuriste Observatory. The reduced geomagnetic element annual mean values accuracy was (Butchvarov et al., 1984):

$$\sigma_D = 0.5', \quad \sigma_H = 4.2 \text{ nT}, \quad \sigma_F = 3.5 \text{ nT}.$$

In 1990 measurements at 15 secular change stations were made. All data from the 1978 – 1980 survey in were reduced to epoch 1990.0. The maps of declination, D, (Figure 39), horizontal intensity, H, (Figure 40), and total intensity, F, (Figure 41) were drawn. The hatched areas on the maps are anomalous zones (Butchvarov and Cholakov, 1994).

## 7. Normal and anomalous geomagnetic fields in Bulgaria

The normal and anomalous geomagnetic fields were determined for epochs 1960.0, 1965.0, 1970.0 and 1980 (Butchvarov and Kostov, 1981). The method used to find these fields is described below.

The normal geomagnetic field was obtained by the method of regression analysis by "sifting out", using an approximating polynomial of 2nd degree (the model) of the geographical coordinates $\varphi$ and $\lambda$

$$\boldsymbol{E}(\Delta\varphi, \Delta\lambda) = a_1 + a_2\,\Delta\varphi + a_3\,\Delta\lambda + a_4\,\Delta\varphi^2 + a_5\,\Delta\lambda^2 + a_6\,\Delta\varphi\Delta\lambda\phi\,, \quad (1)$$

where $\boldsymbol{E}(\Delta\varphi, \Delta\lambda)$ denotes any geomagnetic element, $a_j$, $j = 1,2,...,6$, are the unknown coefficients of the polynomial and $\Delta\varphi = \varphi - \varphi_0$, $\Delta\lambda = \lambda - \lambda_0$ ($\varphi$ and $\lambda$ in '). $\varphi$ is the latitude, $\lambda$ is the longitude. $\varphi_0 =$ 42°30'N and $\lambda_0 = $25°00'E. Using the classical least square method, estimations of the unknown polynomial coefficients $a_j$, the regression residual variance $\sigma^2$, and confidence intervals of the coefficients $a_j$ are determined. A short description of the method is presented below.

Let $\mathbf{X}$ be a rectangular matrix with dimension $6 \times N$ and each row of it has the following form:

$$1, \Delta\varphi_i, \Delta\lambda_i, \Delta\varphi_i^2, \Delta\lambda_i^2, \Delta\varphi_i\Delta\lambda_i,$$

where $\Delta\varphi_i, \Delta\lambda_i$ (defined above) are the geographical coordinates of the measured stations on the Earth surface and $i = 1,2,..., N$, $N$ is the number of these stations. The matrix $\mathbf{X}$ is called "matrix of the observational equations".

The linear unbiased point estimate with minimum variance $\tilde{a}_1, \tilde{a}_2, \tilde{a}_3, \tilde{a}_4, \tilde{a}_5, \tilde{a}_6$ of the $a_1, a_2, a_3, a_4, a_5, a_6$ is given by the following (Wilks, 1967; Butchvarov, 1977; Butchvarov and Cholakov, 1992):

$$\tilde{\mathbf{A}} = \mathbf{C}^{-1}\mathbf{X}^T\mathbf{E}_r,$$

where $\tilde{\mathbf{A}}$ is a vector-column with components $\tilde{a}_1, \tilde{a}_2, \tilde{a}_3, \tilde{a}_4, \tilde{a}_5, \tilde{a}_6$, $\mathbf{X}^T$ is the transposed matrix of the matrix $\mathbf{X}$, $\mathbf{E}_r$ is a vector-column with components $E_{r1}, E_{r2}, ..., E_{rN}$ – the reduced annual mean values to the respective epoch of the geomagnetic elements at the measured stations, $\mathbf{C}^{-1} = (\mathbf{X}^T\mathbf{X})^{-1}$. $\mathbf{X}^T\mathbf{X}$ is the so called "matrix of the normal equations".

The linear unbiased point estimate with minimum variance $\tilde{\sigma}^2$ of the residual variance $\sigma^2$ is given by the formula (Linnik, 1962; Wilks, 1967; Butchvarov, 1977; Butchvarov and Cholakov, 1992).

$$\tilde{\sigma}^2 = \frac{1}{N-6}(\mathbf{E}_r^T\mathbf{E}_r - \mathbf{E}_r^T\mathbf{X}\mathbf{C}^{-1}\mathbf{X}^T\mathbf{E}_r).$$

100 $\gamma$ – % confidence interval $I_{a_j}$ of given coefficient $a_j$ is given according to the expression (Linnik, 1962; Wilks, 1967; Butchvarov, 1992)

$$I_{a_j} = \{a_j : \tilde{a}_j - t_{N-6,\gamma}\,\tilde{\sigma}\sqrt{c_{jj}^{-1}} < a_j < \tilde{a}_j + t_{N-6,\gamma}\,\tilde{\sigma}\sqrt{c_{jj}^{-1}}\},$$

where $t_{N-6,\gamma}$ is the Student coefficient corresponding to the confidence probability $\gamma$ at $N-6$ degrees of freedom, and $c_{jj}^{-1}$ is the $j^{\text{th}}$ diagonal element of the matrix $\mathbf{C}^{-1}$.

The estimation of the normal annual mean values of the geomagnetic field element $\tilde{\boldsymbol{E}}(\Delta\varphi, \Delta\lambda)$ at an arbitrary point with geographical coordinates $\Delta\varphi, \Delta\lambda$ on the Earth's surface is calculated formally changing the coefficients $a_j$ by the estimates $\widetilde{a}_j$ in Eq. (1).

The "sifting out" method consecutively applies the above presented regression analysis over the geomagnetic data, removing every point with a reduced annual mean value that deviates more than a chosen norm from the obtained model $\widetilde{\boldsymbol{E}}(\Delta\varphi, \Delta\lambda)$ (Butchvarov and Cholakov, 1992).

The elements of matrix $\mathbf{C}$ were determined for all observation data (473 measured stations), and the estimates $\widetilde{\mathbf{A}}$ of the regression equation's coefficients for the geomagnetic elements F, Z, and D were found out. Then the estimates $\widetilde{\boldsymbol{E}}(\Delta\varphi_i, \Delta\lambda_i)$ of the normal values for all points of observation were calculated and the differences $\left|\Delta\widetilde{\boldsymbol{E}}_i\right| = \left|E_{ri} - \widetilde{\boldsymbol{E}}(\Delta\varphi_i, \Delta\lambda_i)\right|$ were defined. Further when the difference for a given point was greater than a chosen norm, $\mathbf{N}$, the point was eliminated (sifted out). After the elimination of all points with $\left|\Delta\widetilde{\boldsymbol{E}}_i\right| > \mathbf{N}$, the estimates $\widetilde{\mathbf{A}}$ were calculated again and new estimates $\widetilde{\boldsymbol{E}}(\Delta\varphi_i, \Delta\lambda_i)$ of the normal values were calculated, and the points with $\left|\Delta\widetilde{\boldsymbol{E}}_i\right| > \mathbf{N}$ were eliminated. This procedure was continued till no points were eliminated. Then the estimates $\widetilde{\mathbf{A}}$ were analyzed and the coefficients $a_j$ for which $\left|\widetilde{a}_j\right| < t_{N-6,\gamma}\,\widetilde{\sigma}$ were ignored. Once more, the estimates $\widetilde{\mathbf{A}}$ of the truncated regression equation and estimates $\widetilde{\boldsymbol{E}}$ of the normal values were calculated and the differences $\left|\Delta\widetilde{\boldsymbol{E}}_i\right|$ were defined and compared with the norm $\mathbf{N}$. The procedure iterated until no point or regression coefficients were eliminated. The final received model $\widetilde{\boldsymbol{E}}(\Delta\varphi_i, \Delta\lambda_i)$ was used for drawing the maps of the normal fields.

The chosen deviation norms were 80 nT for F and Z, and 8' for D. The confidence probability $\gamma = 0.95$ was used. The presumption to use these norms was to eliminate the points with reduced annual means deviating from the model $\widetilde{\boldsymbol{E}}(\Delta\varphi_i, \Delta\lambda_i)$ with values grater than the step between the isopleths of the maps (Figure 39, Figure 40, Figure 41) and 2 – 3 time greater then the estimate $\widetilde{\sigma}$ of the residual standard deviation. Some characteristics of the normal geomagnetic field and of the regression equation coefficients and its confidence intervals are given in Butchvarov and Cholakov (1992).

The described procedure was proposed in the past (in the 1960's) by KAPG – the Academy Commission of Planetary Geophysics of former, so-called, socialist countries.

The normal fields of D, Z, and F are presented in Figure 42, Figure 43 and Figure 44.

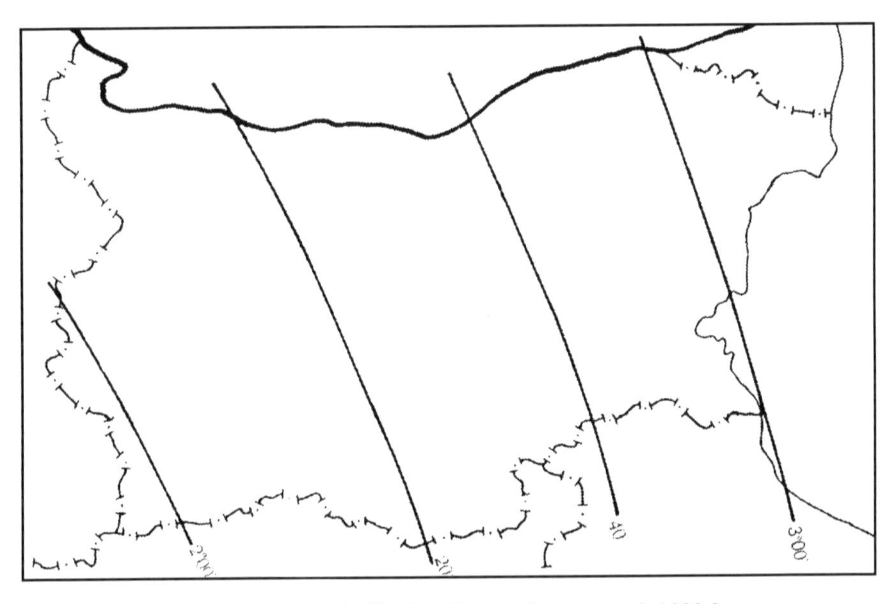


Figure 42. The normal field of the declination, D, in Bulgaria, epoch 1980.0.

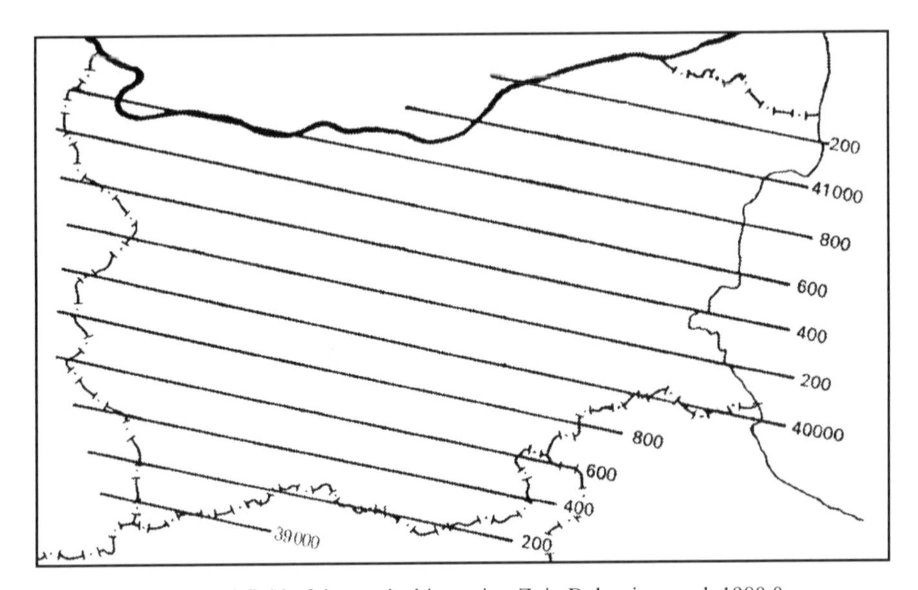


Figure 43. The normal field of the vertical intensity, Z, in Bulgaria, epoch 1980.0.

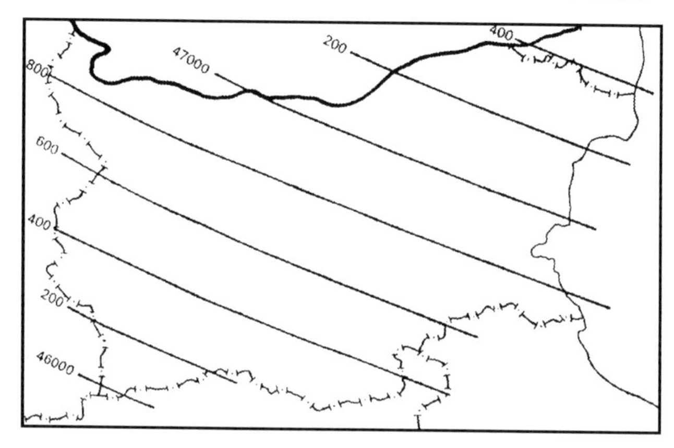


Figure 44. The normal field of the total intensity, F, in Bulgaria, epoch 1980.0.

Remarks:

- We tried using a polynomial with higher than 2nd degree, but the coefficients of the regression equation were unrepresentative (i.e. $\left|\tilde{a}_j\right| < t_{N-6,\gamma}\,\tilde{\sigma}$ ) and were being eliminated.
- The morphologies of the normal fields for all processed epochs are practically identical. We think this is so because of the small size of Bulgaria.

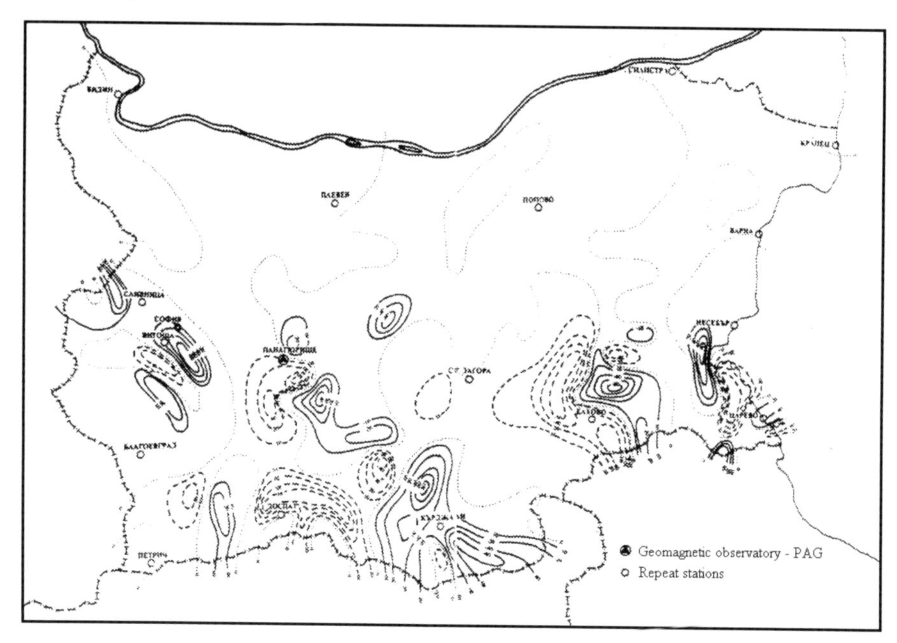


Figure 45. The anomalous field of the declination, D, in Bulgaria, epoch 1980.0.

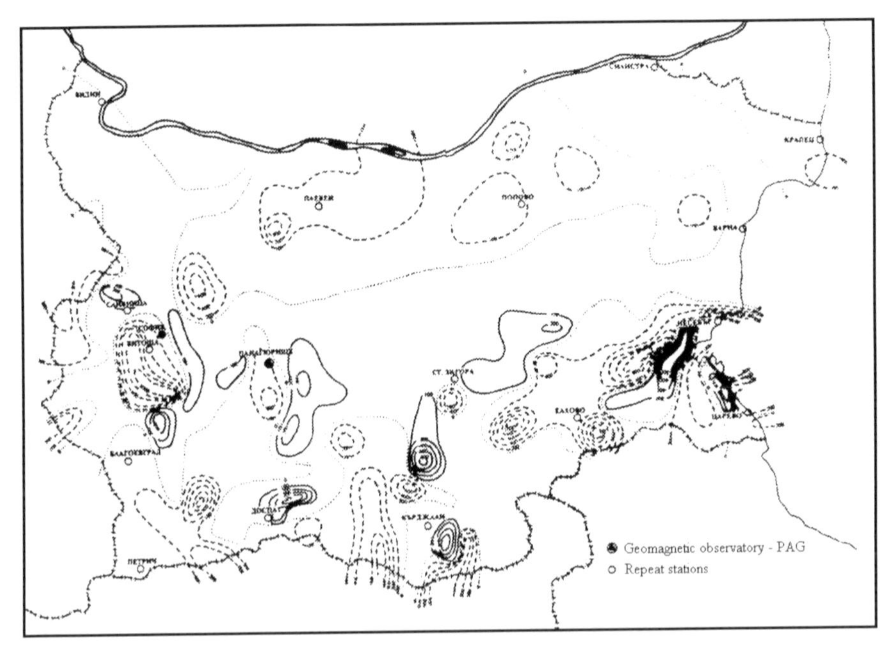


Figure 46. The anomalous field of the vertical intensity, Z, in Bulgaria, epoch 1980.0.

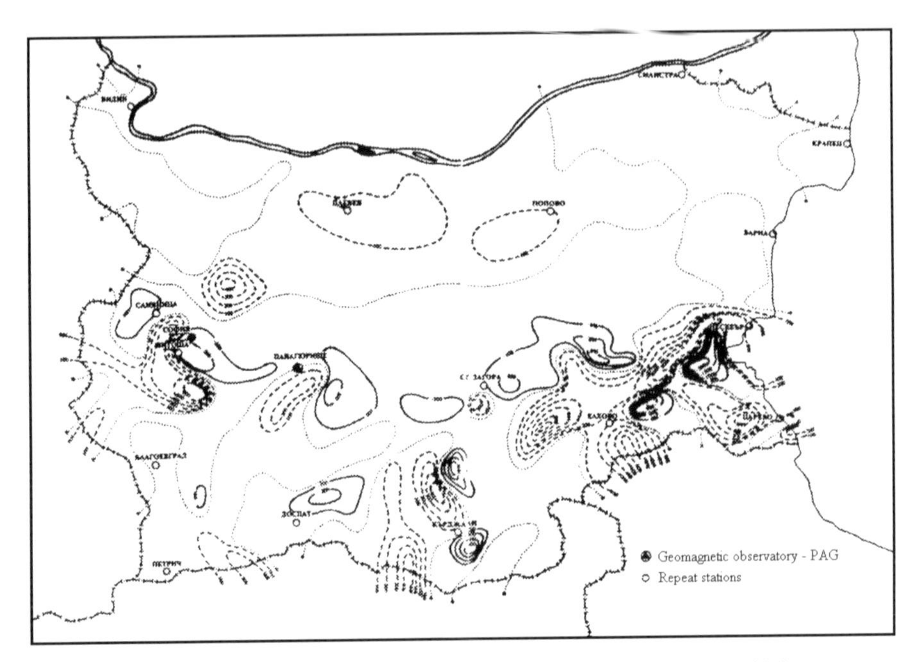


Figure 47. The anomalous field of the total intensity, F, in Bulgaria, epoch 1980.0.

The anomalous field is obtained by extracting the normal field $\widetilde{\boldsymbol{E}}$ from the reduced annual mean values of the geomagnetic element to the respective epoch at the measured stations $E_{r\,i}$ (Butchvarov and Cholakov, 1995). The maps of the anomalous fields of D, Z, and F are presented in Figure 45, Figure 46 and Figure 47.

## 8. Secular change measurements in Bulgaria

The repeat stations network in Bulgaria started in 1934 with 8 stations and was supplemented with 7 more in 1964. All points were investigated, stabilized, and later duplicated with extra stations and secured with lasting azimuth marks. Up to 1980 the measurements were made every three years. After 1980, because of the small secular variations measurements were made every five years. Isoporic maps for different periods were developed (Georgiev et al., 1986). Secular measurements were not made after 1990 for lack of funds.

## 9. Processing and organization of the data obtained from the analog magnetograms of Panagjuriste Geomagnetic Observatory

### 9.1. BRIEF HISTORY OF THE DATA PROCESSING IN THE PAST

At first, the diurnal mean, monthly mean, and annual mean values of the geomagnetic elements were calculated and used only for reduction of the field geomagnetic measurements to the common epoch, and they were not published. The first publication of the data was released in 1965 under the guidance of Dr. D. Zidarov, head of the Section of Geomagnetism and Gravimetry of the Geophysical Institute. The Geomagnetic Yearbooks of the hourly mean, monthly mean, annual mean, and s. o. values were published according to the standards of the International Association of Geomagnetism and Aeronomy (IAGA).

The hourly mean values and their maximum and minimum values (in mm) were read from the magnetograms manually. The calculation of the geomagnetic elements in respective units and the calculation of the diurnal, monthly, and annual means were carried out with a very primitive calculator. The Geomagnetic Yearbooks were typewritten and printed by the Military Topographic Service. They were made in this way from 1956 - 1975.

From 1975 - 1983 all geomagnetic data mentioned above were processed on an IBM 360, and Geomagnetic Yearbooks were printed according to the IAGA standards. All Yearbooks are available in the Panagjuriste Observatory and in the WDCs.

## 9.2. THE DATA PROCESSING AND DATA ORGANIZATION AT PRESENT

Unfortunately, it has not been possible to buy digital equipment so geomagnetic variations continue to be recorded on photographic paper*. The geomagnetic element hourly mean and extreme values (in mm) continue to be read manually. However the data processing is now on a PC.

At the end of 1980's a package of programs (programming software) in TURBO PASCAL 5.5 code, under DOS, for processing of the geomagnetic data and organizing the data into a database was developed by I. Butchvarov. The hourly and annual mean values in machine readable form are sent to the WDC's through the INTERNET. The hourly mean values (standard IAGA files) from 1984 to 2004 are available in the WDCs. The annual mean values from 1948 to 2004 are available in WDCC1 in Murchison House, Scotland, UK.

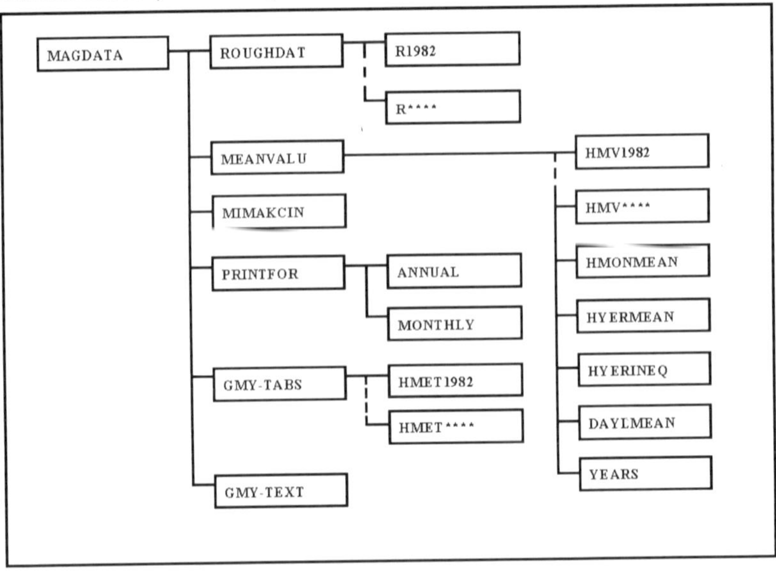


Figure 48. The organization of the geomagnetic data in folders on a PC.

The geomagnetic database structure is presented in Figure 48. The organization of the geomagnetic data in folders on a PC and the contents of all folders and files are in Table 6 (all files are text files are in ASCII-code). We shall introduce the following symbols and conditions to make the folder and file description simpler:

1) **** – defines a year – 1989, 2000 and s. o.

* German colleagues from the Adolf Schmidt geomagnetic observatory, Niemegk, under the guidance of Dr. Lindte, mounted a digital geomagnetic variation station in Panagjuriste observatory during the preparation of the present paper.

2) +++ – defines a month – JAN, FEB, …, DEC.

3) GMFE – defines "GeoMagnetic Field Element(s)".

4) & – defines the GMFE symbols: D – declination, F – total intensity, H – horizontal intensity, I – inclination, X – north component, Y – east component or Z – vertical component, in the names of the files containing the tables of respective elements; and E – in the names of the files containing the tables of extreme values, K and C indices, and the temperature of the "Relative house" in the same form as the old Geomagnetic Yearbook that was edited as a hard copy.

5) "Rough data" – defines the values read from the magnetograms and quantity received after elementary manual processing.

The folder and files in the geomagnetic database of Panagjuriste Observatory and their brief description are given in Table 6. They are described in more details in (Butchvarov, 2005). A CD-ROM with detailed description of all available information in the Panagjuriste Observatory will be produced in the near future. It will be sent to the interested institutions. A WEB page will be created for it.

Table 6. Comparative geomagnetic measurements PAG – NGK.

| Folder | Description |
|---|---|
| MAGDATA | Main folder. Contains all other folders. There are no files in it. |
| ROUGHDAT | Contains all folders R**** of the rough data. There are no files in it. |
| R**** | They contain the files R****+++.PAG of the rough data. There is 1 folder of 12 files for every year – each one for *ONE MONTH* . |
| MEANVALU | Contains the folders HMV**** of the hourly mean values for *ONE MONTH* and the folders: HYERMEAN, HYERINEQ, HMONMEAN, DAYLMEAN and YEARS described below. Contains also files: MONTMEAN.PAG of the monthly mean values, ANNMEANS.PAG of the annual mean values and MDIMONTH.PAG of the average mean diurnal inequalities for unlimited number of years. The last three files are unique in the Observatory. The first two of them are only for 100 years. |
| HMV**** | They contain the files W****+++.PAG of the hourly mean values for *ONE MONTH*. There is 1 folder of 12 files for every year – each one for 1 month. |
| HYERMEAN | Contains the files MDVS****.PAG of the average hourly mean values for *ONE YEAR* and for the *EQUINOX*, *SUMMER* and *WINTER*. |

| Folder | Description |
|---|---|
| HYERINEQ | Contains the files MDIM****.PAG of the mean diurnal inequalities for *ONE YEAR* and the files MDIS****.PAG of the average mean diurnal inequalities for *ONE YEAR* and for the *EQUINOX*, *SUMMER* and *WINTER*. |
| HMONMEAN | Contains the files HMMV****.PAG of the hourly monthly means values. There is 1 file for every year. |
| DAYLMEAN | Contains the files DM****.PAG of the daily mean values. There is 1 file for every year. |
| YEARS | Contains the files PAG****.WDC of the hourly mean values for *ONE YEAR*. There is 1 file for every year. |
| MIMAKCIN | Contains the files EXKC****.PAG of D, H and Z extreme values, K and C indices, and the temperature in the "Relative house". There is 1 file for every year. |
| PRINTFOR | Contains the folders: ANNUAL – of the annual mean value files, and MONTHLY – of the files containing simultaneously annual and monthly mean values. There are no files in it. |
| ANNUAL | Contains the files PAG-ANN.ALL and PAG-ANND.ALL of the annual mean value tables (easy and well arranged for printing). In the first one the declination is presented in ° (degrees) and ' (minutes), and in the second one – in degrees and tenth of degrees. There is only 1 file of the two types (2 files only) for all years. The file PAG-ANN.ALL is sending to the WDCC1 in Edinburgh. |
| MONTHLY | Contains the files PAG****.ALL of the annual and monthly mean value tables simultaneously (easy and well arranged for printing). There is 1 file for every year. |
| GMY-TABS | Contains the folders HMET**** of the files &****+++.PAG described below. There are no files in it. |
| HMET**** | They contain the files &****+++.PAG of the GMFE tables and the tables of extreme values, etc. in the same form as the old Geomagnetic Yearbook that was issued like a hard copy. There is 1 folder of 8 files for every year. |
| GMY-TEXT | Contains different files for the history of the Observatory, some of its characteristics, etc. |

## References

Butchvarov, I., 1977, An application of the regression analysis for processing of geophysical fields and their mapping, Bulg. Geophys. J. **III** (4): 74-81 (in Russian).

Butchvarov, I., 2005, Processing and organization of the data obtained from the analog magnetograms of Panagjuriste Geomagnetic Observatory. Bulg. Geophys. J (in print).

Butchvarov, I. and Cholakov, I., 1992, Normal geomagnetic field in Bulgaria, epoch 1980.0, Bulg. Geophys. J. **XVIII** (4): 65-73 (in Bulgarian).

Butchvarov, I. and Cholakov, I., 1994, Geomagnetic field on the territory of Bulgaria – epoch 1990.0, Bulg. Geophys. J. **XX** (4): 57-66 (in Bulgarian).

Butchvarov, I. and Cholakov, I., 1995, Anomalous geomagnetic fields in Bulgaria, epoch 1980.0, Bulg. Geophys. J. **XXI** (4): 59-72 (in Bulgarian).

Butchvarov, I. and Kostov, K., 1981, Presentation of the geomagnetic field in Bulgaria for the epochs 1960.0, 1965.0 and 1970.0 with a polynomial of second degree of the geographical coordinates, Bulg. Geophys. J. **VII** (4): 51-62 (in Russian).

Butchvarov, I., Kostov, K., Cholakov, I., Georgiev, H. and Chekardjikov, L., 1984, Conclusions on the precision of the absolute geomagnetic survey on the territory of People's Republic of Bulgaria in 1958 – 1961 and 1978 – 1980, Bulg. Geophys. J. **X** (3): 117-126 (in Bulgarian).

Cholakov, I. and Butchvarov, I., 1995, Absolute geomagnetic level of the Panagjuriste Geomagnetic Observatory in the period 1934 – 1993, Bulg. Geophys. J. **XXI** (4): 109-114 (in Bulgarian).

Georgiev H., C., Kostov, K. and Cholakov, I., 1986, Isopors of the declination for the 1940 – 1980, Bulg. Geophys. J. **XII** (3): 40-48 (in Bulgarian).

Kalfin, C., 1939, Magnetic measurements in Bulgaria, Yearbook of State Geogr. Inst., 1937-1938: 58-73 (in Bulgarian).

Kostov, K. and Butchvarov, I., 1969, Normal geomagnetic field in Bulgaria, epoch 1960.0, Bull. Geophys. Inst. **XV**: 33-47 (in Bulgarian).

Kostov, K., 1971, Magnetic measurements in Bulgaria 1958 - 1961, Bull. Geophys. Inst. **XVII**: 87-98 (in Bulgarian).

Kostov, K. and Nozharov, P., 1987, Absolute geomagnetic measurements in Bulgaria 1787 – 1987 (Archives Pangjuriste Geomagnetic Observatory, in Bulgarian).

Kostov, K., Cholakov, I. and Butchvarov, I., 1991, The geomagnetic field on the territory of Bulgaria – epoch 1980.0, Bulg. Geophys. J. **XXII** (2): 42-53.

Linnik, U., 1962, The method of least squares and basis of mathematical-statistical theory of observation treatment, Fizmatgiz, Moscow (in Russian).

Mintchev, M., 1979, Analysis of the Azimuthal measurements with gyrotheodolite, Geodesy, Cartography and Land regulation, **XIX** (6): 77-93 (in Bulgarian).

Wilks, S., 1967, Mathematical Statistics, Nauka, Moscow (in Russian).

### DISCUSSION

Question (Bejo Duka): Did you determine or detect any "bias" effect at your observatory?

Answer (Ivan Butchvarov): No. We have no "bias" effect at our observatory. The magnetic level of Panagjuriste observatory during the period 1963 – 2004 is practically equal to the level of Niemegk Geomagnetic Observatory.

Question (Jurgen Matzka): How large was secular variation of declination the last 10 years?
Answer (Ivan Butchvarov): The secular variation of the declination in Panagjuriste Observatory for the period 1995 – 2004 is 36.4'.
Question (Marjan Delipetrov): Is it possible to make one big map of the Balkans compiled from our magnetic measurements? Could all Balkan countries work together to make a good precision map using measurements carried out at the repeat stations? How do you like this idea?
Answer (Ivan Butchvarov): It will be wonderful to work out a big common geomagnetic map of the Balkans. I think all Balkan countries have very good specialists in that field. There are still some administrative hindrances which will hopefully be overcome in the near future, but the most important is to provide financing for carrying out geomagnetic measurements on the Balkan Peninsula.
Question (Jean Rasson): You mentioned declination measurements on the continental shelf of Black Sea. What technique was used for making those measurements?

**Résumé**

**SUR LA DÉCLINAISON DE L'AIMANT LE LONG DE LA CÔTE DE LA MER NOIRE DE BOURGAS À ACHTOPOL.**

**Par le prof. Dr. Kyrille Popoff, le Dr. ing. Michail Venedikoff et le Dr. Rasoume Andreitchine.**

Sur la démande de la Direction de Navigation au Ministère des Chemins de fer et des Ports nous avons étudiés les variations de la déclinaison de l'aimant le long de la côte de la Mer Noire de Bourgas à Achtopol. Le Bateau de port „Rakovsky" de 100 tonneaux qu'on a mis à notre disposition a été muni d'une boussole marine à amortisement liquide, dont la déviation, après la compensation, ataignait la valeur maximum de 2°. La boussole a été soigneusement étudiée dans les différents cours et dans des conditions des plus variées.

La région éxplorée de la mer avait la forme d'une bande large de 15 à 20 kilomètres, suivant la côte, de Bourgas à Achtopol. Cete région a été couverte de 422 stations d'observation, mesurées ordinairement trois fois de suite. La détermination des coordonnées des points d'observation et la direction du méridien astronomique ont été faites par des methodes géodésiques en employant comme points de repaire les pyramides construites dans la lande à l'usage du service géodésique du pays et dont les coordonnées rectangulaires dans le plan de projection *Gauss-Krüger*, adopté par le service géodésique du pays, nous ont été communiquées par l'Institut Géographique au Ministère de la Guerre.

Dans le tableau de la page 141 nous reproduisons les moyens des mesures dans chacun des 422 points d'observation ainsi que les écarts des mesures individuelles du moyen adopté. Tous ces moyens sont réduits à l'époque *Juin 1938*. Pourtant dans les régions d'anomalies considérables, le bateau ayant été emporté par les courants et les vents, nous avons préferé de donner les mesures individuelles sans former les moyens.

Les isogonnes, traduisant les résultats de nos mesures, réduits à l'époque Juin 1938 ont été portés sur la carte marinne. Les isogonnes des régions fortement perturbées autour de l'île Ste Anastasie ont été portées sur une carte à part. Les resultats des mesures dans la baie du Diable, où nous avons constaté des grandes anomalies, sans avoire le temps pour faire un nombre suffisant de mesures, permetant à désigner la carte des isogonnes, sont simplement indiqués sur la figure 8.

Figure 49. Publication about measurements on the Black Sea shelf. The title reads: "About the magnet's declination along the Black Sea's coast from Bourgas to Achtopol".

Answer (Ivan Butchvarov): The measurements on the Bulgarian Black Sea shelf were carried out during 1937, 1938 and 1939. The iron (magnetic) ship was used. The measured profiles were perpendicular to the coast and measured from the sea towards it. The coordinates and the astronomical azimuth were determined by geodetic way and the magnetic azimuth – by a marine compass. 422 points were measured. The precision of a single measurement of the declination is 0.5'. On Figure 49 a summary of the publication of the staff who made the measurements.

Question (Spomenko J. Mihajlovic): Your experience about digital variometer "BOBROV" - built by M. Beblo?

Did you try to do digitalization of analog magnetograms semi-automatically?

Answer (Ivan Butchvarov): The data received by the "BOBROV" variometers digitalized by Dr. Beblo are relative. In this sense they are inconvenient for receiving absolute values of the geomagnetic field elements. I think that they can be used mainly for illustrating the behavior of the geomagnetic field element variations qualitatively.

We didn't try to digitalize the analog magnetograms semi-automatically.

Question (Angelo de Santis): Regarding your presentation I have a comment and a question.

In your analysis, to reject possible outliers, you applied what you call the "sifting out" method, which looks automatic in its application. We also apply a similar criterion in our analysis of the repeat station data. However we found that sometimes this method can affect any possible model of secular variation you want to find from your data. As example, please consider the case when a repeat station is slightly anomalous in the magnetic field but very good in secular variation. Applying the rejection criterion you could reject the data of this station in two epochs, so loosing an important information about the correct secular variation in that region.

I saw that you have a good series of data from your observatory at Panagiuriste. Do you estimate and publish K-index too?

Answer (Ivan Butchvarov): Indeed, the "sifting out" method is not convenient for receiving quantitative information especially applied over a small number of measured points (for example over the repeat stations). We used it over about 500 measured points and we removed about 20 % of it only with the purpose to obtain a qualitative picture, i.e. to determine approximately the regions in Bulgaria where the geomagnetic field is anomalous.

About the K-indices of the Panagjuriste Geomagnetic observatory, yes we determine it. It is published in the hard copies of the Geomagnetic Yearbook of the Observatory from 1956 till 1983. The hard copies of the Geomagnetic Yearbook are available in the WDC's and in many others institutions all over the world including your Istituto Nazionale di Geofisica, Roma, Osservatorio Geofisico l'Aquila and Istituto Geofisico Geodetico, Genova. For the period 1984 – 2004 the K-indices are in our database and they are available from the Panagjuriste Observatory or directly from me.

# ON THE MODELLING OF A GEOMAGNETIC REFERENCE FIELD FOR THE BALKAN REGION

BEJO DUKA[6]
*Department of Physics, Faculty of Natural Sciences, University of Tirana*

ANGELO DE SANTIS

LUIS R. GAYA-PIQUÉ
*Istituto Nazionale di Geofisica e Vulcanologia*

**Abstract.** This paper presents the situation of the geomagnetic measurements over Albanian territory, distinguishing the campaigns of 1942.0, 1961.0, 1990.4, 1994.75 and 2004.7 epochs. The results of the latest model of the geomagnetic field for Southern Italy and Albanian territory are therefore shown. This model was created using SCHA (Spherical Cap Harmonic Analysis) technique, applied on all ground data from three different epochs (1990.4, 1994.5 and 2003.7) and a selected dataset from Oersted satellite mission. The ground data were confined in a small Spherical Cap that we expect to enlarge with ground data from adjoining countries in order to compose an accurate Balkan Geomagnetic Reference Field Model.

**Keywords:** geomagnetic field; geomagnetic observatory; Geomagnetic Reference Field; Spherical Cap Harmonic Analysis; repeat stations.

## 1. The geomagnetic field measurements over Albania

### 1.1. 1942.0 EPOCH

We have no information about old geomagnetic field measurements over Albania. The earliest geomagnetic observations we know of are those from 1942, when a German team carried out absolute measurements of

[6] To whom correspondence should be addressed at: Fakulteti i Shkencave Natyrore, Bulevardi Zog I, Tirane, Albania; e-mail address:bduka@fshn.edu.a



*J.L. Rasson and T. Delipetrov (eds.), Geomagnetics for Aeronautical Safety*, 83–95.

declination $D$, horizontal component $H$ and inclination $I$ at 65 points in Albanian territory. Their results were inserted into the European Magnetic Declination Atlas for the 1944.5 epoch (Bock, 1948). We do not have the original values from their measurements, but the values were interpolated to the 1961 epoch according to the secular variation given by M. Rossinger (Bolz, 1963) using the Niemegk Observatory as reference point. Inverting this transformation, the observed values were recovered and the Normal Geomagnetic Field for the 1942 epoch was calculated (Duka and Bushati, 1991):

$$D_{1942..0} = -1.596^{\circ} + 0.038^{\circ} \cdot \Delta\varphi + 0.327 \cdot \Delta\lambda \quad (1)$$

$$H_{1942..0} = 24064nT - 537.3 \cdot \Delta\varphi + 53.1 \cdot \Delta\lambda \quad (2)$$

$$Z_{1942..0} = 37281nT + 670.7 \cdot \Delta\varphi + 94.5 \cdot \Delta\lambda \quad (3)$$

Where $\Delta\varphi$ and $\Delta\lambda$ (in degrees) are respectively the geographic latitude and longitude deviation from the reference point "Tirana 1" ($\varphi$ = 41.374°, $\lambda$ = 19.876°).

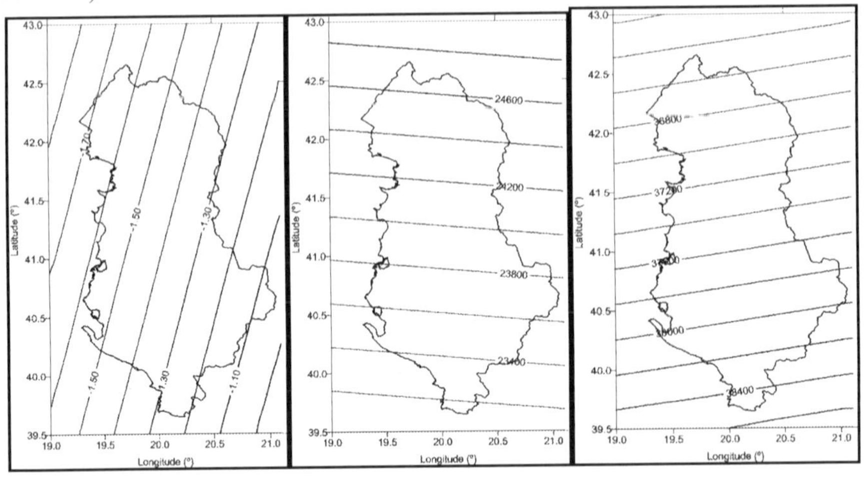


Figure 50. Contour maps of NGF (epoch 1942.0), respectively: a) Declination D, b) Horizontal Component H, c) Vertical Component Z.

The normal geomagnetic field (NGF) was calculated as a polynomial in $\Delta\varphi$ and $\Delta\lambda$. We were content with a first order polynomial, since Albania is a small country, and with such little data the values of the higher order polynomial coefficients can be smaller than their errors. In Figure 50, the isolines of these components of NGF for this epoch are shown.

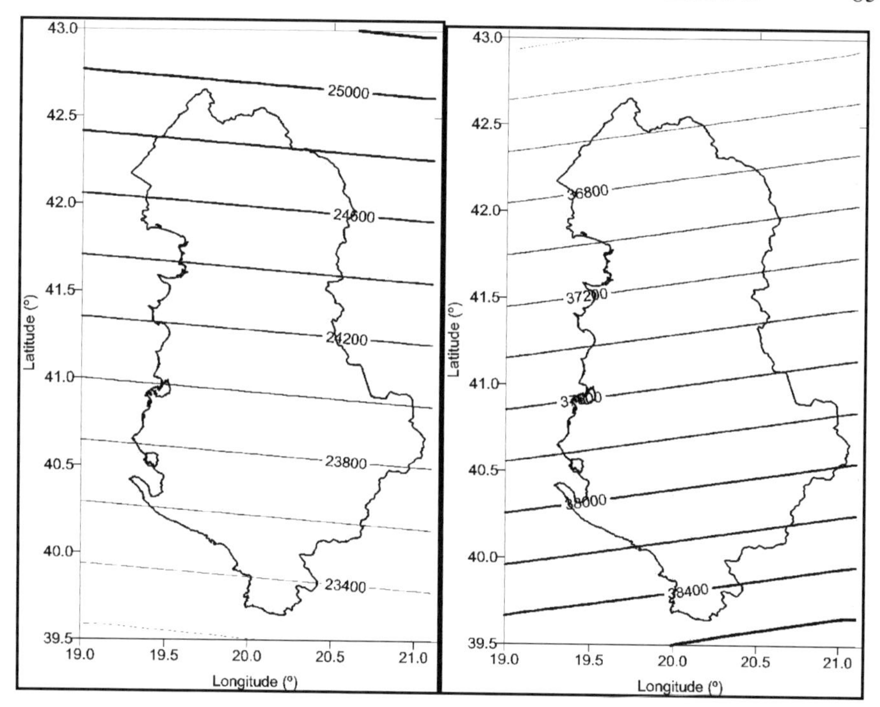


Figure 51. Contour maps of NGF (epoch 1960.0), respectively for: a) Horizontal Component H, b) Vertical Component Z.

## 1.2. THE 1961.0 EPOCH

During 1961, an expedition from the Potsdam Geomagnetic Institute performed geomagnetic surveys over a limited area of Albania (absolute measurements of *H* and *Z* components at 13 points were made). In order to enlarge the measurement area, we also used Bolz's field values (Bolz, 1963) reduced from 1942 to 1961 and the following NGF was found:

$$H_{1961..0} = 24248.71nT - 565.86 \cdot \Delta\varphi + 42.23 \cdot \Delta\lambda \tag{4}$$

$$Z_{1961..0} = 38041.84nT + 650.87 \cdot \Delta\varphi + 121.87 \cdot \Delta\lambda \tag{5}$$

The NGF at this epoch has less accuracy than for other epochs; the respective contour maps of this epoch are shown in Figure 51.

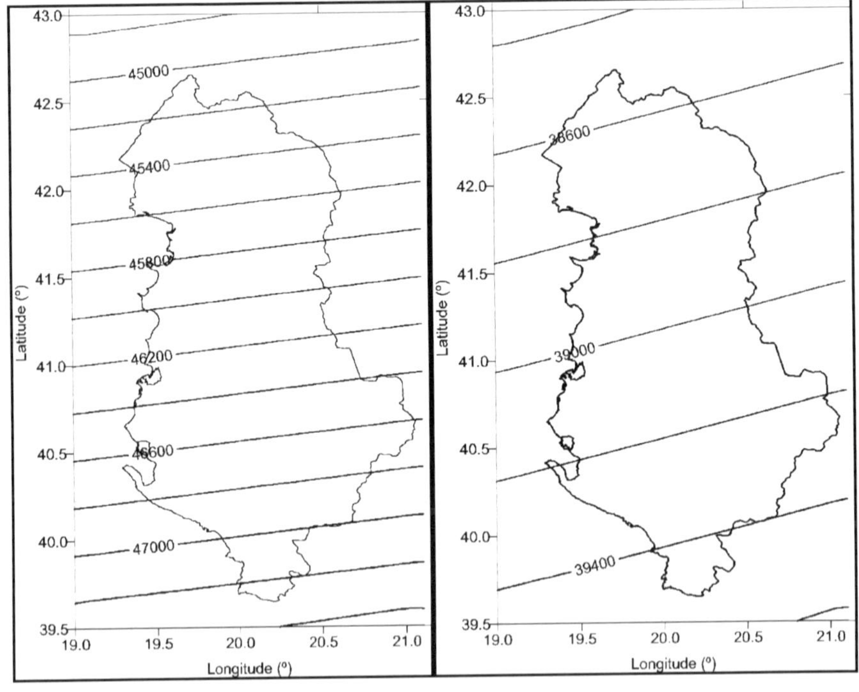


Figure 52. Contour maps of NGF (epoch 1990.4), respectively for: a) Total Field F, b) Vertical Component Z.

## 1.3. THE 1990.4 EPOCH

Trying to setup an Albanian Geomagnetic Network, a group from Tirana Geophysics Enterprise performed geomagnetic surveys throughout Albanian territory (34 points) at the 1990.4 epoch. The group measured the relative vertical component (referring to "Tirana 1" station) with a Flux-Gate magnetometer (1 nT accuracy) and the absolute value of this component at Tirana 1 station. They also measured, at all the points, the absolute value of the total field *F* with an MP-2 magnetometer (1 nT accuracy). Using all these data, the NGF for the 1990.4 epoch was calculated (Duka and Bushati, 1991):

$$F_{1990..4} = 45989.72nT + 321.79 \cdot \Delta\varphi + 77.57 \cdot \Delta\lambda \tag{6}$$

$$Z_{1990..4} = 38925.94nT + 738.42 \cdot \Delta\varphi + 78.56 \cdot \Delta\lambda \tag{7}$$

Isolines of NGF of this epoch are shown in Figure 52.

## 1.4. THE 1994.75 EPOCH

The first complete three-component geomagnetic survey covering all of Albania was carried out between September and October 1994 by an Italian team from the Istituto Nazionale di Geofisica (Roma, Italy) in collaboration with the Institute for Geochemistry and Geophysics of Tirana and the Department of Physics of Tirana University. They measured absolute values of *I*, *D*, and *F* elements at 10 points using a proton precession magnetometer for *F*, a DI fluxgate theodolite for inclination and declination, and a gyroscope theodolite for geographic north determination. In order to reduce the data to the 1994.75 epoch, data from the Geomagnetic Observatory of L'Aquila (Italy), which is 500 km from Tirana, was used. In the following equations, the NGF for the three elements *F, I, D*, are shown. The latitude and longitude of the NGF of the 1994.75 epoch represented here are 41°N, 20°E (see: Chiappini *et al.*, 1997):

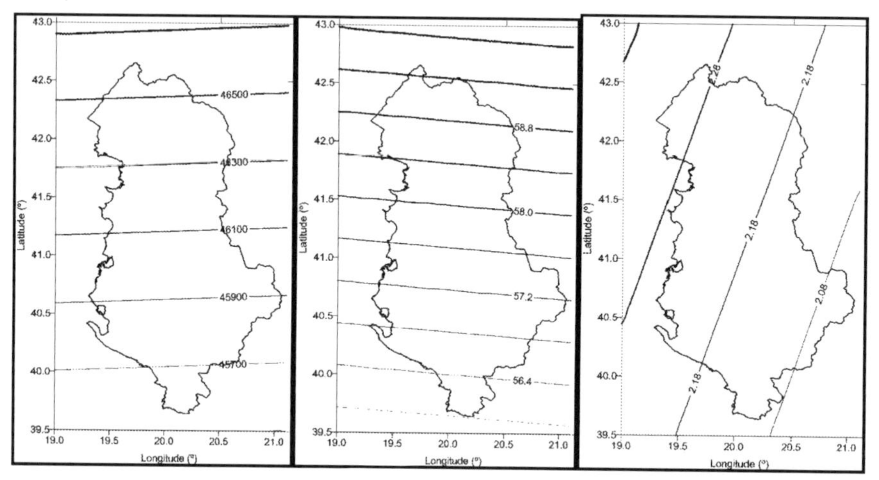


Figure 53. Contour maps of NGF (epoch 1994.75), respectively for: a) Total Field F, b) Inclination I, c) Declination D.

$$F_{1994.75} = 46028.8\,nT + 5.757\cdot\Delta\varphi - 0.215\cdot\Delta\lambda \tag{8}$$

$$I_{1994.75} = 57.489° + 0.0184\cdot\Delta\varphi - 0.001\cdot\Delta\lambda \tag{9}$$

$$D_{1994.75} = 2.1848° + 0.00075\cdot\Delta\varphi - 2.016\cdot\Delta\lambda \tag{10}$$

Isolines of NGF of this epoch are shown in Figure 53.

1.5. THE 2004.7 EPOCH

The same institute (now named Istituto Nazionale di Geofisica e Vulcanologia) that built up the first Albanian repeat station network (October 1994), carried out in September 2004 (in collaboration with Albanian colleagues from the Center for Geophysics of Tirana, the Department of Physics, and the Academy of Sciences) the second campaign of absolute measurements at all the 1994 points, adding a new repeat station in Berat. They measured the same elements *D, I,* and *F* in all 11 points and reduced the data using the automatic registration (every 30'') of the diurnal variation of these elements at a fixed point (near the residence of the Academy of Sciences of Tirana) not far from Tirana 1 station. We present here the NGF of the measured elements and their respective isolines, from Tirana 1 station:

$$F_{2004.7} = 46446.9\ nT + 352.95 \cdot \Delta\varphi - 9.847 \cdot \Delta\lambda \qquad (11)$$

$$I_{2004.7} = 58.037^\circ + 1.09^\circ \cdot \Delta\varphi + 0.0626^\circ \cdot \Delta\lambda \qquad (12)$$

$$D_{2004.7} = 2.991^\circ + 0.0251^\circ \cdot \Delta\varphi - 0.082^\circ \cdot \Delta\lambda \qquad (13)$$

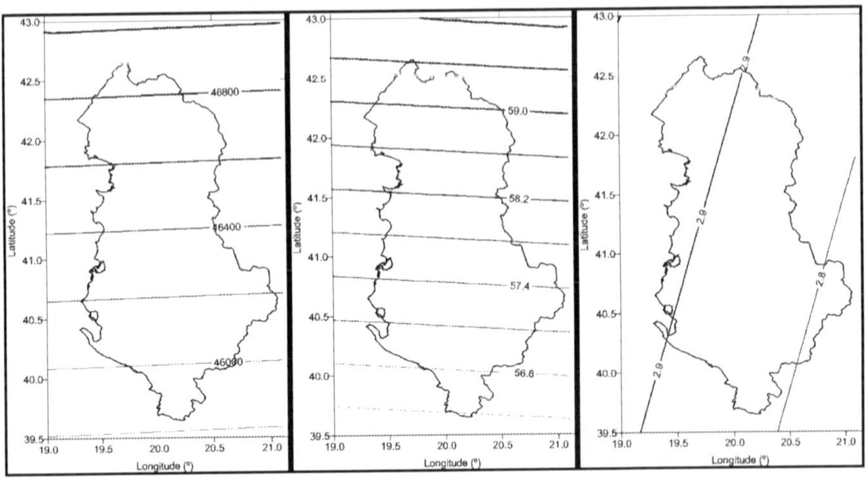


Figure 54. Contour maps of NGF (epoch 2004.7), respectively for: a) Total Field F, b) Inclination I, c) Declination D.

## 2. A geomagnetic reference model for Albania, Southern Italy, and the Ionian Sea from 1990 to 2005

Using the measurements from the Albanian and Italian magnetic repeat station networks since 1990, as well as from a selected set of Ørsted satellite total field measurements, a geomagnetic reference model for Albania, Southern Italy, and the Ionian Sea, was recently published (Duka *et al.*, 2004). Scalar values from 1990.0 and 2003.6 from 31 and 8 Albanian stations, respectively, together with vector and scalar measurements reduced to epoch 1995.0 from a total of ten Albanian repeat stations of epoch 1994.75 were used to develop the reference model. In order to define a model not only for Albanian territory but also for the Ionian Sea and the southeastern part of Italy, measurements at seven locations from the Italian magnetic repeat station network (Coticchia *et al.*, 2001) at epochs 1990.0., 1995.0, and 2000.0 were also taken into account.

A selected subset of total intensity magnetic field measurements from the Ørsted satellite was also included. The inclusion contributes to a homogeneous coverage of the studied region, especially in the sea areas. Moreover, the satellite data act as a boundary to develop a three-dimensional model, valid not only at sea level but also at any distance between the surface and the satellite height. A total of 30 scalar values measured between 1999.5 and 2002.5 were selected according to procedures to reduce the presence of external magnetic fields in the data. The data chosen are distributed in a height range between 650 and 850 km above the Earth's surface.

The reference model was obtained applying Spherical Cap Harmonic Analysis (SCHA; see Haines, 1985). This choice represents an improvement with respect to the previously presented polynomial models for the region (see: Chiappini *et al.*, 1997), due to the fact that an SCHA model allows computation of the field component values through expressions that satisfy Laplace's equation. Moreover, the radial variation of the magnetic field is implicitly described, without need to assume a dipolar continuation of the field as with polynomial models. Although a similar approach was used by Chiappini *et al.* (1999), their model was valid for a fixed epoch only, without secular variation modeling.

The solution of Laplace's equation for the magnetic potential due to internal sources over a spherical cap in spherical coordinates ($r$, $\theta$, $\lambda$) can be written as an expansion of non integer spherical harmonics (Haines and Torta, 1994):

$$V = a\sum_{k=0}^{K}\sum_{m=0}^{k}\left(\frac{a}{r}\right)^{n_k(m)+1} P^m_{n_k(m)}(\cos\vartheta)\cdot\sum_{q=0}^{Q}\left[g^m_{k,q}\cos(m\lambda)+h^m_{k,q}\sin(m\lambda)\right]\cdot t^q \tag{14}$$

where the polynomial time dependency is included. The spherical cap harmonic coefficients $g^m{}_{k,q}$ and $h^m{}_{k,q}$ are those which determine the model. The number of coefficients depends on the maximum spatial and temporal indices of the expansion, $K$ and $Q$ respectively. The associated Legendre functions $P_{n_k(m)}(\cos\theta)$ that satisfy the boundary conditions (a zero of the potential or its derivative with respect to colatitude at the border of the cap; Haines, 1985) have integer order $m$ but a generally non-integer degree $n_k(m)$, where $k$ is the index used to order the different roots for a given order $m$. Legendre functions were computed using a procedure proposed by Olver and Smith (1983), since it seems to provide more reliable values than the original approach suggested by (Haines,1985) when the cap is rather small (Thébault *et al.*, 2004). The magnetic components are obtained as gradient components of the potential (14) in spherical coordinates. The fact that vector measurements are combined with total field measurements introduces a nonlinearity in the equations involved to obtain the coefficients of the model. To avoid this problem, a first order Taylor expansion of the total magnetic field intensity, as a square root function of the $X$, $Y$, and $Z$ components, was used.

After many tests, the parameters that defined the best model in terms of fit to the input data and spatial and temporal behavior corresponded to $K$=2 and $Q$=1 ($q$ = 0, 1) covering the period between 1990.0 and 2005.0. The coefficients were obtained through a least squares regression procedure. The cap over which the model was originally defined had a semi angle of 3°. Nevertheless, some problems arose because such small cap was considered, as typically happens when SCHA is applied to small caps. To avoid these problems, the cap was enlarged up to 8° half angle, in order to cover the most significant harmonics. The final model has a total of 18 coefficients (Table 7).

Table 7. Coefficients of the Geomagnetic Reference Model developed by using SCHA.

| $k$ | $m$ | $n_k(m)$ | $g^m_{k,0}$ | $h^m_{k,0}$ | $g^m_{k,1}$ | $h^m_{k,1}$ |
|---|---|---|---|---|---|---|
| 0 | 0 | 0.0000 | 45.735 | | -219.959 | |
| 1 | 0 | 16.7209 | 6.221 | | 7.429 | |
| 1 | 1 | 12.7139 | -.564 | 16.519 | 5.745 | -40.357 |
| 2 | 0 | 26.9471 | -2.824 | | -3.997 | |
| 2 | 1 | 26.9471 | -1.448 | -2.800 | -1.972 | 8.893 |
| 2 | 2 | 21.4163 | -5.127 | -4.654 | -.315 | .257 |

The RMS between the observed and modeled field values for $X$, $Y$. $Z$, $F$ components were respectively 26.8 nT, 53.5 nT, 32.7 nT, and 36.4 nT for ground data, and 4.3 nT for satellite data (only $F$). In Figure 55 and Figure 56[7], the contour lines of the same values for different components and different epochs are shown.

[7] The Figure 55, Figure 56 and Figure 57 are reproduced from (Duka *et al.*, 2004)

The secular variation for all the magnetic elements for epoch 1995.0 (Figure 57), obtained from the SCHA model as the field differences between epochs 1995.5 and 1994.5, confirms that the region under study has low values for the temporal variation of the geomagnetic field for this period.

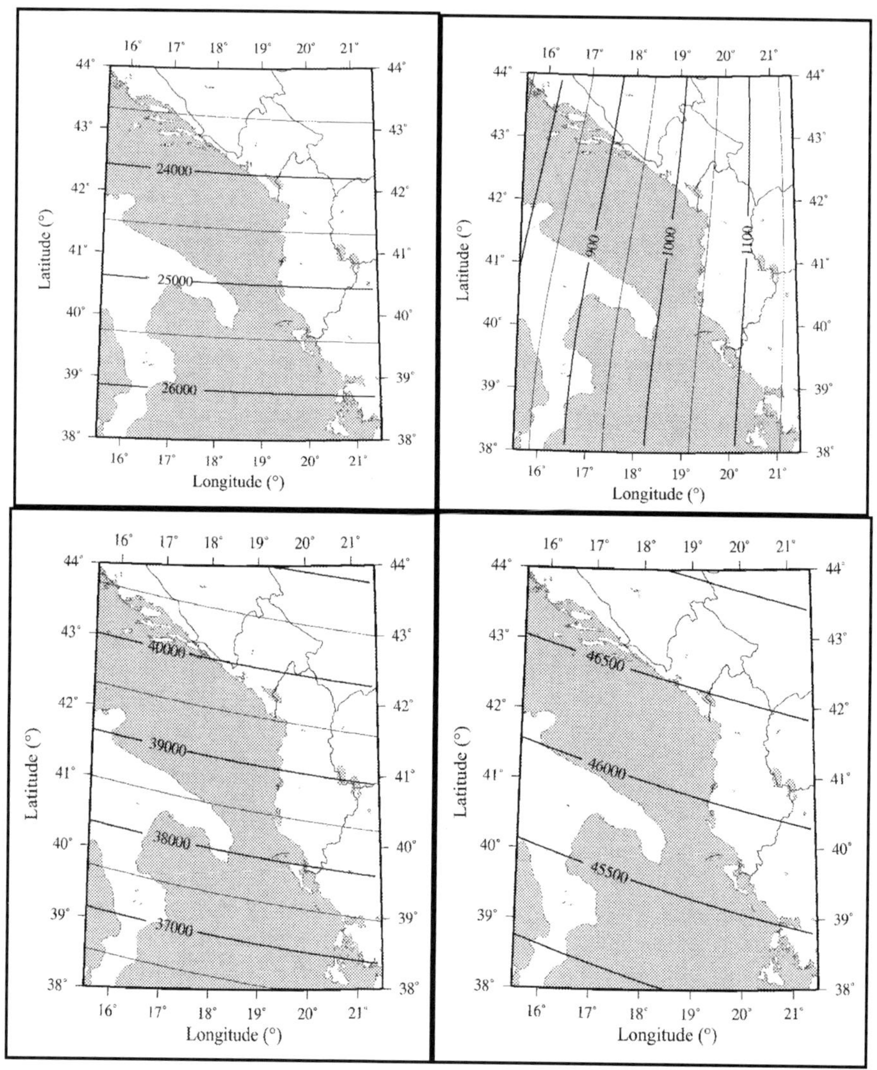


Figure 55. Contour Maps (epoch 2000.0) in nT, respectively for: a) X Component, b)Y Component, c) Z Component, and d) Total field F at sea level obtained from anSCHA model developed on a 8° semi angle cap.

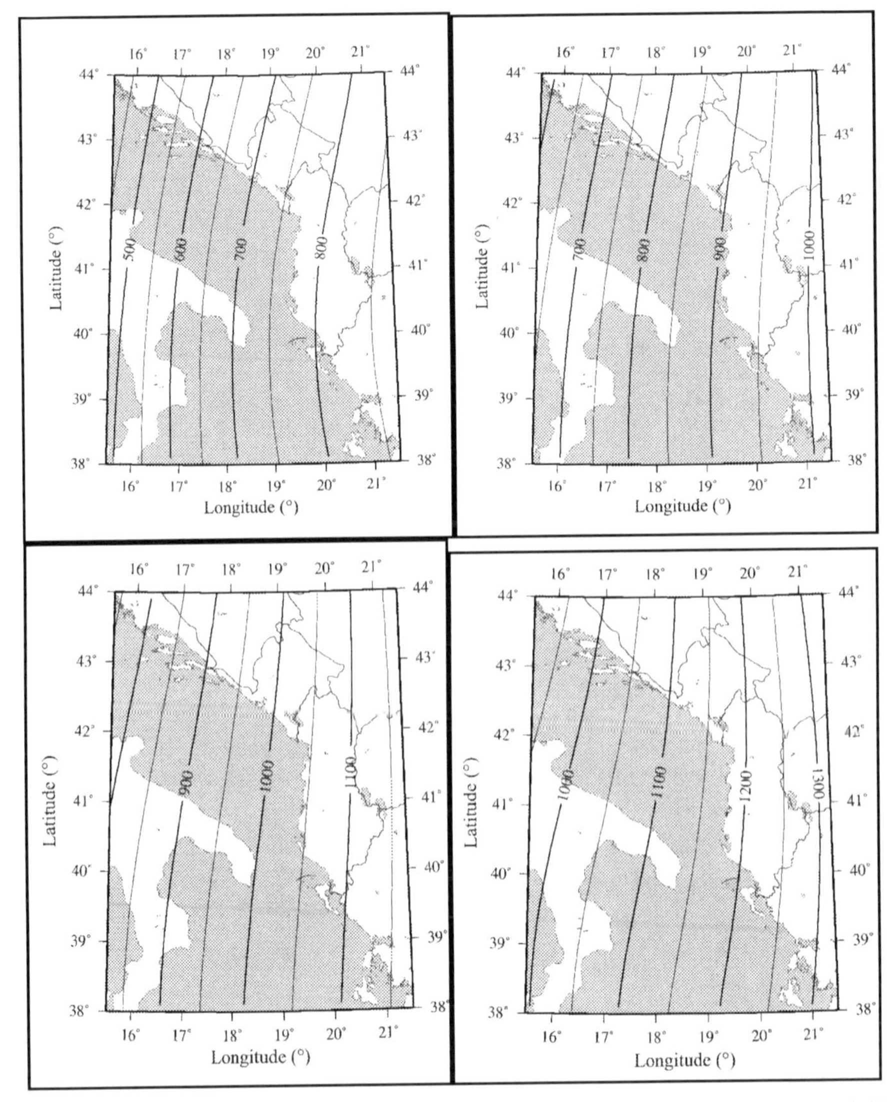


Figure 56. Contour Maps in nT, respectively for: a) Y Component (epoch 1990.0), b)Y Component (epoch 1995.0) , c)Y Component (epoch 2000.0) , and d) Y component (epoch 2005) at sea level obtained from an SCHA model developed on a 8° semi angle cap.

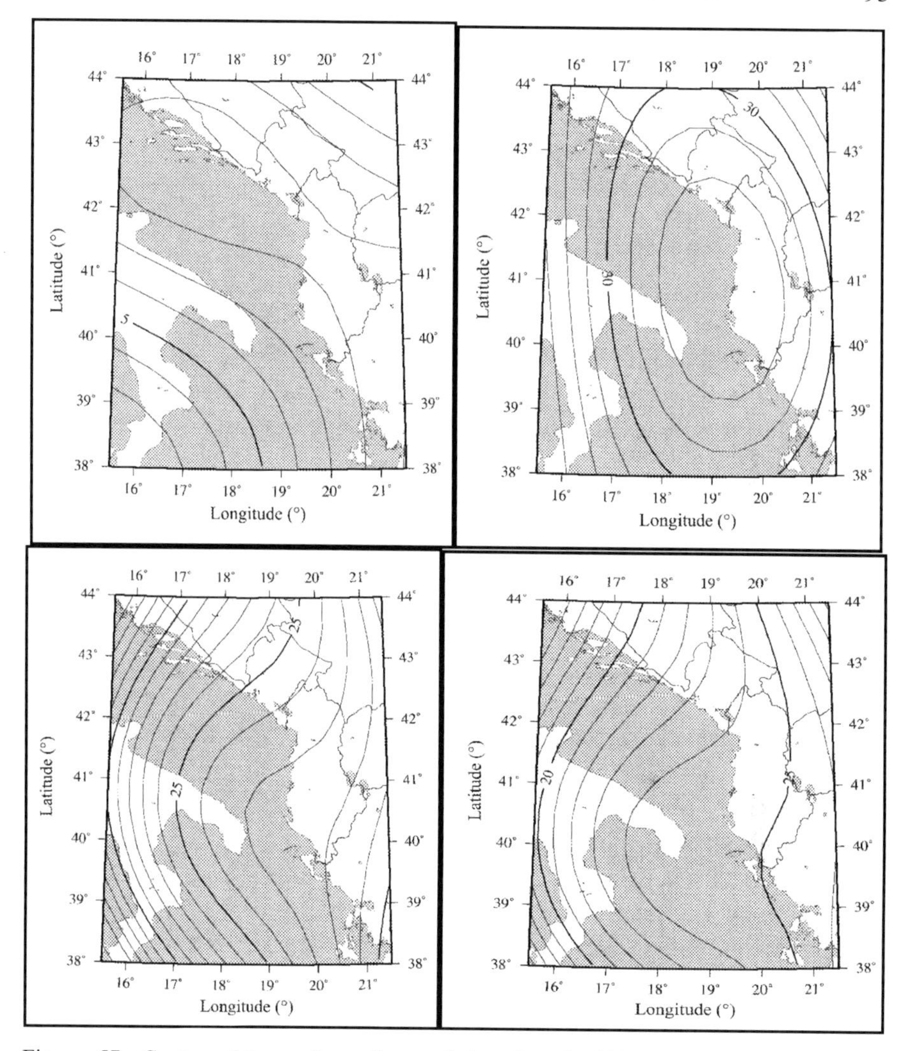


Figure 57. Contour Maps of secular variation (epoch 1995.0) , respectively: a) X Component, b) Y Component , c) Z Component and d) F total field at sea level obtained from SCHA mode. Contour lines are at 1 nT/year interval.

## 3. On the possibility of a reference geomagnetic field model for the Balkan region

For economical and political reasons, the Balkan countries did not collaborate on geomagnetic measurements and geomagnetic studies in the past. The situation is still worse than in other parts of Europe. The NATO workshop "New data for the magnetic field in the Republic of Macedonia for enhanced flying and airport safety" represents the starting point for good collaboration between Balkan countries. Countries need this collaboration not only to connect their national repeat station networks, but also to build up a common Reference Geomagnetic Field Model for Balkan countries.

There are several methods for regional geomagnetic modelling, like Polynomial Approximation, Rectangular Harmonic Analysis, Spherical Cap Harmonic Analysis, etc. Several studies (see for example: De Santis *et al.*, 2003; Duka, 1998) demonstrate the advantages of the latest method (SCHA) especially for a region like the Balkan Peninsula. For that reason we would suggest that the geomagnetism representatives from Balkan countries collaborate to build up a Regional Geomagnetic Field Model using the SCHA technique. This technique is well standardized, the algorithm is well known, and the respective programs run well on every computer.

The success of such an endeavor depends mainly on the amount and quality of the data available. It is very important to have as much data as possible from the different countries and epochs. These data should have good accuracy and should be well distributed over the area of study. It is also important to have data in a standard format. In order to enrich the database, especially for the area not covered by ground measurements, satellite data over the Balkan area would be a great help.

In previous sections, we have presented Albanian magnetic measurements and modeling data. As far as we know, the situation is better in neighboring countries like Greece, Bulgaria, ex-Yugoslavia, Romania and Turkey, where geomagnetic observatories are present: Pendeli, (PEG, Greece), Grocka (GCK, Serbia), Panagyurishte (PAG, Bulgaria), Surlari (SUA, Romania), Istambul (ISK, Turkey). Apart from the data that these observatories supply to the Geomagnetic World Data Center and that can be accessed through Internet, we are expecting to receive from these countries other data from their repeat station networks, especially their latest measurements.

Publishing all these data on the Internet would allow every geomagnetic scientist from the Balkan countries to have access to the data, and would allow them to compare the models made by different groups so that they could select the best geomagnetic reference model for the Balkans for a given epoch or from an interval of epochs. Having reference models for the geomagnetic field at different epochs, one can study the secular variation of the geomagnetic field at a regional scale. Regional studies would improve the global knowledge on the fluid flow of outer Core and Core-Mantle boundary. The Reference Geomagnetic Field Model would also be useful in reducing the number of local magnetic surveys carried out in the region for geological exploits or for aeronautic studies.

## References

Bock, R. (1948): Atlas of magnetic declination of Europe for epoch 1944.5, Washington, *Army Map Service.*

Bolz, H., (1963): Ergebnisse geomagntetischer Feldmessungen in Albanien, *Gerlands Beiträge zur Geophysik*, 72, 266-300.

Chiappini, M., O. Battelli, S. Bushati, G. Dominici, B. Duka and A. Meloni (1997): The Albanian geomagnetic repeat station network at 1994.75, *J. Geomagn.Geoelectr.*, 49, 701-708.

Chiappini, M.,A. De Santis, G. Dominici and J.M. Torta (1999): A normal reference field for the Ionian Sea area, *Phys. Chem. Earth A*, 24 (5), 433-438.

Coticchia, A., A. De Santis, A. Di Ponzio, G. Dominici, A. Meloni, M. Pierozzi and M. Sperti (2001): Italian magnetic network and geomagnetic field maps of Italy at year 2000.0, *Boll. Geod. Sci. Affini*, 4, 261-291.

De Santis, A., L.R. Gaya-Piqué, G. Dominici, A. Meloni, J.M. Torta and R. Tozzi (2003): Italian Geomagnetic Reference Field (ITGRF): update for 2000 and secular variation model up to 2005 by autoregressive forecasting, *Ann. Geophysics*, 46 (3), 491-500.

Duka, B., Gaya-Piqué, L. R., De Santis, A., Bushati, S., Chiappini, M. and Dominici, G.: A geomagnetic reference model for Albania, Southern Italy and Ionian Sea from 1990 to 2005, *Annals of Geophysics*, Vol. 47, N. 5 (2004), 1609-1615.

Duka, B. and S. Bushati (1991): The normal geomagnetic field and the IGRF over Albania, *Boll. Geofis. Teor. Appl.*,XXXIII (130/131), 129-134.

Duka, B. (1998): Comparison of different methods of analysis of satellite geomagnetic anomalies over Italy, *Annali di Geofiscica*, Vol. 41, 49-61).

Haines, G.V. (1985): Spherical cap harmonic analysis, *J. Geophys. Res.*, 90 (B3), 2583-2591.

Haines, G.V. and J.M. Torta (1994): Determination of equivalent currents sources from spherical cap harmonic models of geomagnetic field variations, *Geophys. J. Int.*, 118, 499-514.

Haines, G.V. and L.R. Newitt (1997): The Canadian geomagnetic reference field 1995, *J. Geomagn. Geoelectr.*, 49, 317-336.

Olver, F.W.J. and J.M. Smith (1983): Associated Legendre functions on the cut, *J. Computat. Phys.*, 51, 502-518.

## DISCUSSION

Question (Spomenko J. Mihajlovic): Who made the measurements in Albania in 1961?
What is the minimal area for which SCHA method can be applied?
Answer (Bejo Duka): 1) There was a German expedition from the Potsdam Geomagnetic Institute that carried out geomagnetic measurements in Albania on 1961.
2) Theoretically, the minimum area for which the SCHA method can be applied is determined by the minimum of values of noninteger degree $n_k(m)$ of spherical cap harmonic expansion, where $k = 0, 1, 2 \ldots K_{max}$ and $m = 0, 1, \ldots k$. For a given $K_{max}$ , the smaller cap the greater is the maximum value of $n_k(m)$ and there is a lack of lower harmonics in the spherical cap harmonic expansion. In practice, the problem appears for the caps smaller than 4°, so we can say that the SCHA method can be used for all caps greater than 4°.

# REPEAT SURVEYS OF MACEDONIA

JEAN L. RASSON[8]
*Institut Royal Météorologique de Belgique*
MARJAN DELIPETROV
*Faculty of Mining and Geology*

**Abstract.** We present our geomagnetic repeat station work in the Republic of Macedonia performed during the years 2002, 2003 and 2004. A total of 15 stations were established. New stations were created as the localization data of the old stations were not available at the time. The paper will describe the measuring and localization equipment used, the observation techniques applied and the way the data were reduced to obtain the annual means of the components of the geomagnetic field vector over the area. As a result of this work, isogonic maps are available for the benefit of aircraft navigation with compass in Macedonian airspace.

**Keywords:** Title, Magnetic measurement, Declination, Repeat station, Geomagnetism

## 1. Short history

### 1.1. PAST REPEAT STATIONS

The website with the most extensive list of repeat station data is presently the page maintained by the British Geological Survey (BGS) with the URL: http://www.geomag.bgs.ac.uk/gifs/surveydata.html. Worldwide data from 1900 until present are available. Macedonian repeat station data are easily obtained from this database Data collected prior to 1900 are not relevant to this study.

Figure 58 shows data retrieved from the BGS site plotted on a map with latitude/longitude indications. The data can be grouped by period and country:

[8]To whom correspondence should be addressed at: IRM/CPG, Rue de Fagnolle, 2 Dourbes, B-5670 Viroinval, Belgium ; e-mail : jr@oma.be



*J.L. Rasson and T. Delipetrov (eds.), Geomagnetics for Aeronautical Safety*, 97–114.

- Measurements obtained in Macedonia between 1911 and 1918. (The oldest data on the map.)
- Measurements in Bulgaria between 1930 and 1931. More information on Bulgarian data can be found in the paper by Butchvarov and Cholakov in this Volume.
- Measurements obtained in Albania during 1942.
- Measurements obtained in Greece reduced to 1944.5.
- Measurements obtained in Albania between 1961 and 1995. More information on Albanian data is in the paper by Duka in this Volume

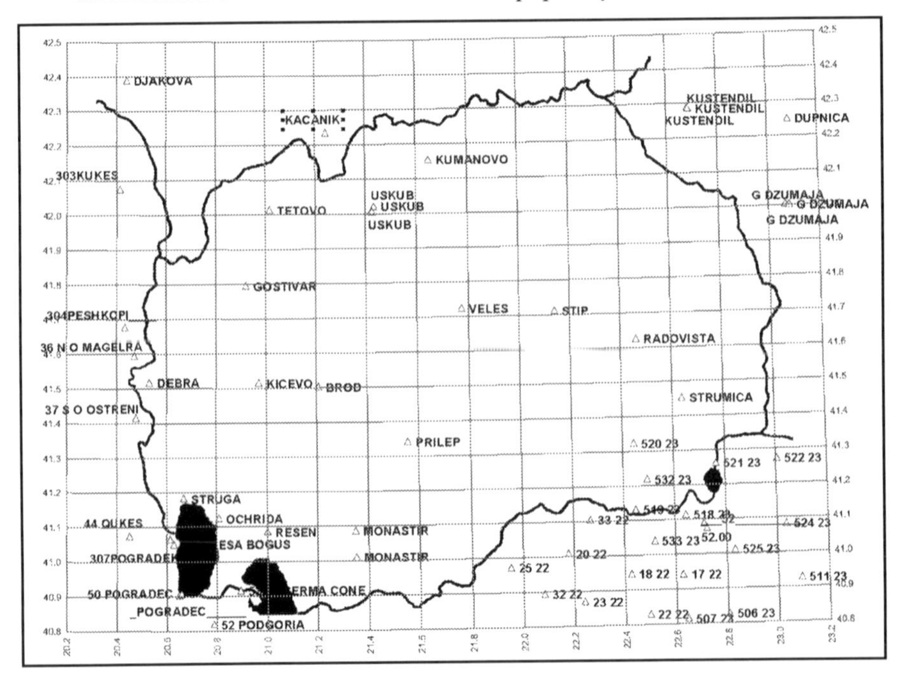


Figure 58. Map of the Republic of Macedonia and neighboring countries showing the magnetic repeat stations occupied in the past.

## 1.2. THE INITIATIVE OF THE YOUNG REPUBLIC FOR CREATING AN OBSERVATORY

Very early after its creation in 1992, the Macedonian Republic felt it was necessary to gain knowledge of the geomagnetic field in its territory. This responsibility was borne by the Geophysical Institute in Belgrade before 1992. Jean Rasson, the first author of this paper was contacted in the year 2000 to provide expertise, guidance, and help in a project entitled "Establishing a Geomagnetic Observatory in the Territory of the Republic of Macedonia". From then on, a collaboration started between Macedonia (represented by the Faculty of Mining and Geology, Štip) and Belgium (represented by the Royal Meteorological Institute, Centre de Physique du Globe, Dourbes) with the goal of establishing a magnetic observatory. Establishing a regular magnetic repeat survey of the country was also planned.

### 1.3. THE BILATERAL AGREEMENT WITH BELGIUM

Following the decision described in section 1.2, an approval process was initiated for signing an official agreement between Macedonia and Belgium. Due to delays caused by political events, the bilateral agreement was not approved and signed until the year 2002, and a first meeting between the parties was organized in April 2002 at the Dourbes magnetic observatory.

The agreement, valid for 3 years and renewable, set-out the tasks (work program), conditions, benefits, and obligations for both parties.

### 1.4. THE TEMPUS PROJECT (2003)

Fortunately, it was also possible to have a project approved under the EU TEMPUS Cards Education and Culture program. This was a joint undertaking entitled "Geomagnetic Measurements and Quality Standards" with the same parties as in section 1.3 but with the addition of the Austrian colleague, Dr. G. Duma from the Zentralanstalt für Meteorologie und Geodynamik (ZAMG) in Vienna, Austria. The duration of the project was set for 3 years.

This important project not only provided funding and involved a large number of Macedonian academics and students but provided:

- Education - of a core of Macedonian experts in geomagnetic measurements, magnetic pollution basics, and electro-smog detection
- Geomagnetic instrumentation – such as DIflux, magnetic variometers, data acquisition systems (for the purpose of training students) and other instruments for magnetic environment characterisation
- Funds - for international travel to train magnetic experts and for management visits.

## 2. Looking for a repeat network and establishing the stations

### 2.1. THE FIRST MEASUREMENT POINTS IN THE YEAR 2002

At the onset of our collaborations, we overlooked the database described in section 1.1 and the repeat station locations from the former Yugoslavia were not available. In absence of information about past repeat networks, our first repeat measurements network had to be planned from scratch. We focussed on an optimal geographical distribution of stations and on the necessity to perform magnetic tests in order to find a location for the future magnetic observatory. For the latter we had to take into account not only geological and magnetic considerations but also what sites were practical, available, and affordable. The three station network pictured on Figure 59, was established. The quality of the measurements was not always optimal, and the D measurement in Ohrid afterwards proved to be bad. Hence, the data collected in 2002 should be regarded as tentative at best.

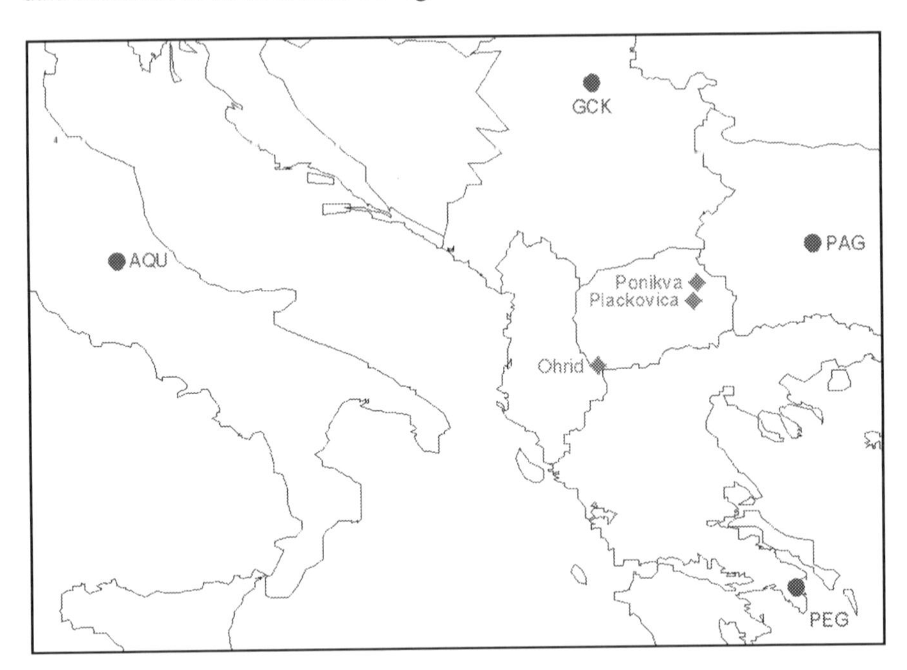


Figure 59. The first repeat stations of the Republic of Macedonia (diamonds), created in 2002. Also represented are nearby magnetic observatories (dots).

## 2.2. THE CAMPAIGN OF THE YEAR 2003

Building on the first 3 stations mentioned in section 2.1, we planned a second campaign for 2003 with the goal of covering the whole country with an even distribution of stations. It was felt that 15 stations would be adequate and affordable considering the available means and time. The resulting network of stations is shown in Figure 60.

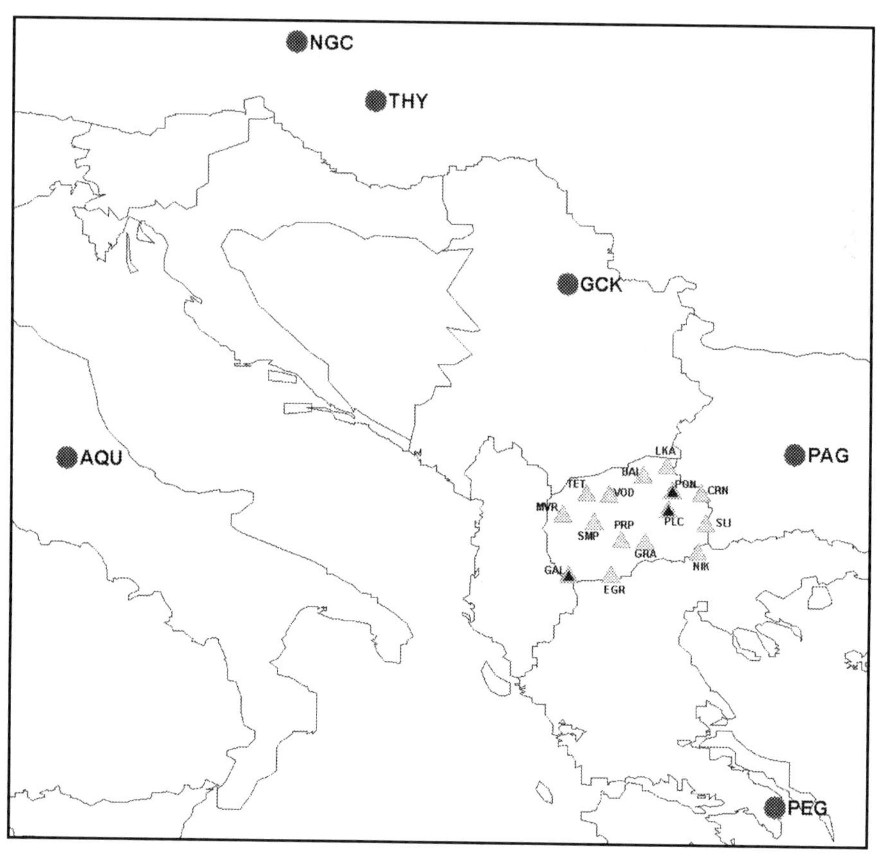


Figure 60. Map of the repeat station network created in 2003.

## 2.3. REOCCUPATION OF THE NETWORK IN 2004

The entire network mentioned in section 2.2 was reoccupied during August 2004. It was a short time span between the occupations but this allowed us to check the network and to strengthen the recently acquired skills of the Macedonian team.

## 3. Measuring the stations

### 3.1. LOGISTICS

The Faculty of Mining and Geology in Štip handled repeat station logistics. Transport between the stations and to the base locations was by car. Because the country of Macedonia is so small, its people know each other well. Consequently good contact persons were available throughout the country. The Faculty of Geology has alumni in every city in Macedonia, so we always found support and a warm welcome in the immediate vicinity of our repeat stations.

Base locations were in Štip, Kavadarci, and Ohrid. Typically we measured at one station in the morning and at another one in the afternoon. This was possible because of the short distances between stations. It was possible to measure the whole network of 15 stations in 3 weeks time, including 1/2 week for preparation and 1/2 week for preliminary data processing.

### 3.2. INSTRUMENTS

We used the instruments from the Dourbes Observatory fieldwork pool. The equipment were thoroughly checked and calibrated before leaving Dourbes. We had the following instrumentation list:

- Proton magnetometer G816. We favour this device because it is simple to use yet robust and accurate. The electrical supply is provided by D-cell batteries which are readily available.
- Zeiss 010 Diflux (Mingeo demagnetisation) with Pandect fluxgate sensor. This is our preferred DIflux because of its 1" accuracy and convenient telescope which allows easy and fast sunshots for azimuth determination.
- Zeiss tripod (non-magnetic)
- Adapter bracket to mount proton magnetometer sensor on tripod. This bracket is necessary to ensure that the total field measurement and the D and I measurements are made at the same location.
- Level for tripod rough levelling. This level comes in handy for positioning the tripod over the station mark with its bearing plate roughly levelled.
- Telescope with internal compass display (KVH DATASCOPE). This instrument facilitates finding targets when setting up the stations or

when revisiting them. The KVH is convenient, fast, and provides approximate magnetic azimuths. See Figure 61.

- Coudés (90° eyepieces) for Zeiss 010 Telescope and Microscope (non-magnetic). Coudés are necessary to sight elevated targets, like the Sun at midday. See Figure 61. If coudés are not available (they are hard to find nowadays) then sighting of the Sun can be done in the morning after sunrise or in the late afternoon before sunset.
- Solar filter for sunshots. See Figure 61. This small optical device is needed to make sunshots with the Zeiss 010 theodolite. An experienced observer will take about the same amount of time for sighting the Sun as for sighting the target. Therefore, this method is much cheaper and faster than other gyro or GPS-based methods, and have similar accuracy - provided the Sun is visible.

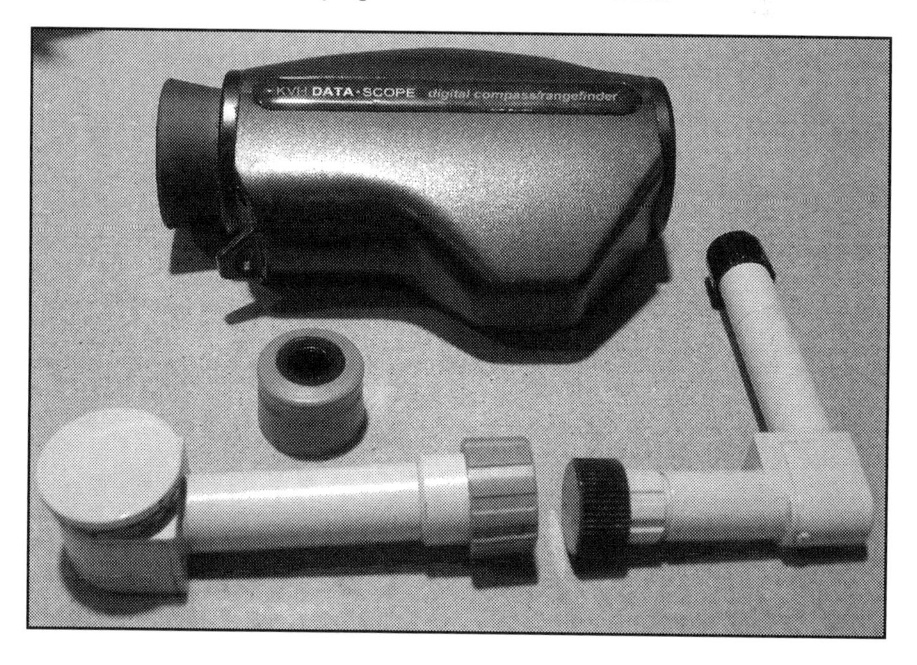


Figure 61. The KVH DATASCOPE Telescope with internal magnetic compass displays the magnetic azimuth of the telescope's optical axis. Also shown are the coudé eyepieces and the solar filter. Essential components for successful sunshots.

- FLM3/A fluxgate electronics with GPS receiver and battery. This device, looks like a small suitcase (Figure 62). It was designed and built by Dourbes instrumentation lab and has 4 functions:
    - Holds the fluxgate electronics and digital readout.
    - It contains a GPS receiver giving UTC time, lat/long, and elevation information.

- It has a battery for powering the above mentioned devices.
- It has a carrying case for the ZEISS 010 DIflux. The FLM3/A can be operated closed with the readouts and switches accessible. The cover provides waterproof protection for the battery and electronic circuits during measurements.

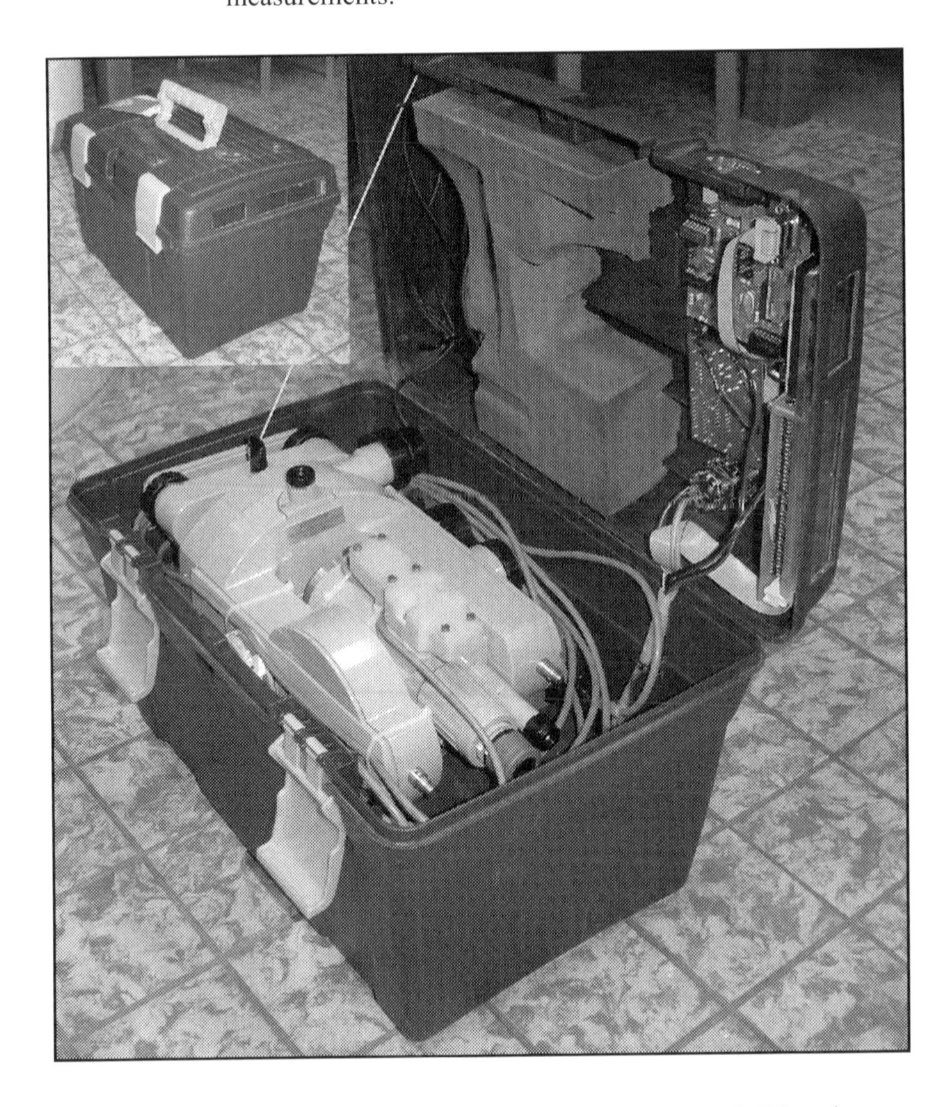

Figure 62. The ZEISS 010B DIflux with FLM3/A fluxgate electronics and GPS receiver.

- Aluminium Markers. Permanent markers should be set into the ground at each repeat station. The line between the permanent marker and the azimuth mark will have a known azimuth after the sunshot measurements have been calculated. Markers should be nonmagnetic. They should be easy to install but difficult to remove. They should be easily found in overgrown vegetation, yet should not attract too much attention. Finally, they should be durable but not made of expensive materials. The best markers are ground level pillars made of nonmagnetic stone or concrete with engravings on the emergent surface. Unfortunately these monuments are cumbersome and difficult to install. For our network we used 25 mm diameter, aluminium rods which were driven into the ground with a hammer and painted blue. The problem with using these was that people removed them to get the valuable aluminium.

- Sledge-hammer. This is a necessary tool for driving markers into the ground. In Macedonia the ground is not soft and is often stony. We used a 10 kg hammer successfully, but we had to be careful not to leave it in the vicinity when measuring the magnetic field!

### 3.3. OBSERVATION TECHNIQUES

Based upon past experience in measuring at repeat stations, we have refined our procedures for the Macedonian network.

- With only 2 weeks to make measurements, we wanted to streamline our operations without compromising accuracy. Therefore we opted for geodetic azimuth measurements by sunshots. (Macedonia is sunny in the summer so we expected the clear skies required for sunshots.) In fact we were only twice deprived of them. Also, we limited the magnetic measurements to a maximum of 4 independent sessions, unless a problem developed during the session at the station and we had to do more.
- Measurements were done in early morning and/or late afternoon, because these times favour elimination of the daily variation differential between the station and observatory used for the reduction. Also, accuracy of sunshot measurements are improved since theodolite levelling errors are reduced and sunshots can be made without coudés (90° eyepieces) since the sun is low in the sky.
- A preliminary proton magnetometer survey was performed before occupying each station to check the magnitude of the horizontal and vertical gradients to rule out magnetic pollution at the site.
- The standard DIflux 12 step session (protocol): 2 x target, 4 x D, 2 x targets, 4 x I was used. This protocol is very strict and eliminates almost all DIflux and theodolite dimensional and mechanical errors.

- We selected distant (>5 km) targets. Far away targets provide high accuracy orientation in space for declination measurements. The only drawback to them is that they are difficult to observe in hazy weather; therefore, an additional, closer target should be established as a back-up.

An example of a sunshot measurement session follows for the target at the repeat station in Ponikva. The target is the centre of the left tower of the right building of a Bulgarian border post at 35° magnetic azimuth. The following text must be entered in the basic program for processing sunshot data (Rasson 2005). Note that the Zeiss 010A DIflux has a circle graduation in grads. A total of 6 independent sunshots sets with circle left/right are given so that a standard deviation can be calculated.
Time of the shot (hh.mmssd): 06.0950
Horizontal circle reading of sunshot (grades): 176.5044
Horizontal circle reading of target sighting (grades): 108.7860
Latitude of sunshot (sdd.mmss): 42.0135
Longitude of sunshot (sddd.mmss) (! sign: - E longitude, + W): -22.2130
GW sidereal time @ 00:00 UT: 21.46474
Right Ascension for Sun on the sunshot day @ 00:00 UT: 9.50384
Right Ascension for Sun on the next day @ 00:00 UT: 9.54214
Declination for Sun on the sunshot day @ 00:00 UT: 13.0331
Declination for Sun on the next day @ 00:00 UT: 12.4400
*** Sun azimuth: 096 ° 13 ' 28 " *** Target azimuth: 35°16'40 "
*** Sun azimuth: 096 ° 29 ' 39 " *** Target azimuth: 35°16'44"
*** Sun azimuth: 098 ° 33 ' 11 " *** Target azimuth: 35°16'57"
*** Sun azimuth: 098 ° 49 ' 06 " *** Target azimuth: 35°16'45"
*** Sun azimuth: 100 ° 36 ' 44 " *** Target azimuth: 35°16'28"
*** Sun azimuth: 100 ° 59 ' 21 " *** Target azimuth: 35°16'24"
*** Sun azimuth: 103 ° 50 ' 21 " *** Target azimuth: 35°16'50"
*** Sun azimuth: 104 ° 11 ' 7 " *** Target azimuth: 35°16'22"
*** Sun azimuth: 106 ° 41 ' 22 " *** Target azimuth: 35°16'33"
*** Sun azimuth: 106 ° 55 ' 39 " *** Target azimuth: 35°16'42"
*** Sun azimuth: 108 ° 41 ' 03 " *** Target azimuth: 35°16'39"
*** Sun azimuth: 109 ° 00 ' 48 " *** Target azimuth: 35°16'33"
Target azimuth final result (mean): 35°16'38" standard deviation = 10”

## 4. Calculations

### 4.1. REDUCTION TECHNIQUE

Magnetic observatories are in operation in neighbouring countries, so we decided to use their data to reduce our field instantaneous data to annual mean data. We had to wait until the observatories published their definitive data, about typically 6 months after the year end.

The data reduction procedures and formulae follow:
Given that the repeat survey gives us a measurement of $D$ in station *Stat* at the epoch $t$:

$$D_{Stat}(t)$$

and that a nearby Observatory *Obs* supplies us with a measurement in the observatory at the same epoch $t$ :

$$D_{Obs}(t)$$

as well as an annual mean in the observatory for epoch $a$:

$$\overline{D_{Obs}}(a)$$

We want to calculate the annual mean in the station *Stat* for epoch $a$:

$$\overline{D_{Stat}}(a)$$

We postulate that:

$$\overline{D_{Stat}}(a) - \overline{D_{Obs}}(a) = D_{Stat}(t) - D_{Obs}(t) \tag{1}$$

Hence we find the annual means at epoch $a$ at station *Stat* for the component $D$:

$$\overline{D_{Stat}}(a) = D_{Stat}(t) - D_{Obs}(t) + \overline{D_{Obs}}(a) \tag{2}$$

The validity of the above postulate (1) depends mainly on the differential in daily variation between *Stat* and *Obs* and hence is influenced by:

- Distance in longitude and latitude between *Stat* and *Obs*
- Geomagnetic Field activity
- Time in the day
- t – a

We performed this reduction for each measured component. Data from several observatories were used to reduce repeat station measurements:
L'Aquila, Italy (AQU),
Pedeli, Greece, (PEG),
Grocka, Serbia, and Montenegro (we could not get the data for 2003),
Tihany, Hungary, (THY),
Nagycenk, Hungary (NGC) and
Panagyurishte, Bulgaria (PAG).

Each reduction provides an annual mean. In an ideal situation where the postulate is correct and there are no measurement errors, the same annual means should be calculated regardless of which observatory was used to reduce the data. But of course this is not the real situation. We believe that inspection of the magnitude of the differences between means will be a good evaluation of the quality of the measurements and in how far the postulate (1) is valid.

Table 8. Summary of the Declination measurements at the repeat station Santa-Maria Precesna on august 9th 2003 and the differences with the synchronously measured declination at neighboring observatories.

| Time UTC | Declination measurement at location: | | | | | | Difference in Declination between *Stat* and Observatory | | | | |
|---|---|---|---|---|---|---|---|---|---|---|---|
| | *Stat* | AQU | PEG | NGC | THY | PAG | Stat-AQU | Stat-PEG | Stat-NGC | Stat-THY | Stat-PAG |
| 13:42 | | 1.784 | 3.198 | 2.600 | 2.779 | 3.184 | | | | | |
| 13:43 | | 1.785 | 3.200 | 2.602 | 2.781 | 3.184 | | | | | |
| 13:45 | | 1.782 | 3.198 | 2.598 | 2.778 | 3.184 | | | | | |
| 13:46 | | 1.782 | 3.198 | 2.598 | 2.777 | 3.184 | | | | | |
| Mean: | 3.035 | 1.783 | 3.199 | 2.600 | 2.779 | 3.184 | 1.252 | -0.163 | 0.436 | 0.257 | -0.149 |
| Session #2 | | | | | | | | | | | |
| 14:18 | | 1.804 | 3.220 | 2.637 | 2.818 | 3.218 | | | | | |
| 14:19 | | 1.806 | 3.222 | 2.638 | 2.820 | 3.218 | | | | | |
| 14:21 | | 1.806 | 3.222 | 2.637 | 2.819 | 3.218 | | | | | |
| 14:22 | | 1.805 | 3.222 | 2.637 | 2.819 | 3.218 | | | | | |
| Mean: | 3.062 | 1.805 | 3.221 | 2.637 | 2.819 | 3.218 | 1.257 | -0.159 | 0.425 | 0.243 | -0.155 |
| Session #3 | | | | | | | | | | | |
| 14:50 | | 1.805 | 3.220 | 2.640 | 2.820 | 3.218 | | | | | |
| 14:53 | | 1.809 | 3.222 | 2.643 | 2.824 | 3.218 | | | | | |
| 14:58 | | 1.807 | 3.222 | 2.643 | 2.823 | 3.218 | | | | | |
| 15:00 | | 1.809 | 3.223 | 2.645 | 2.825 | 3.218 | | | | | |
| Mean: | 3.067 | 1.808 | 3.222 | 2.643 | 2.823 | 3.218 | 1.260 | -0.155 | 0.424 | 0.244 | -0.150 |
| Session #4 | | | | | | | | | | | |
| 15:23 | | 1.814 | 3.223 | 2.652 | 2.831 | 3.218 | | | | | |
| 15:25 | | 1.811 | 3.223 | 2.647 | 2.827 | 3.218 | | | | | |
| 15:26 | | 1.811 | 3.223 | 2.647 | 2.827 | 3.218 | | | | | |
| 15:27 | | 1.812 | 3.225 | 2.648 | 2.829 | 3.218 | | | | | |
| Mean: | 3.068 | 1.812 | 3.224 | 2.648 | 2.828 | 3.218 | 1.256 | -0.156 | 0.420 | 0.240 | -0.149 |
| | | | | | | | | | | | |
| **Final result, average of all sessions:** | | | | | | | **1.256** | **-0.158** | **0.426** | **0.246** | **-0.151** |

AQU: l'Aquila (IT), PEG: Pedeli GR), NGC: Nagycenk (HU), THY: Tihany (HU), PAG: Panagyurishte (BG)

The differences in annual means at the station *Stat* can be statistically processed to deliver the final value as an average, and a standard deviation is also computed. The standard deviation was our principal evaluation tool torank the quality of each station.

We believe this procedure is better than computing the standard deviation of the successive field measurements reduced to one observatory only. But, our method is only possible if many observatories encircle the repeat stations and it involves more work.

To illustrate our reduction method we include tables of the August 2003 measurements from the Macedonian repeat station "Santa Maria Precesna" in the mountains North of Makedonski Brod and the subsequent reduction data from the neighbouring observatories.

In Table 8, four sessions of declination measurements are listed for the Santa Maria Precesna repeat station. The declination values observed for the same epoch at the neighbouring observatories are listed. In the last five columns, the differences between repeat station and observatory observations are indicated and the mean of each session is also given. Those values will be added to the observatory annual mean to produce the Station annual mean.

In Table 9 the repeat station annual means, at the stated epoch, are calculated. To assess the quality of the observations we inspected the column headed by "At Station by reducing on", there we found essentially the same annual mean values of the declination. The low standard deviation (0.006°) shows the excellent reduction we obtained using the various neighboring observatories and confirms the good quality of our measurements and validity of the postulate (1). The standard deviations for the reduction to annual means of Inclination and total field were 0.004° and 2.2 nT respectively.

Table 9. Final result for the annual mean of magnetic declination at the station Santa-Maria Precesna for epoch 2003.5.

| 2003.5 Declination Annual Means [°] | | | | | | | | | | | |
|---|---|---|---|---|---|---|---|---|---|---|---|
| At Observatories | | | | | At Station by reducing on | | | | | **At Station** | |
| AQU | PEG | NGC | THY | PAG | AQU | PEG | NGC | THY | PAG | **Average** | StanDev |
| *1.833* | *3.233* | *2.660* | *2.843* | *3.231* | *3.089* | *3.076* | *3.086* | *3.089* | *3.081* | ***3.084*** | *0.006* |

Not all reductions were so successful; however, observations from the Egri repeat station were made close to local noon during a period of high diurnal variation of the magnetic field at the Pedeli and l'Aquila observatories (Figure 63). The data reduction is not as good as from Santa-Maria Precesna (where observations were made when the diurnal activity was lower). At the Egri repeat station, the standard deviation for the annual means reduction to the neighboring observatories was 0.037°, 0.014° and 7.5 nT, respectively, for D, I and F.

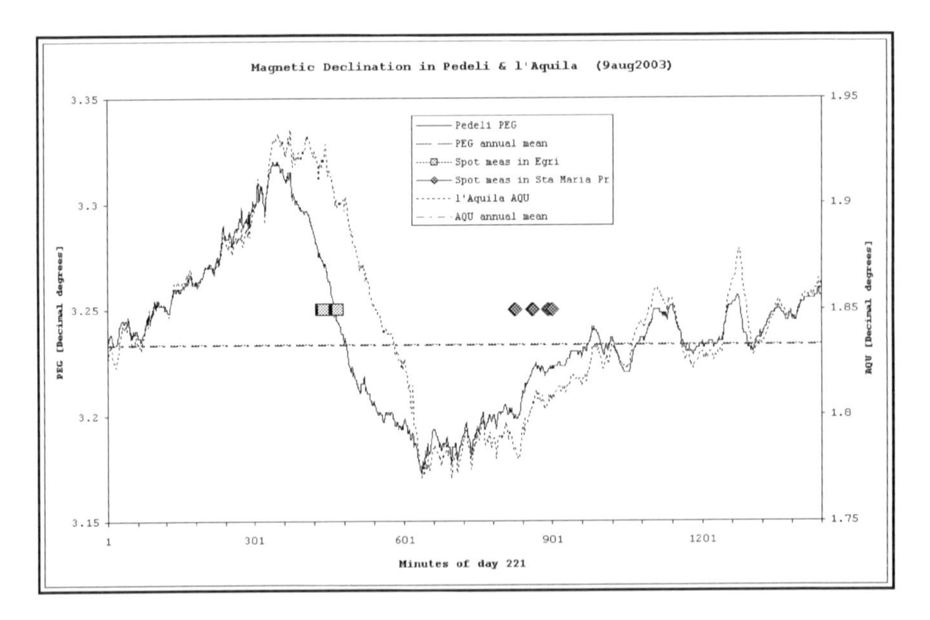


Figure 63. Diagram showing the diurnal variation as measured at l'Aquila and Pedeli. The squares and diamonds indicate the epoch (not the value) of the measurements made at Egri and Santa-Maria Precesna repeat stations in Macedonia.

## 4.2. FIRST RESULTS

Our goal was to obtain the three magnetic elements D, I and F for all the magnetic repeat stations, evaluating the accuracy of the measurements using the method explained above. Results are given in Table 10.

Table 10. Final results for the Macedonian repeat stations magnetic elements reduced to annual means for 2003.5. The standard deviation calculated from the annual means reduction to the neighboring observatories AQU, PEG, NGC, THY, and PAG is also shown.

| Repeat Station @ 2003.5 | | Coordinates WGS84 | | **Total Field** [nT] | | **Inclination** [°] | | ***Declination*** *[°]* | |
|---|---|---|---|---|---|---|---|---|---|
| Code | Locality | Long | Lat | Mean | st dev | Mean | st dev | Mean | *st dev* |
| BAI | BAILOVCE | 42.221 | 21.921 | **46723.5** | 3.2 | **59.248** | 0.013 | **2.918** | *0.026* |
| CRN | CRNA SKALA | 41.995 | 22.791 | **46886.4** | 4.8 | **58.883** | 0.005 | **3.188** | *0.017* |
| EGR | EGRI | 40.966 | 21.448 | **46399.2** | 7.4 | **57.748** | 0.014 | **3.004** | *0.037* |
| GAL | GALICICA | 40.956 | 20.814 | **46264.7** | 2.8 | **57.694** | 0.005 | **2.881** | *0.027* |
| GRA | GRADOT island | 41.388 | 21.952 | **46414.2** | 5.0 | **58.079** | 0.015 | **3.535** | *0.025* |
| LKA | LUKA | 42.344 | 22.275 | **47015.4** | 4.2 | **59.387** | 0.006 | **3.265** | *0.011* |
| MVR | MAVROVO | 41.716 | 20.727 | **46533.8** | 4.2 | **58.564** | 0.010 | **2.977** | *0.015* |

| Repeat Station @ 2003.5 | | Coordinates WGS84 | | **Total Field** [nT] | | **Inclination** [°] | | ***Declination*** *[°]* | |
|---|---|---|---|---|---|---|---|---|---|
| Code | Locality | Long | Lat | Mean | st dev | Mean | st dev | Mean | *st dev* |
| NIK | NIKOLIC | 41.265 | 22.743 | **46569.5** | 3.2 | **58.197** | 0.002 | **3.078** | *0.006* |
| PLC | PLACKOVICA | 41.795 | 22.304 | **46645.9** | 3.1 | **58.618** | 0.004 | **3.162** | *0.010* |
| PON | PONIKVA | 42.026 | 22.358 | **46799.8** | 1.9 | **58.987** | 0.003 | **2.821** | *0.015* |
| PRP | PRILEP lake | 41.403 | 21.609 | **46634.6** | 5.6 | **58.275** | 0.009 | **3.042** | *0.013* |
| SLI | SLIVNICA | 41.615 | 22.863 | **46665.4** | 4.7 | **58.499** | 0.006 | **3.381** | *0.008* |
| SMP | ST MARIA PRE | 41.627 | 21.193 | **46532.1** | 2.2 | **58.444** | 0.004 | **3.084** | *0.006* |
| TET | TETOVO | 41.986 | 21.079 | **46717.7** | 2.4 | **58.757** | 0.003 | **3.109** | *0.005* |
| *VOD* | *VODNO* | *41.978* | *21.416* | ***46712.3*** | *2.6* | ***58.789*** | *0.002* | ***3.199*** | *0.009* |

To extrapolate the data to cover he entire Macedonian territory, the normal field must be calculated and a gridding program such as Surfer used. Isogonal maps can then be drawn and used for aeronautical applications. These maps can be found elsewhere in this Volume.

Stations have been reoccupied regularly and measurements are now available for 2002 and 2004. The secular variation of the components can now be studied in the absence of an active observatory. Once the network is well established and the Macedonian observers are familiar with procedures, a reoccupation schedule of 2 years should be adopted.

## 5. Quality checks

Checking the quality of the data is important, especially for a new network, where extrapolation of the past data cannot be used to spot errors. Therefore any time a quality check can be done it should be done. We have identified the following ways to detect errors:

- Check true north determination. An error in the determination of True North is difficult to spot, unless it is large. If the sunshot method is used, care should be taken to make several fully independent measurements well spaced in time. During post-processing of the sunshots, any error will show up as a wandering direction for supposed true North. For instance, an error in time or in longitude will show up as a shifting value of the target's azimuth with the Sun's apparent motion. In the first years of the network, sunshots should be repeated and results checked for similarity with previous year's determinations. The use of the KVH compass telescope also guards additionally against large azimuth errors.

- Examine Diflux collimation errors. The so-called DIflux collimation and magnetisation errors should remain constant, if the DIflux is treated carefully and is not subjected to shocks and/or sensor position adjustment. A collimation error outside the norm indicates a bad declination or inclination measurement. Because we make a minimum of 4 complete measurements, it is possible (at reduction time) to simply discard a bad value and use only good values.
- Examine the field differences between the Station and the nearby Observatory. These differences should remain fairly constant, especially if the Observatory is close (distance < 500km) and at the same longitude because the diurnal variation will then be similar. This is a powerful error finder. We were able to detect errors in the *nearby Observatory* data, because we were sure of the correctness of repeat station measurements when checked using other observatories!
- Examine secular variation. If data is available at the station for previous years, use it to compute the secular variation. Detection of anomalies in the latter may point to measurement errors.
- Examine reduction data and standard deviation. If values are significantly different and standard deviation large, errors may exist. This is probably the most sophisticated way in detecting anomalous measurements. Measurement errors of all types will dramatically increase the standard deviation like the one displayed in Table 10.

## 6. The Future

The repeat station network in Macedonia should be connected to networks in neighbouring countries. During this Advanced Research Workshop this goal is being addressed. Further, the network should be integrated into the recently created MagNetE project, which plans the coordination of European magnetic repeat stations and the creation of a European database.

The network data should be used in providing magnetic services and products to Macedonian airports and other aeronautical agencies.

Finally, construction of a Macedonian magnetic observatory at the Plackovica repeat station site should improve reduction procedures and provide important geomagnetic data.

## Acknowledgements

We gratefully acknowledge the Belgian Army and Air-Force ComOps for air transport of equipment and staff between Brussels and Skopje. The TEMPUS-Cards Joint European Project IB_JEP-17072-2002 provided a powerful incentive for our observation tasks. But most of all we are very much indebted to the wonderful Macedonian people who made the hard work look like a holiday!

## References

Rasson J.L., 2005, About Absolute Geomagnetic Measurements in the Observatory and in the Field, Publication Scientifique et Technique No **040**, Institut Royal Meteorologique de Belgique, Brussels, 1-43

## DISCUSSION

Question (Angelo De Santis): Your opinion is that in Macedonia it is not relevant to use a variometer to reduce repeat station data to nighttime. This is because of the vicinity of observatories from neighboring countries and because of the possible temperature drift of a fluxgate magnetometer that could affect the diurnal variation recorded. Have you considered the possibility to install a proton magnetometer continuously recording in a fixed centered site of the country in order to have a stable check at least of the reduction of total intensity?
Answer (Jean Rasson): Yes it is true that the mediocre performance of a variometer after its first days of installation make me think it is not worth the effort of the additional instrumentation cost and logistics. However a proton magnetometer should not be affected by this drawback.

But the results obtained show that the reduction of our total field measurements to the proton magnetometers of 5 nearby observatories agree within a few nT (stan.dev. of 2.2nT mentioned in 4.1). The use of a dedicated proton magnetometer recording in the center of Macedonia would only marginally improve this deviation, and is hence not deemed necessary.
Question (Valery Korepanov): Your opinion: is it better to make repeat surveys with variometers or without?
Answer (Jean Rasson): My opinion is to do repeat surveys without a local variometer. I believe it is better in terms of accuracy, speed, simplicity and cost to measure in conditions where the diurnal variation is absent or negligibly small at both the repeat station and the Observatory(ies) serving for data reduction. This may imply measurements at dusk or dawn or even night measurements.
Question (Alan Berarducci): What is distance to observatories used for data reduction?
Answer (Jean Rasson): The approximate distances from the center of the Republic of Macedonia to the various observatories (see Figure 60) used in the repeat station data reductions are:

| | | |
|---|---|---|
| PAG | 225 | km |
| PEG | 435 | km |
| GCK | 335 | km |
| THY | 655 | km |
| NGC | 770 | km |
| AQU | 700 | km |

Question (J. Miquel Torta): Have you found different deviations from the mean depending on the distance between the repeat station and the observatory used to reduce the data?
Answer (Jean Rasson): No, we did not see any obvious correlation.

# NATURE OF EARTH'S MAGNETIC FIELD AND ITS APPLICATION FOR COMMERCIAL FLIGHT NAVIGATION

RUDI ČOP*
*University of Ljubljana, Faculty of Maritime Studies and Transportation*
DUŠAN FEFER
*University of Ljubljana, Faculty of Electrical Engineering*

**Abstract.** This article presents properties of the Earth's magnetic field and its impact on the commercial flight navigation. Scientific studies of the Earth's magnetic field in Slovenia during the last thirty years are presented. This paper deals with theoretical research regarding the magnetopause that protects life on Earth from lethal radiation from space. Changes in the Earth's magnetic field have been measured and are presented. The magnetic compass, which operates uses the Earth's magnetic field, is the oldest navigation instrument. Through centuries it has been the most important aid to navigation on the sea, helping seafarers to steer the right courses. Other physical measurements are not as easy or reliable as measurements by magnetic compass. Today, the magnetic compass is still the basic navigation instrument on ships and aircraft. It is used in exceptional cases when radio navigation systems do not operate reliably.

**Keywords:** Earth's magnetism, magnetic compass, aeronautical navigation

## 1. Description of the Earth's Magnetic Field for Future Navigators

The Earth's magnetism has been researched longer than any other natural phenomenon and has had the greatest number of differing explanations. Ancient nations used a magnetic needle, which could rotate freely around its vertical axis, for navigation and geodesy. The ancient Chinese people

*Address for correspondence: ddr. Rudi Čop, Faculty of Maritime Studies and Transport, Pot pomorščakov 4, 6320 Portorož, Slovenia; e-mail: rudi.cop@fpp.edu



*J.L. Rasson and T. Delipetrov (eds.), Geomagnetics for Aeronautical Safety,* 115–126.

were acquainted with magnetic inclination. In their opinion this inclination was the connection between the earth and the sky[1]. An explanation of how and why the Earth's magnetic field is generated got its final form in modern times when the astronomical and geophysical researchers resolved how sunspots and the Sun's magnetic field came into being. Their theory was verified with considerable help from spacecraft in the second half of the last century.

William Gilbert started modern research into the Earth's magnetism and published his findings in 1600[2]. Measuring the Earth's magnetism was done mainly by seafarers during the seventeenth and eighteenth centuries (Figure 64). They measured the directions and size of the magnetic forces that originated from the Earth. They first drew compass maps for the entire surface of the Earth with all the anomalies they noticed. They indicated the central magnetic field source in the form of a magnetic dipole. The same conclusions were reached by Carl Friedrich Gauss and Wilhelm Weber[3], whose findings were based on measurements and theoretical research in 1836 – 1841. Analyses of volcanic rocks from the edges of tectonic plates in the oceans, done during the 1960s, showed that the Earth's magnetic dipole reverses its polarization each half million years on average[4]. Why the Earth's magnetic poles reverse has not yet been determined.

Dutch physicist Pieter Zeeman found in 1897 that a strong magnetic field may change spectral light lines[5]. Such changes grew in proportion to the density of the magnetic field. This conclusion resulted in the enhancement of measuring methods, so that it became possible to measure the magnetic field of sunspots and the Sun. The Sun's magnetic field changes its direction cyclically about every eleven years.

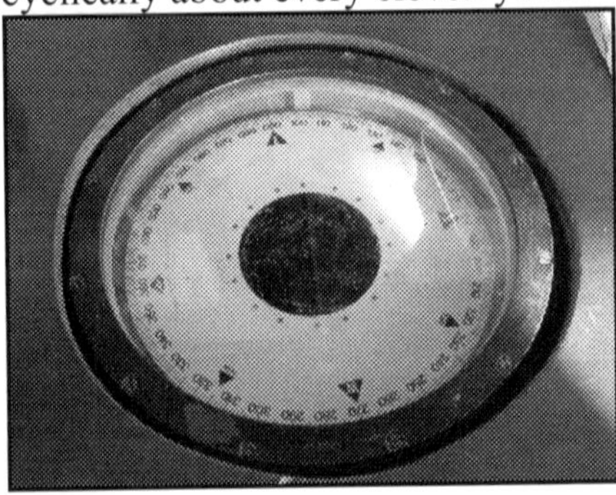

Figure 64. Liquid magnetic compass on the vessel.

In 1919, Sir Joseph Larmor explained how sunspots developed. Based on an electric dynamo principle, according to his explanation, an electric current emerges in the Sun's plasma. The electric current generates a magnetic field which, in turn drove the electric current. The discovery, by astronomers, that the Sun does not rotate as a solid was important. The

Sun's equator rotates fastest, and the velocity of rotation decreases towards its poles. Therefore, the Sun can act as a self-stimulated electric dynamo. A useful theoretical explanation on the basis of these findings was introduced by Stanislav Braginsky in 1964. He assumed that the magnetic field was axially symmetrical with small asymmetries[6]. Because of the heat, the plasma flowed towards the surface. The origin of the heat has not been completely explained. The straight-line movement of the plasma towards the surface turned into rotation because of the Coriolis acceleration. It appeared because the reference frame was moving and therefore the whirling was stronger in the direction of rotation. Such asymmetries are also expected in the Earth's fluid core. Electric currents, which cause the Earth's magnetic field, are the result of the different speeds of stratum rotation of the three basic layers: solid core, fluid magma and solid outer crust.

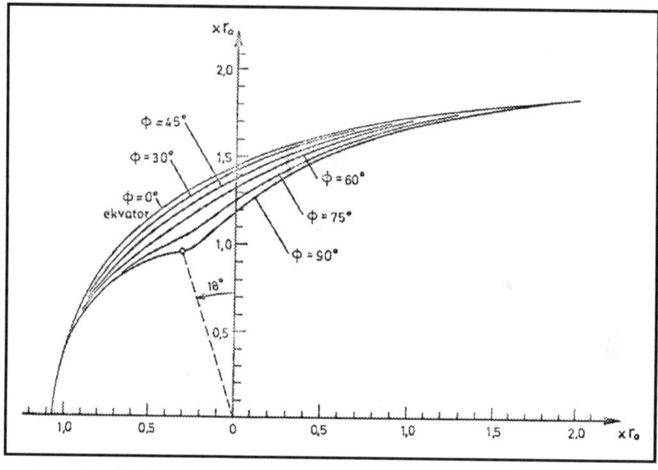


Figure 65. Results of calculation of Earth magnetopause cross-section shape at different angles Φ with regard to the half-meridian plane Φ = 90°.

## 2. Influence of the Solar Wind on the Earth

In 1896, Norwegian physicist Kristian Birkeland proved, on the basis of laboratory tests, that the Northern Lights are the results of electrified solar particles colliding with the Earth's magnetic field. The phenomenon of there being less space radiation when the Sun's activity is greater was explained in 1952 by Swedish physicist, Hannes Alfvén[7]. He proposed that the Sun's magnetic field was in direct proportion to the density of its radiation. The stronger the Sun's magnetic field, the more efficient it was in deflecting galactic space radiation. In 1953 Eugen Parker, an English physicist, found that the Sun's corona was not a static phenomenon but that it spread through space in the form of the solar wind[8].

The solar wind, composed of hydrogen nuclei and electrons, spreads through space at a speed that increases as the wind's distance from its source increases. When it reaches the Earth, it is deflected by the Earth's magnetic field and therefore splits into two flows, a proton flow and an electron flow, which will whirl in opposite directions. They make a curved crossing almost parallel to the Earth's surface, and then disperse and fly off into space[9]. The currents that appear around the Earth during this phenomenon make the Earth's magnetic field stronger in the direction towards the Earth and weaker in the direction away from the Earth. The stratum where the magnetic field decreases to zero is on the sunny side at a height of 100 km above the Earth and is called the Magnetopause. This stratum isolates the Earth from the destructive impact of space radiation and by doing so protects life on the Earth.

## 3. Research into the Earth's Magnetism in Slovenia

In the 1970's the Faculty of Electrical Engineering in the University of Ljubljana carried out comprehensive theoretical research on the magnetosphere[10]. Based on the fact that the solar wind always blows comets' tails away from the Sun, it was assumed that the magnetopause was teardrop shaped with its tail turned away from the Sun. The research started with a magneto-hydrodynamic equation for the pressure of the solar wind (1.1).

$$grad\,\vec{p} = \vec{J}\;x\;\vec{B} \tag{1.1}$$

where:
p = solar wind pressure
J = density of the solar wind flow
B = density of the Earth's magnetic field

In the Earth's magnetopause it is possible to use the equation (1.1) of uniform part-flow in a form where the mechanical pressure of the solar wind is equal to the Earth's magnetic-field effect. By using an iteration calculation procedure according to the Newton – Raphs method, researchers came to the conclusion that the bipolar magnetic field on the sunny side of the Earth is rather deformed (Figure 65). Based on additional results, which were reached at the same time by other research groups, the magnetopause was given a shape like a teardrop with a tail, on the opposite of the sunny side of Earth. This shape was confirmed by measurements carried out by spacecraft. Theoretical research and measurements verified that the Earth's magnetic field does not spread into space.

In the last fifteen years, the Laboratory for Magnetic Measurements at the Faculty of Electrical Engineering in the University of Ljubljana has systematically researched changes in the Earth's magnetic field in different settings. First, researchers were interested in Changes to the Earth's magnetic field caused by plants. Measurements verified that plants impact the local magnetic field, at a micro-level in different ways depending on their states and inner processes[11], and that the strength of the magnetic field impacts plant growth, both in plantations and in forests (Figure 66)[12].

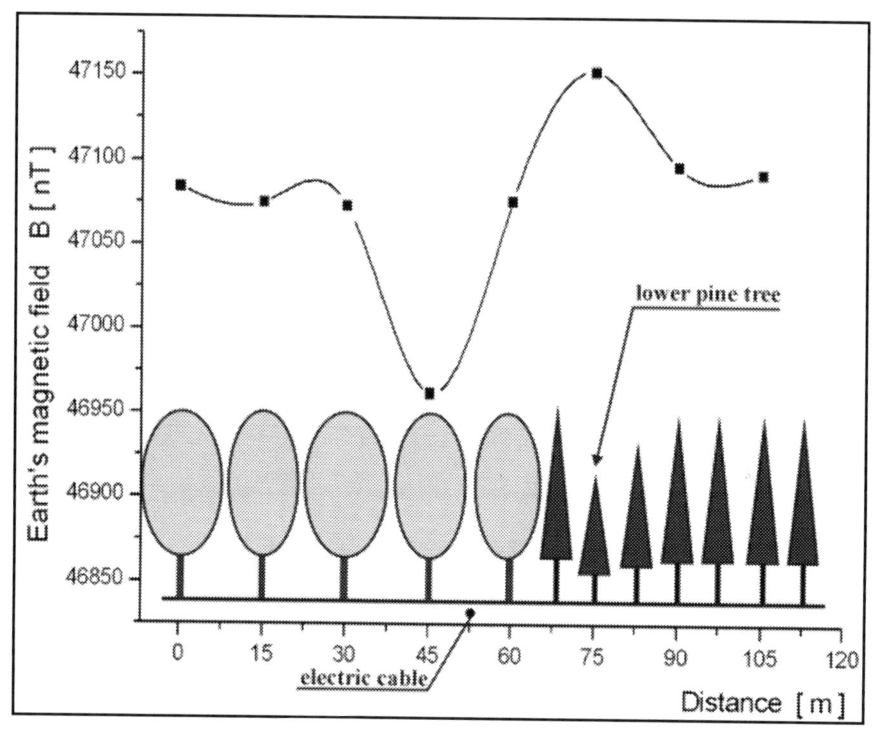


Figure 66. Impact of increase in gradients of the Earth's local magnetic field on the growth of young pine trees.

Changes in the Earth's magnetism in residential areas and at archaeological sites were also of interest. In both cases, there were large-scale changes of gradients, particularly in the vicinity of ferromagnetic materials. Especially large deviations from normal levels of magnetism were measured indoors, in residential buildings constructed of reinforced concrete with branched electrical installations[13]. However all values were below the approved limits in accordance with the valid regulations. Measurements indicating changes in the Earth's magnetic field in non-residential areas were also of interest. Changes were observed during weather changes and during different kinds of storms (Figure 67).

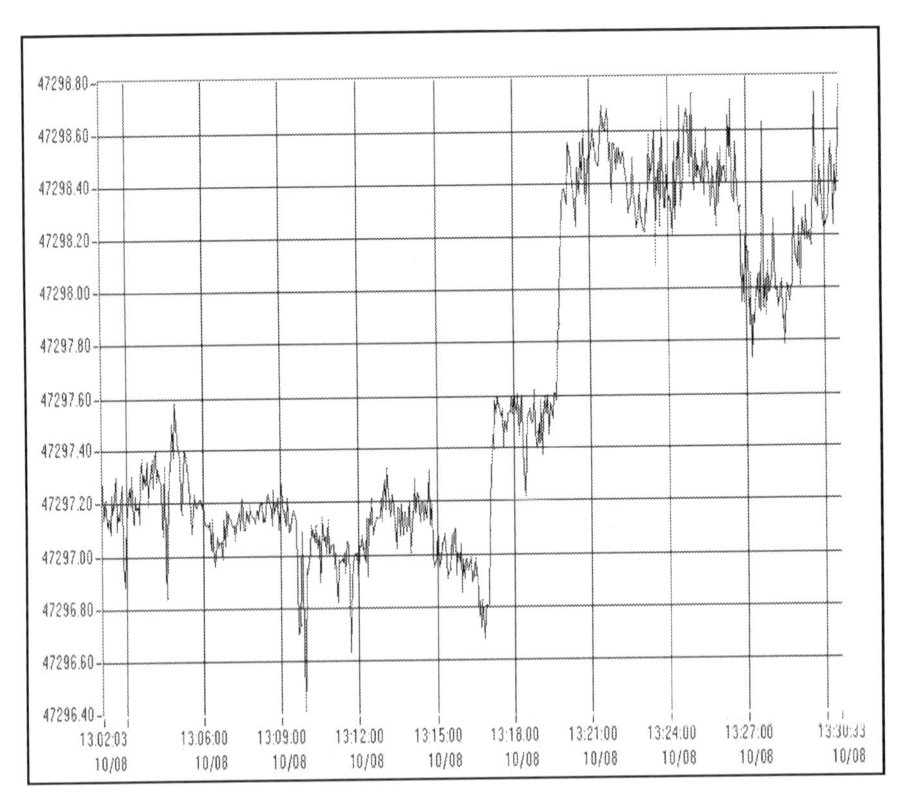


Figure 67. Changes in the Earth's magnetic field at a location with small deformations while the weather was changing and before the rain started.

## 4. Aeronautical Navigation by the Use of the Magnetic Field

Navigation is a method for planning a journey from one place to another safely and in accordance with given conditions. In its basic meaning, navigation is the art and science of planning, controlling, and piloting a vessel from its starting place to its destination. Man improved navigation methods several thousand years ago, some sources have even said 8,000 years ago, to such a level that it became a science [14].

Navigation means recognising the place where one finds oneself and determining the time necessary to reach a particular destination. The most basic type of navigation is called Dead Reckoning. It is also the most demanding. Ancient Phoenicians used this type of sea navigation when visibility was poor. Today, it is still used as an independent method of navigating, but most often it is used as an additional instrument to other methods of navigation (terrestrial, astronomic, radio-navigation) and as a check on other types of navigation. New positions can be defined on the basis of the known direction and speed of movement as well as on the basis of the position previously reached. For a safe journey, the following navigation parameters must be calculated: current position, distance to destination, travelling speed and direction, current fuel consumption and quantity of fuel needed, and the time left for the rest of the journey[15]. Because of the number of unknowns in such a method of calculation, the accuracy range for defining a position is ±1/8 of the route already travelled. Before the introduction of the satellite navigation system, Dead Reckoning navigation was the only way to navigate on the open sea or in the air where all other methods failed.

The magnetic compass is a useful navigation instrument that must be calibrated prior to use, as its frame may cause magnetic deviation. The magnetic declination should be added to the adjusted magnetic heading (the deviation of magnetic North from true North), at any single position on the globe. Easterly declination is a negative quantity and westerly declination is a positive quantity. The necessary correction can be made by adding the declination to True North. The North Pole is the appropriate reference point for maps, but finding its precise position is rather difficult. So it can only be a useful reference point for course control together with the magnetic pole. However, in the case of aeronautical navigation, a true flight course can be defined only with additional consideration of the wind correction angle (Figure 68). The heading is obtained when wind correction angle is applied to a True Course. When wind is blowing from the left a negative wind correction is added and when wind is blowing from the right a positive correction is added. For a flight plan worksheet, all these influences are taken into consideration as a matter of course, and the flight log is drawn up in accordance with them (1.2).

In exceptional cases, the Dead Reckoning method may be used for IFR (Instrument Flight Rules) in civil air navigation. In the cases of flights over very remote places and seas, where all other ways of navigation are unusable, unavailable, or do not operate reliably, the pilot must be prepared to use this kind of navigation. Before taking off, the flight speed and wind correction angle have to be carefully calculated. After take off, the wind influence in the upper air strata should be closely estimated. Dead Reckoning is used rarely, but it is the most appropriate kind of navigation for covering gaps between areas that use different kinds of aeronautical navigation.

$$TC + WCA + VAR + DEV = CH \tag{1.2}$$

| **Sign** | **Meaning** | **Influence** |
|---|---|---|
| TC | True Course, line drawing on a map | |
| WCA | Wind Correction Angle | left -, right + |

| Sign | Meaning | Influence |
|---|---|---|
| VAR | Variation | East -, West + |
| DEV | Deviation | -, + |
| CH | Compass Heading | |

The last resort for a pilot who has lost electric power or has had a fire in the cockpit is Dead Reckoning. In these cases he has only a magnetic compass, a clock, and an airspeed indicator. Errors in Dead Reckoning depend on the accuracy of available data before beginning calculations. During emergency situations a pilot will follow the magnetic compass heading to the point of destination. For a new course and speed of flight he should also coordinate already known data on wind direction. He should assure himself that there are no high-rising barriers in his course. The experienced pilot will try to reach good visibility quickly, to reduce the possibility of collision. The pilot plans a new flight course with the following conditions; beginning the visible part of the route with the best conditions for landing, the flight time and the time before fuel runs out. The pilot should observe flight speed and time simultaneously, and not panic. To prevent flying around in circles, the MPP (Most Probable Position) position should be observed at regular intervals. The MPP observation is the only way to prevent circling in emergency situations.

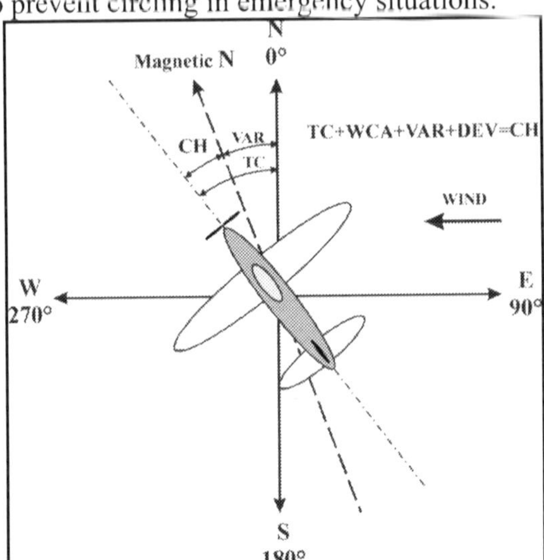


Figure 68. The wind correction angle.

Gliders and light motor aircraft must be equipped with a radio as well as the basic navigation instruments before being registered[16]. Basic instruments are: airspeed indicators, altimeters and magnetic compasses. The simple magnetic compass on an airplane operates the same way as one on a ship (Figure 64). It has a circular scale on which is fixed a permanent magnetic bar. The direction indicator floats in liquid (liquid magnetic compass) (Figure 69). The gimballed magnetic compass is fixed to its base by joint connections. Such magnetic compasses are incomparably more accurate and less sensitive to changes of flight direction than magnetic compasses with a simple construction, although they are much more expensive.

## 5. Conclusion

The oldest and easiest way to set a course is to use a magnetic compass. Compass measurements are easily made and reliable. Compasses are unsuitable for navigation in the four following cases:

1. If the frame is magnetic and has too much influence on the magnetic compass.
2. If natural or artificial objects have an unpredictable influence on the Earth's local magnetic field.
3. If the local vertical component of the Earth's magnetic field is larger than the horizontal component.
4. If it is impossible to take reliable measurements with a magnetic compass because the frame is accelerating too fast.

Figure 69. Liquid magnetic compass on a light aircraft.

The modern electronics industry enables the construction of digital magnetic compasses which are precise and inexpensive devices for setting courses. They are also accurate in most cases where traditional magnetic compasses are not reliable. Electronic magnetic compasses were introduced for aeronautical navigation during the Second World War. In the 1970's microprocessors were included in electronic magnetic compasses. Such digital magnetic compasses are able to measure a local Earth magnetic field with high accuracy and reliability[17]. These compasses have an electric output signal and, therefore, are easily handled. They can also display data in multiple locations simultaneously And can be used in a closed control loop, where they can be automatically controlled, even if the precise course with reference to the magnetic North Pole is unknown[18].

A normal magnetic field strength in Slovenia[19] is somewhat greater than 47,000 nT and changes is no more than 100 nT (1 nano-Tesla = $10^{-9}$ T) in a full day. Measuring the change in field strength is important for aeronautical and shipping navigation. Such measurements also enable predictions of changes connected with the Sun's activity, which affect the space around the Earth: its weather, density of the atmosphere, influence of solar wind on electric power-lines and transformers, and changes in Earth satellite orbits[20].

Digital observatories IMO (INTERMAGNET Magnetic Observatories) must be equipped with a magnetometer for measuring the three-components of the Earth's magnetic field and with a scalar magnetometer for measuring the absolute value of the Earth's magnetic field. Such instruments should take measurements each second with a resolution better than 1 nT. The measurement results should first be registered and then processed each minute and during the following minute sent to a GIN (Geomagnetic Information Node). In Europe this information is sent via METEOSAT, the geostationary satellite for observing the Earth, or via Internet[21].

Two of the six GIN world centres for collecting geo-magnetic measurements are in Europe and operate in near real time. More than 80 geo-magnetic observatories through out the world are connected to INTERMAGNET (International Real-time Magnetic Observatory Network) via these centres. INTERMAGNET has been active since 1991 and makes it possible to collect and store measurements of changes in the Earth's magnetic field. The collected data are communicated to all IMO's and are available to other users as well.

In order to join the International Project INTERMAGNET, the Republic of Slovenia would have to find an appropriate place to construct a digital geomagnetic observatory with low magnetic pollution and a low Earth magnetism gradient (< 1nT/m). The location must not have any direct electric currents such as those associated with electric railways and industries with electroenergetic sources. Finally, a location for a magnetic observatory must be guaranteed to stay magnetically clean in the future. These conditions can only be ensured by an appropriate national institution with expert staff supported by the governmental administration.

## References


1. S. Silverman, *Compass, China, 220 BCE* in Virtual Museum of Ancient Inventions, Smith College, 1997-2004 (March 03, 2004); http://www.smith.edu/hsc/museum/-ancient_ inventions/home.htm.
2. W. Gilbert, *De Magnete*. (Dover Publications, New York, 1991).
3. J. J. O'Connor, E. F. Robertson, *Johann Carl Friedrich Gauss* in The MacTutor History of Mathematics archive, University of St Andrews, 1996 (March 17, 2005); http://www-groups.dcs.st-and.ac.uk/~history/Mathematicians/Gauss.html.
4. W. Klous, T. Jackquelyne and R. I. Tilling, *This Dynamic Earth: The Story Of Plate Tectonics*. (USGS Information Services, Denver, 1996).
5. *Pieter Zeeman – Biography* in The Official Web Site of the Nobel Foundation. (March 16, 2005); http://nobelprize.org/physics/laureates/1902/zeeman-bio.html.
6. D. P. Stern, *The Great Magnet, The Earth*, Nasa Goddard Space Flight Center, Greenbelt, 2002 (March 15, 2005); http://pwg.gsfc.nasa.gov/earthmag/dmgmap.htm.
7. A. L. Peratt, *Dean Of The Plasma Dissidents*, in: The World & I, May 1988, P. 190-197.
8. A. Hood, *The Solar Wind*, in Alan Hood's Personal Home Page, University Of St Andrews 1995 (March 21. 2005); http://www-solar.mcs.st-andrews.ac.uk/~alan/-sun_course/ Introduction/Solar_wind.html.
9. *Hubble Provide Complete View of Jupiter's Auroras*. Space Telescope Science Institute, Baltimore, 1998 (March 21, 2005); http://hubblesite.org/newscenter/newsdesk/archive/releases/1998/04/.
10. B. Bottin, *Zemljino Magnetno Polje V Sončnem Vetru: The Earth Magnetic Field In Solar wind*, in: Disertation, mentor prof.dr. F. Avčin and prof.dr. M. Željeznov (Fakulteta za elektrotehniko, Ljubljana, 1978).
11. R. Thompson, F. Oldfield, *Environmental Magnetism* (Unwin Hyman, London, 1986).
12. D. Mancini, *Zemeljsko magnetno polje: The Eart's magnetic field*, in: Diplom of university graduate engineer, mentor prof.dr. A. Jeglič and doc.dr. D. Fefer (Fakulteta za elektrotehniko in računalništvo, Ljubljana, 1991).
13. *Merjenje magnetnih anomalij zemeljskega magnetnega polja v okolici Ljubljane: The measurement of anomalies of Earth's magnetic field in the suburbs of Ljubljana*, in: Porocilo - Report, mentor prof.dr. D. Fefer (Fakulteta za elektrotehniko, Ljubljana, 2004).
14. N. Bowditch, *American Practical Navigator: An Epitome Of Navigation* (Defense Mapping Agency: Hydrographic and Topographic Center, Washington, 1984).
15. D. J. Clausing, *The Aviator's Guide To Modern Navigation*. (TAB Books, Blue Ridge Summit, 1990).
16. J. Brezar, *Jadralno Letalstvo: Glidering*, editor Moškom Marjan. (Letalska zveza Slovenije, Ljubljana, 1995).
17. *Overview of compass technology*. Technical documents (KVH Industries Inc., Middletown, 1998).

18. *Compass heading using magnetometers.* AN-203. (Honeywell International's Solid State Electronics Center (SSEC), 1995).
19. K. Weyand, *Eine Neues Verfahren zur Bestimmung des Gyromagnetischen Koeffizienten des Protons* (PTB, Braunschweig, 1984).
20. D. Kerridge, *Intermagnet: Worldwide Near-Real-Time Geomagnetic observatory data*, in: International Real-time Magnetic Observatory Network - INTERMAGNET (March 24, 2005); http://www.intermagnet.org/Publications_e.html.
21. *INTERMAGNET technical reference manual*, edited by Benoît St-Louis (U.S. Geological Survey, Denver, 2004).

## DISCUSSION

Question (Ivan Butchvarov): What is the difference between the magnetic and gyrotheodolite in use?

Answer (Rudi Cop): When a theodolite is oriented by magnetic compass, the magnetic variation and deviation must also be taken into consideration when carrying out measurements regarding true North. Magnetic variation changes with time and by changing the location. Magnetic deviation is caused by the influence of the frame of a magnetic compass on the accuracy of the measuring instrument.

A gyrocompass enables direct theodolite orientation regarding true North. Its operation does not rely on the Earth's magnetism.

In both cases, when using navigation instruments for defining a position, it is necessary to take into consideration the accuracy of the measuring instrument.

Question (Angelo de Santis): Your presentation focuses our attention on an important matter: even if the geomagnetic community can give information to airports / heliports about the mean value of declination, this value can change rapidly and significantly during magnetic storms that can last for a few days. This is something to be clearly said to operators of aircraft. A possible solution is to ask these operators to look at websites when magnetic recordings from observatories or space weather forecasts are shown (see, for example, the ingv website: www.ingv.it).

Answer (Rudi Cop): Small changes in the Earth's magnetic field over smaller or greater areas do not only influence the accuracy of the navigation instruments but, as the measurements taken in Slovenia and elsewhere in the world have proved, there is also a mutual connection between the state of the magnetic field and the reactions of living beings in it. There is considerable work in this field to be done by researchers, among other things, to determine the ability of the crew to operate the aircraft in the case of short term changes in the Earth's magnetic field caused by the solar wind.

# GEOMAGNETIC OBSERVATIONS IN TURKEY

C. CELIK[9], M.K. TUNCER, E. TOLAK, M. ZOBU, O. YAZICI-ÇAKIN, B. CAĞLAYAN, AND Y. GÜNGÖRMÜŞ
*Bogaziçi University, Kandilli Observatory and Earthquake Research Institute*

**Abstract.** Systematic geomagnetic observations in Turkey started in 1947 at Istanbul-Kandilli (ISK) Observatory, which is located about 15 km out of the city center of Istanbul. Increasing population and demand for residential, commercial, and industrial zones caused the city to expand and with time ISK Observatory will be engulfed by the city. Consequently, geomagnetic observations are affected by artificial noise generated by the surrounding city. Despite the decrease in the data quality, ISK Observatory submits its data to World Data Centers on a daily basis (near-real-time). In order to improve the data quality, a new observatory was set up in Iznik, about 100 km away from Istanbul. Iznik is a site of great geophysical interest because a fault segment of the North Anatolian Fault Zone is nearby. Since 1986, 9 continuous geomagnetic total intensity stations have been running to observe the tectonomagnetic field in the region. All stations will greatly improve the geomagnetic field models and will facilitate the study of the geomagnetic field distribution and variation in Turkey.

**Keywords:** Geomagnetic Observatory, Turkey, ISK observatory, INTERMAGNET

## 1. Introduction

For the study of the geomagnetic field and field models, geomagnetic observatory data are very important. There are several factors affecting the data quality of Turkey's Istanbul Kandilli Geomagnetic Observatory. The most important factor is the location of the observatory. Istanbul Kandilli geomagnetic observatory was originally built well away from the city but

[9] To whom correspondence should be addressed at: celikc@boun.edu.tr



*J.L. Rasson and T. Delipetrov (eds.), Geomagnetics for Aeronautical Safety*, 127–135.

with increasing population the observatory is now nearly in the middle of the city. Naturally, the geomagnetic signal is affected by the artificial noise of the city. The only solution to the problem is to construct a new geomagnetic observatory away from all sources of artificial noise.

## 2. Istanbul-Kandilli geomagnetic observatory (ISK)

There has been an observatory at Kandilli since 1911, when the Istanbul astronomical and meteorological observatory moved there from the centre of the city. The first magnetic measurement at Kandilli was made by the Director, Fatin Gökmen, on March 12, 1927. He used a Chasselon-Brunner magnetic theodolite and a dip circle brought from France in 1926 to make the measurements. Further measurements were made by Osman Sipahioglu between 1936 and 1947, but it was not until 1947 that systematic magnetic measurements were started. Between 1947 and 1996 geomagnetic measurements were made as photographic records. In 1996 two three-component Fluxgate magnetometers and a BGS Flare Data Logger were bought and set into operation at ISK. For absolute measurements, a Scintrex EDA OMNI portable proton magnetometer and an Askania (Schmidt large size) declinometer are now used.

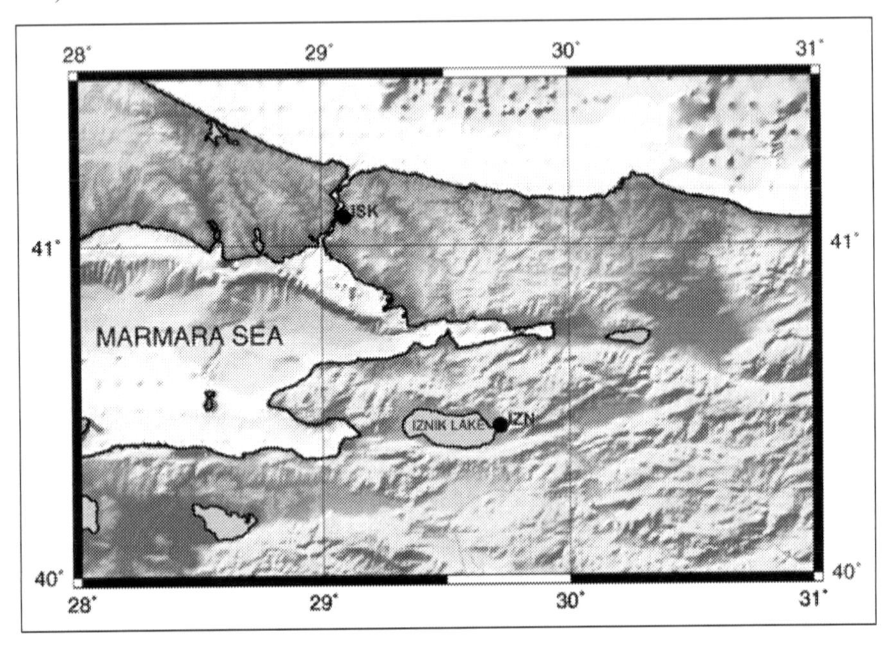


Figure 70. Location of the observatories in Turkey.

a)

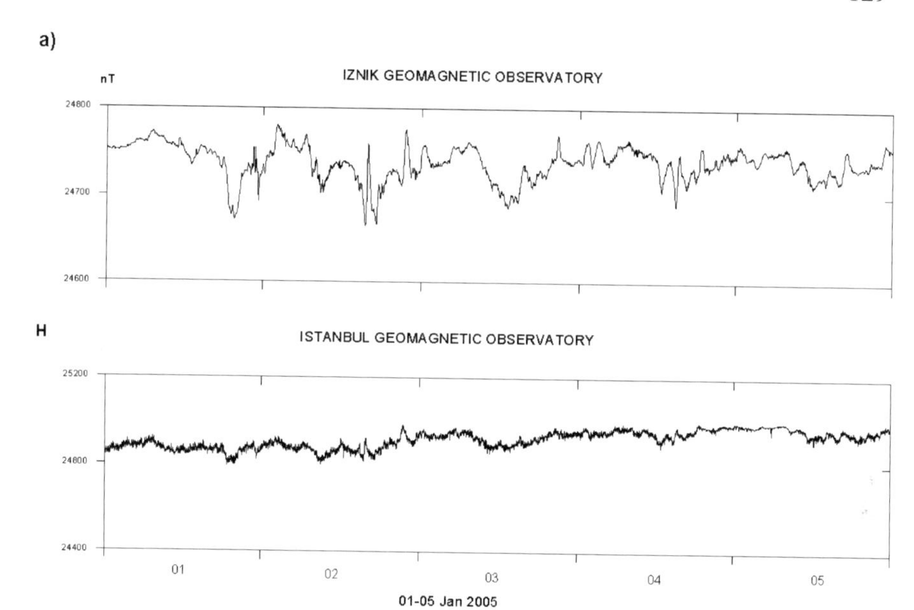
nT
IZNIK GEOMAGNETIC OBSERVATORY
24800
24700
24600
H
ISTANBUL GEOMAGNETIC OBSERVATORY
25200
24800
24400
01
02
03
04
05
01-05 Jan 2005

b)

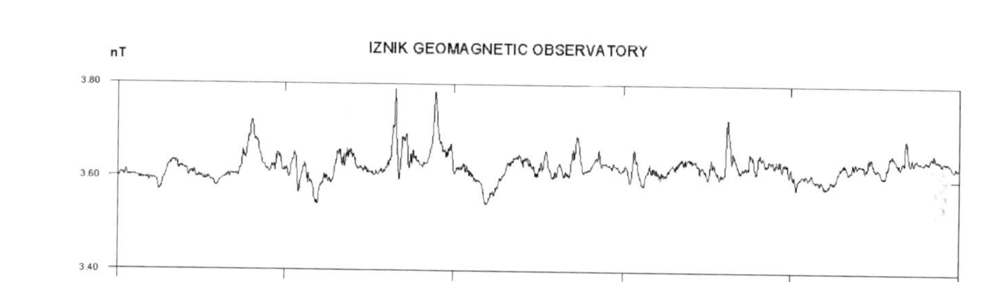
nT
IZNIK GEOMAGNETIC OBSERVATORY
3.80
3.60
3.40

D

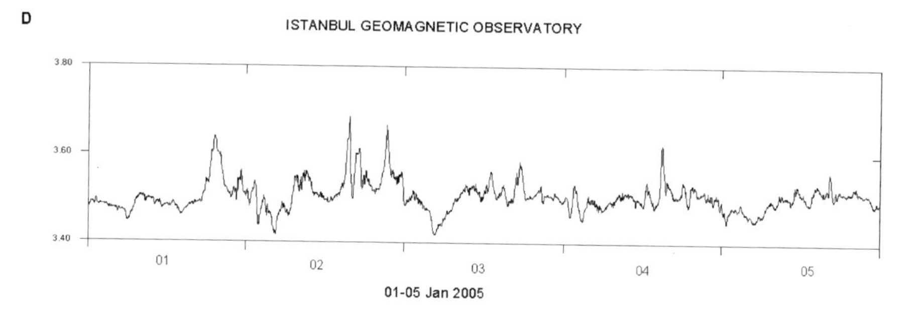
ISTANBUL GEOMAGNETIC OBSERVATORY
3.80
3.60
3.40
01
02
03
04
05
01-05 Jan 2005

c)

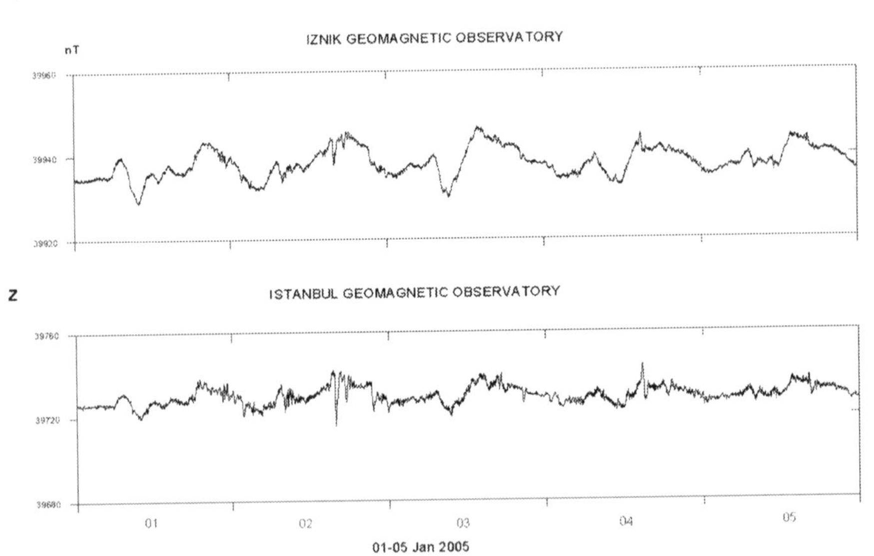


Figures 71a, b, c. *H*, *D*, and *Z* components of the geomagnetic field recorded at ISK and Iznik observatories on 01-05 Jun, 2005, respectively.

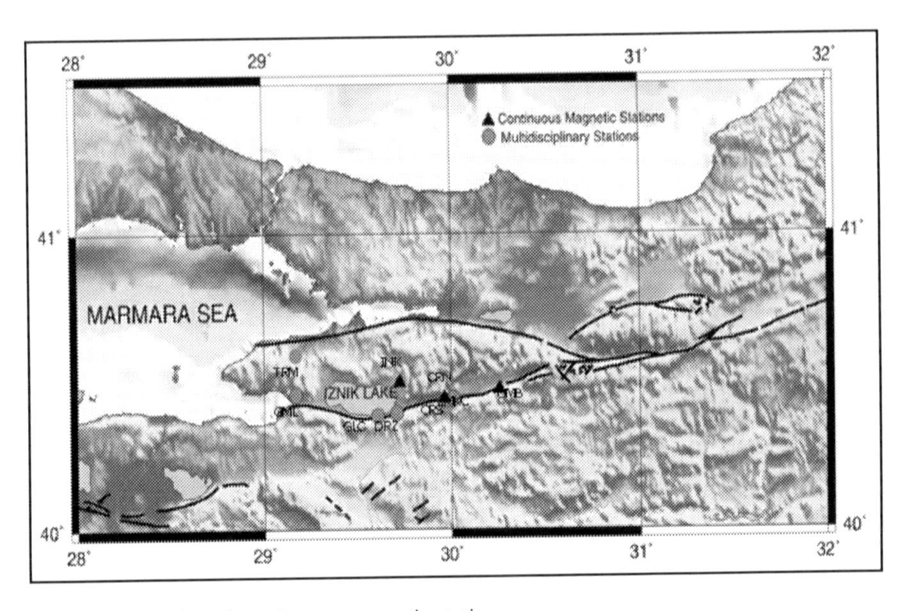


Figure 72. Location of continuous magnetic stations.

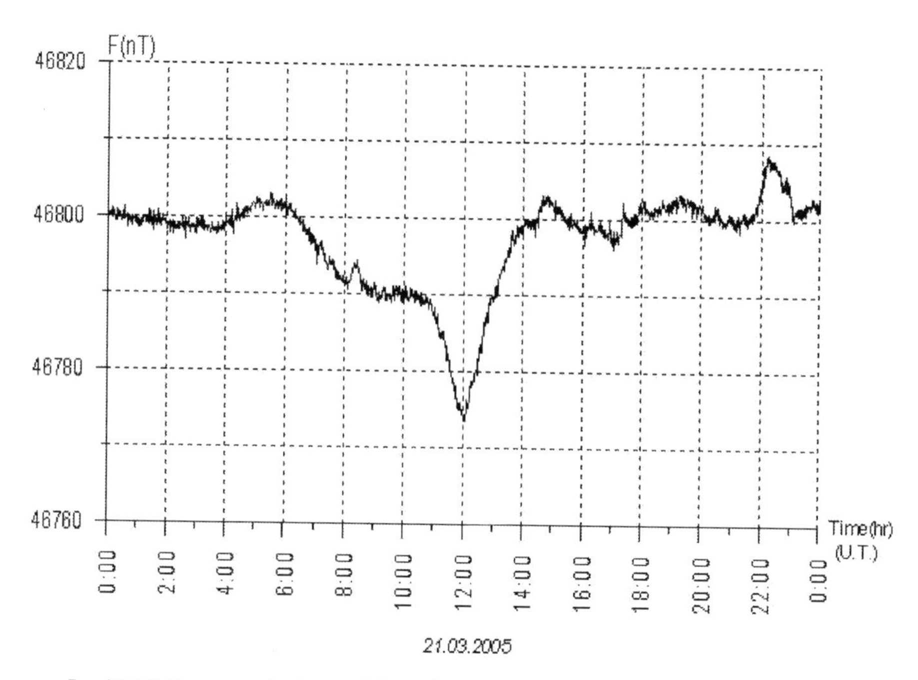


In 1947 the population of Istanbul was about one million and Kandilli was situated out in the country. The population today is about 12 million. The city has expanded and now engulfs Kandilli. Naturally, the geomagnetic observations at ISK are affected by the artificial noise of the surroundings. Despite the low data quality, ISK observatory still submits its data to World Data Centers on a daily basis (near-real-time).

## 3. Iznik geomagnetic observatory

A new geomagnetic observatory began operation in 2004. This new observatory, the Iznik Observatory was built to acquire high quality, quiet data. Figure 70 shows the location of the observatories in Turkey. The new observatory has a three-component Scintrex fluxgate magnetometer recording on a BGS flare data logger (HDZ) and a three-component DMI fluxgate magnetometer variometer installed. For absolute measurements a D/I theodolite and a proton Magnetometer are used.

## 4. Comparison between Iznik and Istanbul geomagnetic observatory data

Figures 71 shows the variation of the H, D, and Z components of the Earth's magnetic field recorded at both the ISK and Iznik observatories on 01-05 Jun, 2005. It is seen clearly from the figures that the data from ISK are very noisy, displaying both short and long period noise. Data from Iznik are comparitively quiet.

## 5. Tectonomagnetic and magnetotelluric observations

Iznik is a site of great geophysical interest because it lies on an active fault segment of the North Anatolian fault zone in a seismic gap, so a major earthquake could occur there any time. The institute has maintained many field sites there since 1986. Nine sites are maintained for the continuous monitoring of the total magnetic field. One of them is situated away from the fault segment and is used as a reference station. Figure 72 shows the locations of the nine magnetic stations. The sampling interval is one minute and data are transferred to the research center in the region using a modem connection. To detect the tectonomagnetic field, a simple difference technique with night-time data is used. Figure 73 shows a record of daily variations and Figure 74 shows the tectonomagnetic field observed in 2004 in the region.

In addition to the tectonomagnetic observations in the region, an MT survey was done to determine the structure of the crust around Iznik. Seven profiles were completed using more than 100 MT sites. MT surveys will continue in the coming years. Figure 75 shows the location of the MT profiles and stations.

## 6. Conclusion

A new geomagnetic observatory was constructed at Iznik, Turkey in 2004 away from all kinds of artificial noises. It was built so that high quality geomagnetic data could be obtained. The Iznik Observatory and the nearby geomagnetic total intensity stations will greatly improve the geomagnetic field models for both Turkey and the Balkan region. Iznik is planning to apply for membership to INTERMAGNET soon.

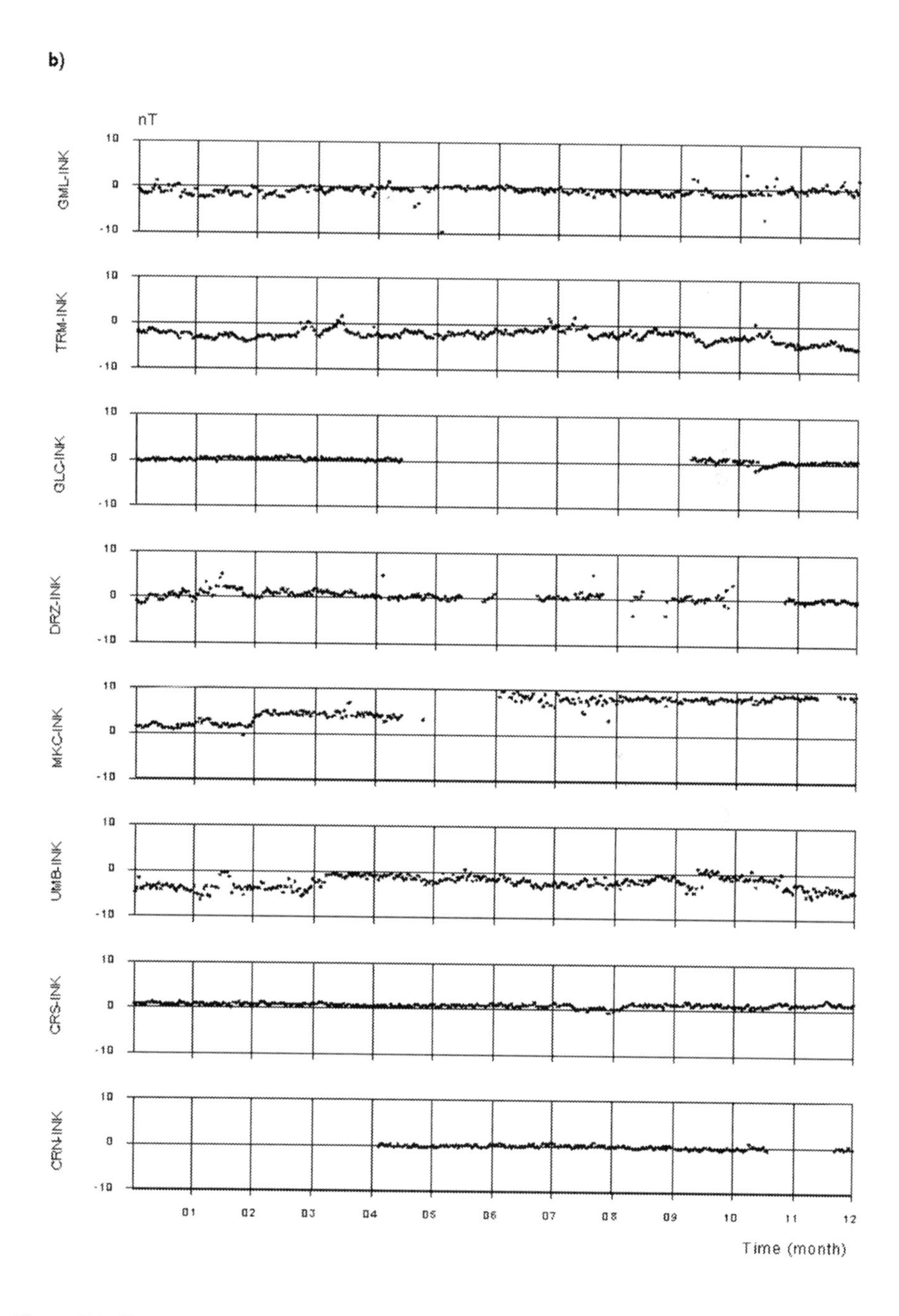


Figure 74. Tectonomagnetic field recorded in the region in 2004. Troubles related to the power supply system caused gaps in the data.

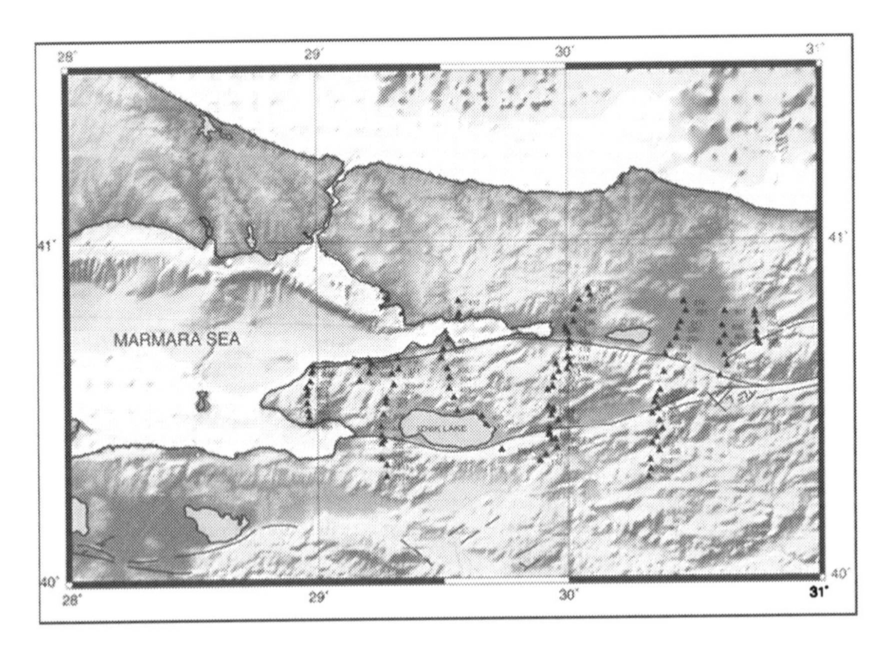


Figure 75. Location of the MT profiles in the region.

**DISCUSSION**

Question (Spomenko J. Mihajlovic): What is space difference between total field (proton magnetometer) stations for tectonomagnetic effect (project)?
Answer (Cengiz Celik): In the western part of the Turkey, we have 9 continuous stations running along the Southern branch of the North Anatolian Fault Zone since 1986. The distance between the stations is about 10 km.
Question (Jean Rasson): For dual display of ISK and IZN there seems to be very long term drift in ISK.
The proton magnetometer is a in-house design?
Answer (Cengiz Celik): Systematic geomagnetic observations in Turkey started in 1947 at Istanbul-Kandilli (ISK) Observatory, which is located about 15 km out of city center of Istanbul. Increasing population and demand for residential, commercial and industrial zones caused enlargement of the city and thus in time ISK Observatory is engulfed by the city. Naturally, the geomagnetic observations are affected by the artificial noise throughout this process. So the ISK data are very noisy.

Since 1986, 9 continuous geomagnetic total intensity stations have been running to observe tectonomagnetic field in the Western part of North Anatolian Fault Zone. This is a cooperative project with Japan. Therefore, the proton magnetometers running in this project are of Japanese design.

Question (Angelo De Santis): For true reference of your magnetometer for seismomagnetic studies do you use GPS time?
Answer (Cengiz Celik): In the western part of the Turkey, we have 9 continuous stations running along the Southern branch of the North Anatolian Fault Zone since 1986. Simple differences technique is used to determine the tectonomagnetic field. So one of the stations which is away from fault segment is used as reference station. Also we use night-time data to detect seismomagnetic effect because night-time data are quieter than day-time data. All stations operate simultaneously. This is very important, so we use GPS time.

# CROATIAN GEOMAGNETIC REPEAT STATIONS NETWORK

MARIO BRKIĆ[10], DANIJEL ŠUGAR, MILAN REZO,
DANKO MARKOVINOVIĆ AND TOMISLAV BAŠIĆ
*Faculty of Geodesy*

**Abstract.** The establishment of the first geomagnetic repeat stations network of the Republic of Croatia is presented. Repeat stations were designed in accordance with the recommendations of the Coordination Committee for Common European Repeat Station Surveys, and of the International Association of Geomagnetism and Aeronomy, as well as experiences of the European countries. Final locations of the repeat stations were determined by field evaluation of all criteria. Following the testing of various materials, rock monuments were selected to mark station locations. The network stations were established by implanting the monument using standard geodetic procedures. Simultaneously with the network setup, the first declination, inclination, and total intensity survey took place in the summer of 2004.

**Keywords:** geomagnetic network setup; repeat station; total magnetic field gradients; geomagnetic survey.

## 1. Milestones

The oldest geomagnetic measurements in Croatian territory date from 1806, when the first declination survey was performed on the Adriatic coast. This survey was followed by numerous geomagnetic surveys on the continent (for a review see e.g. Brkić et al., 2003). In 1928 J. Mokrović put together all the surveys relating to the epochs of 1806 - 1918 and used them in the calculation of the geomagnetic elements for the epoch 1927.5. Mokrović proclaimed the significance of the geomagnetic observatory and pointed out the relationship between horizontal intensity anomalies and the geological structure of the Earth's crust. With no Croatian observatory, a declination

[10] Address for correspondence: University of Zagreb, Faculty of Geodesy, Department for Geomatics, Kačićeva 26, HR 10000 Zagreb, Croatia, e-mail: mario.brkic@geof.hr



*J.L. Rasson and T. Delipetrov (eds.), Geomagnetics for Aeronautical Safety,* 137–143.

survey took place along the Adriatic coast and on islands in 1949 (Goldberg et al., 1952). These declination measurements, reduced to the 1950.0 epoch, represented the last available information of the geomagnetic field in Croatian territory. Prior to the proclamation of the Republic of Croatia in 1991, the geomagnetic observatory in Grocka was in charge of geomagnetic surveys. Thus in 2002, aware of the significance of geomagnetic information, a preliminary study was prepared by the Faculty of Geodesy for the Ministry of Defense (Bašić et al., 2002). Two projects of the Faculty of Geodesy supported the renewal of geomagnetic studies in Croatia today. The first one, "Geomatica Croatica", launched by the Ministry of Science in 2002, provided a Bartington D/I MAG01H fluxgate with MAG Probe A, a Zeiss 010B theodolite, and a GEMSyS GSM-19G gradiometer. The second project, started in 2003, called the "Basic geomagnetic network of the Republic of Croatia – for the purposes of official cartography" was done under contract with the State Geodetic Administration. This project funded the setup of the Croatian Geomagnetic Repeat Station Network. In addition, there is a parallel effort at the Faculty of Science in which a geomagnetic observatory will be built (Vujnović et al., 2004).

## 2. Repeat Stations Network Design

It is advantageous to know the behavior of the geomagnetic field before setting up a repeat station network. Unfortunately, with the exception of the 2003 small scale total intensity measurements done by the Faculty of Science, the available geomagnetic data considered in the network design dated from the mid 20th century or even earlier. Since the exact positions and the descriptions of the Yugoslav geomagnetic repeat stations were unknown, it was decided to make use of existing networks locations. However, trigonometric, gravimetric and GPS points generally do not fulfill the International Association of Geomagnetism and Aeronomy criteria (Newitt et al., 1996, Wienert, 1970, Jankowski and Sucksdorff, 1996), nor are they recommended by the Coordination Committee for Common European Repeat Station Surveys (2003), so a new network was designed.

Before establishing a network of repeat stations, networks from other European countries were studied (see Korte and Fredow, 2001, Coticchia et al., 2001, Kovács and Körmendi, 1999). Then, major anomalous structures were determined. Areas with large total field anomalies (nT) and gradients, which must be avoided, were found by subtracting IGRF model from Mokrović's data for the 1927.5 epoch (Brkić et al., 2005). Ferrous ore is unevenly distributed across Croatia, and exists at numerous sites. Such sites were excluded. Taking advantage of maps and orthophotos, as well as new infrastructure plans, civilization noise sources (like railway, roads, etc.) were identified. Taking into consideration the peculiar shape of Croatian territory, as well as interpolation requirements, the macro locations for repeat stations were proposed.

## 3. Repeat Stations Network Setup

Field know-how is of utmost importance in finding the repeat station locations. By exploring the proposed macro locations, the actual repeat station locations were chosen based on easy access to unused meadows, little geomagnetic noise, prominent landmarks (like churches), and the possibility to establish new reference points. However, the exact locations for repeat stations followed from the thorough field assessment of all the criteria.

### 3.1. GRADIOMETRY

The most important criteria used to select repeat station locations are low total intensity gradients, less then a few nT/m (Newitt et al., 1996). Selected locations were visually checked for sources of noise first. Then, positions of stations were determined by rough gradient measurements in cardinal directions (NS–EW), and by checking PPM short time records. The positions were confirmed by measuring total field gradients and using the appropriate software (developed to utilize only one GEMSyS GSM-19G gradiometer). A quick (up to 2 minutes) series of measurements and an acceptable K-index were required to reduce the data to the measurements at the central station point (Brkić et al., 2005). Gradients were measured (1) above the station; (2) in a 10 m radius of the station; (3) in a 1 m x 1 m, inner grid; and (4) in a 10 m x 10 m, outer grid. Total intensity gradients at the repeat stations were low, typically less then 1 nT/m.

All these findings are documented in a 'Geomagnetic Repeat Stations Parameters' form. The form includes: total intensity gradients (along with the height of the probes), differences between primary and secondary repeat stations, differences between auxiliary and primary stations, D-I-F times, and measurements and their errors, with height of the probes, Kp-indices, reduced measurements with reference to methods and observatories, geological description of the locations, and notes concerning physical condition as well as possible sources of magnetic contamination.

### 3.2. REPEAT STATION MONUMENT ERECTION

After testing of various materials, Istrian hard limestone 'Kanfanar' was selected to mark repeat station locations. Repeat station monuments are

15 cm x 15 cm x 60 cm limestone blocks which weigh 36.45 kg. The blocks have two 15 cm x 15 cm x 5 cm underground centers, each with a weight of 3.04 kg, and have a cross carved on the upper facet. The monument erection procedures and the station description procedures are standard geodetic practice (Brkić et al., 2005).

### 3.3. AUXILIARY AND AZIMUTH REFERENCE POINTS

Besides the repeat stations, other points were set up in order to take declination measurements. The PPM auxiliary (AUX) points were usually set up in the NE corner of the outer grid for gradient measurements (approximately 7 m from the repeat stations) and were marked with wooden stakes. Also, three azimuth reference marks were placed a few hundreds meters from the repeat stations and marked with steel spikes.

For each repeat station a 'Position Description' form was maintained. The main elements of the form are: the repeat station name and coordinates, the name of the county and town, reference to the 1:50000 topographic map, a sketch of the location, a sketch with azimuths to reference marks, and a sketch and photo of the monument.

Table 11. Geomagnetic primary repeat stations names and positions.

| St. Name | Lat. (dec.deg.) | Lon. (dec. deg.) | h (m) |
|---|---|---|---|
| POKUpsko | 45.473 | 15.983 | 105 |
| MEDJimurje | 46.484 | 16.332 | 199 |
| BARAnja | 45.836 | 18.787 | 86 |
| RACInovci | 44.856 | 18.969 | 81 |
| KONAvle | 42.532 | 18.340 | 47 |
| SINjsko Polje | 43.649 | 16.689 | 296 |
| KRBavsko Polje | 44.670 | 15.630 | 648 |
| PONte Porton | 45.356 | 13.735 | 5 |

### 3.4. GPS SURVEY

Taking advantage of the existing '10km' GPS network (having points with 10 km spacing), the coordinates of all the repeat stations and azimuth reference marks were determined by GPS relative static positioning with Trimble 4000 SSI. The resulting positions are presented as $\phi$, and $\lambda$ coordinates on Bessel's ellipsoid (Table 11).

3.5. REPEAT STATIONS NETWORK

The Geomagnetic Primary Repeat Stations Network of the Republic of Croatia consists of eight stations. The minimum distance between the neighboring stations is 93 km, and the maximum distance is 183 km (Figure 76).

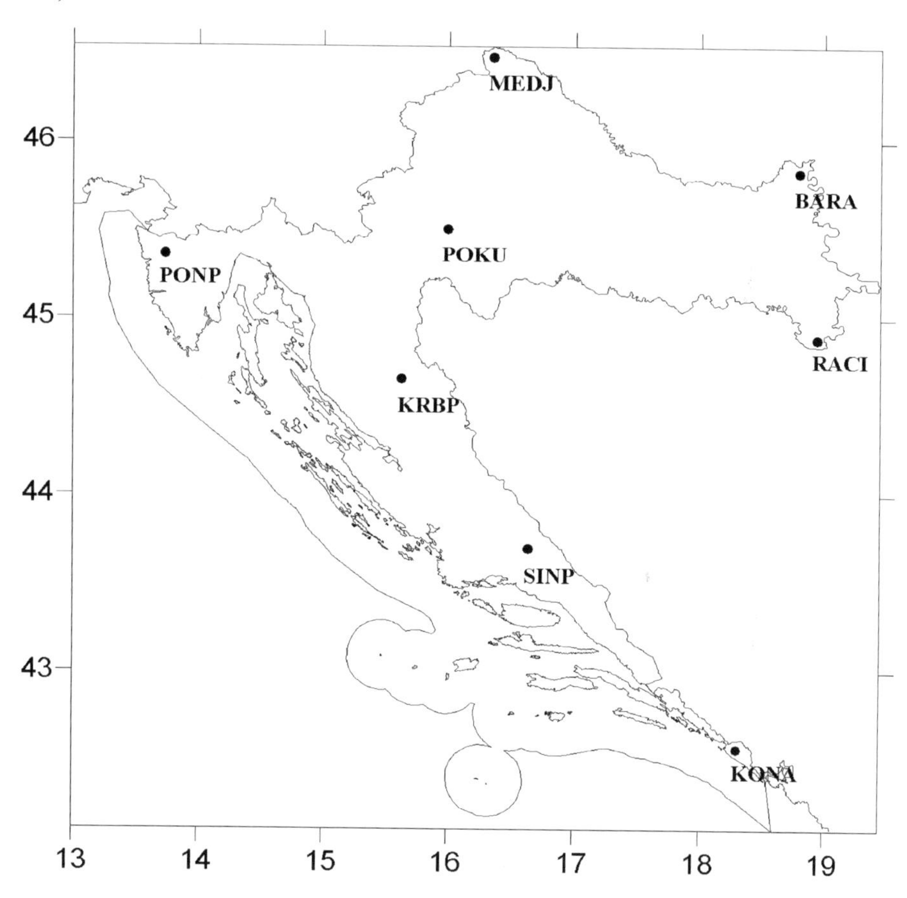


Figure 76. Geomagnetic Primary Repeat Stations Network of the Republic of Croatia.

## 4. Geomagnetic Repeat Stations Survey

The first survey of declination, inclination and total intensity in the Republic of Croatia was carried out in the summer of 2004. The instruments used were a Bartington Mag-01H D/I Fluxgate and a GEM Systems GSM-19G Overhauser Magnetometer. As a rule, two to three declination and inclination measurements were performed in the morning, as well as in the evening. For these measurements the null-method was employed. The total intensity was measured simultaneously at the AUX point. Because of financial and time constraints, only a few measurement sets were performed. These measurements are planned to be reduced to the nearest observatories.

## 5. Future prospects

The suitability of the established repeat stations for monitoring secular variation remains to be verified through periodic measurements of gradients and other geomagnetic elements over a longer time span. In addition to primary repeat stations, the establishment of secondary repeat stations, a few variometer points, and a Croatian geomagnetic observatory are planned. A denser vector field ground survey and airport surveys are also expected in the near future. Exciting times are ahead for geomagnetism studies in Croatia.

## References


Bašić, T., Brkić, M., Hećimović, Ž., Šljivarić, M., Markovinović, D., Rezo, M., Jungwirth, E., Viher, M., Horvat, S., 2002, Basic geomagnetic declination network of the Republic of Croatia, - preliminary study (in Croatian), Ministry of Defense of the Republic of Croatia, Institute for defense studies, research and development, Zagreb.

Brkić, M., Bašić, T., Verbanac, G., 2003, Geomagnetism in Croatia – a Historical Overview, a talk given at the Ludwig-Maximilians-Universität München on 15th May 2003, *Geodetski list*, 57 (80) **3**, 183-194, Zagreb.

Brkić, M., Šugar, D., Rezo, M., Markovinović, D., Bašić, T., 2005, Croatian Geomagnetic Repeat Stations Network (in Croatian), *Geodetski list*, 59 (82) **2**, Zagreb.

Coordination Committee for Common European Repeat Station Surveys, 2003, Recommendations for European repeat magnetic station surveys, *http://www.gfz-potsdam.de/pb2/pb23/GeoMag/eurepstat.html.*

Coticchia, A., De Santis, A., Di Ponzio, A., Dominici, G., Meloni, A., Pierozzi, M., Sperti, M., 2001, La Rete Magnetica italiana e la Carta Magnetica d'Italia al 2000.0, Estratto dal *Bollettino di Geodesia e Scienze Affini*, Rivista dell'Istituto Geografico Militare, Anno LX, N. **4**.

Goldberg, J., Baturić, J., Mokrović, J. , Kasumović, M., 1952, Determination of the magnetic declination in Yugoslav part of Adriatic sea 1949. (in Croatian), *Discussion on materials of the Institute of the history of the Science and Medicine of the Yugoslav Academy of Arts and Sciences*, **1**/2, 13-43, Zagreb.

Jankowski, J., Sucksdorff, C., 1996, *Guide for magnetic measurements and observatory practice*, IAGA, Warsaw, Poland.

Korte, M., Fredow, 2001, *Magnetic repeat station survey of Germany 1999/2000*, Scientific Technical Report STR01/04, GeoForschungsZentrum Potsdam.

Kovács, P., Körmendi, A., 1999, Geomagnetic Repeat Station Survey in Hungary during 1994-1995 and the Secular Variation of the Field between 1950 and 1995, *Geophysical Transactions*, 42, **3-4**, pp. 107-132.

Mokrović, J., 1928, Distribution of the main elements of the Earth's magnetism in the Kingdom of Serbs, Croats and Slovenes (in Croatian), Geophysical Institute in Zagreb, 3-14, Zagreb.

Newitt, L. R., Barton, C. E., Bitterly, J., 1996, *Guide for Magnetic Repeat Station Surveys,* IAGA, Boulder, USA.

Vujnović, V., Verbanac, G., Orešković, J., Marki, A., Marić, K., Lisac, I., Ivandić, M., 2004, Results of the preliminary geomagnetic field strength measurements in the northern part of middle Croatia. *Geofizika*, **21**, 1-13.

Wienert, K. A., 1970, *Notes on geomagnetic observatory and survey practice*, UNESCO.

## DISCUSSION

Question (Spomenko J. Mihajlovic): Did you measure at any repeat station in Croatia?

Answer (Mario Brkic): Yes – in addition to a new network setup in the summer of 2004 – the geomagnetic declination, inclination and total field were measured at all the repeat stations. Immediately after each station setup, two or more measurement sets were performed in the evening, and in the morning, depending on available time. Declination and inclination measurement method was the null-method, utilizing DI fluxgate, while total field was recorded in the nearby auxiliary point by means of Overhauser PPM. These first D-I-F measurements in the Republic of Croatia, are not checked and not reduced yet.

Question (Jean Rasson): There is a tendency when building a new network to put stations on the country border. If everybody does the same, than we will have high concentration of stations on the borders.

Answer (Mario Brkic): That is a fact. The tendency is a result of the network design requirement to cover the whole area of interest with properly spaced stations, along with the need for as much surveying and modelling independence as possible. Still, a high concentration of stations at the border could be seen not as an obstacle to regional field study, but an advantage.

Question (Bejo Duka): Did you ask Grocka observatory of Serbia and Montenegro for the old repeat station information?

Answer (Mario Brkic): Yes, during the 'Workshop on European Geomagnetic Repeat Stations' in Niemegk in 2003, an attempt to withdraw the old survey data regarding the Croatian territory has been made, but without success. However, the question is, are the repeat stations, as well as the azimuth marks, 'alive' so that continuation of measurements is possible? Perhaps not. On the other hand, the growing demands for the update of the geomagnetic information on the maps initiated a new network setup.

# GEOMAGNETIC INSTRUMENTATION FOR REPEAT STATION SURVEY

VALERY KOREPANOV[11]
*Lviv Centre of Institute of Space Research*

**Abstract.** Repeat station survey measurements are important geomagnetic data because they are widely used both for fundamental science (e. g., study of Earth's magnetic dynamo) and for applied purposes (e.g., declination charts for aviation safety). To execute repeat station surveys, normally three types of instruments are used: absolute scalar magnetometers, three-component vector variometers, and theodolite-mounted one-component magnetometers. The modern specifications of each magnetometer are described together with simplified fundamentals of their operation. The recommended set of such devices is given and possible further development of this type of instrumentation is discussed.

**Keywords:** repeat station, flux-gate magnetometer, Overhauser magnetometer, non-magnetic theodolite

## 1. Introduction

Observation of the Earth's magnetic field remains an important branch of scientific research. Moreover, the number of geomagnetic observatories continues to increases – both manned and especially unmanned ones. Observatory data are shared via international networks. INTERMAGNET works with stationary observatories; CANOPUS, IMAGE and some other networks are for unmanned observatories. The same trend is observed with repeat station surveys – more often they are executed by international teams, especially in regions close to state borders.

Repeat station data are obtained by national institutions that are responsible for magnetic surveys with a maximum interval of five years. These measurements are used to determine the so called secular variations

[11] To whom correspondence should be addressed at: Lviv Centre of Institute of Space Research, Naukova 5-A, 79060 Lviv, Ukraine; e-mail: vakor@isr.lviv.ua.



*J.L. Rasson and T. Delipetrov (eds.), Geomagnetics for Aeronautical Safety,* 145–166.

of the Earth's magnetic field – the slow changes of direction and value of the field vector with time. The example (Figure 77), shows that these changes are big and they demonstrate the evolution of the magnetic dynamo in the core of the earth. Permanent magnetic observatories are the most accurate sources of secular variation information, but the present network of magnetic observatories does not adequately cover the globe. Repeat stations provide an important and cost-effective means of supplementing observatory data with the most valuable stations being those remote from observatories. Repeat station data have long been used for producing regional field models and charts.

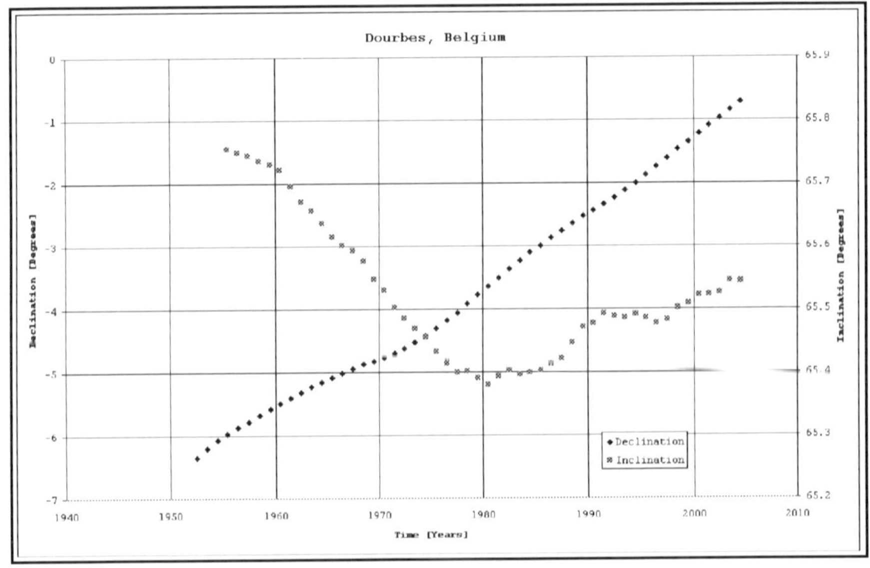


Figure 77. Declination and inclination changes at Dourbes magnetic observatory for 50 years.

Besides the fundamental scientific importance of the secular variations for understanding the Earth's magnetic dynamo mechanism, knowledge of the geomagnetic field elements (i.e. vector components) is of special interest for navigation in general and in particular for aircraft navigation.

The directions of geographic North (True North) and magnetic North do not coincide. The difference is the angle called "Magnetic Declination": to find True North by means of a compass, a correction should be introduced to the direction the compass indicates. Measurement of the declination is especially important at airports. Without information about the declination angle there is real danger that aircraft will have wrong headings when landing. Additionally, airports must provide adequate calibration pads for aircraft compass certification and checks (swings), where the magnetic declination should be precisely known. The magnetic compass is still the primary navigation device on aircraft. In case of failure of other electronic navigation devices (GPS, VOR) the magnetic compass is an important backup tool. The failure to correctly calibrate and operate magnetic compasses represents a security risk to aircraft and airports.

The knowledge of the geographical distribution of magnetic declination allows for the mathematical calculation of magnetic headings which also appear on aeronautical charts.

Declination is not a constant value, but changes rather chaotically due to secular variations of the geomagnetic field. In order to know the declination value precisely, regular magnetic measurements must be made. In the past, these measurements were made every 5 years, but now, most European airports require making such measurements once per year, for the sake of improving flight safety. This work is rather expensive, needs precise magnetic instrumentation, and takes qualified manpower. Additionally, there is no regular production of non-magnetic instrumentation for performing such measurements.

The aim of this paper is to analyze the state of geomagnetic instrumentation and to present recent developments in this field.

## 2. Requirements of the instrumentation

The key to obtaining useful repeat station data is the ability to make accurate corrections for transient field variations so that the secular variation can be determined from the differences between results from successive station occupations. These corrections can be obtained either by using one or more permanent observatories as a reference standard, or by installing a variometer on-site and running the repeat station as a temporary observatory.

Each method has its advantages and limitations. The reference observatory method is quicker, easier, and cheaper and is the natural choice when it can be demonstrated that the transient-field corrections obtained at the reference observatory are applicable at the repeat station site. It is important to investigate and confirm this result, and not simply assume it for convenience, before adopting the reference observatory method. Data from a local variometer help to determine the quiet level of the magnetic field when the repeat station is far from an observatory (Newitt et al., 1996). This is usually so when we apply the repeat station survey techniques for calibration pad certification at airports. A complete set of geomagnetic instrumentation must include three devices: an absolute magnetometer, a three-component variometer, and a theodolite-mounted declinometer-inclinometer. To satisfy the acceptable level of measurements error (~20" for declination and ~10" for inclination) the instrumentation has to fulfill necessary requirements to its metrological parameters. The main parameters are given in Table 12.

Table 12. Main requirements of the instrumentation for repeat station surveys.

| ABSOLUTE MAGNETOMETER | |
|---|---|
| Obligatory parameters | |
| Magnetic field measurement band | ± 60 000 nT |
| Measurement resolution | 0.1 nT |
| Absolute error of measurement | ≤ 0.5 nT |
| Desirable parameters | |
| Stability of precessing liquid | ≥ 5 year |
| THREE-COMPONENT VARIOMETER | |
| Obligatory parameters | |
| Magnetic field measurement band | ± 60 000 nT |
| Measurement resolution | 0.1 nT |
| Linearity | ≤ 0.1 % |
| Temporal drift | ± 2 nT/day |
| Thermal drift | ≤ 0.2 nT/°C |
| Desirable parameters | |
| Power consumption | ≤ 2 W |
| Sensor tilt compensation available | |
| DECLINOMETER-INCLINOMETER | |
| Obligatory parameters | |
| Magnetic field measurement band | ± 1000 nT |
| Magnetometer offset | ≤ ± 10 nT |
| Measurement resolution by field | 0.1 nT |
| Measurement resolution by angle | 1 arc sec |
| Desirable parameters | |
| Autonomous power supply | |

Each of these instruments is described below in detail and some examples are given. Our goal was not to compare and criticize other available instruments – we used some known brands only to describe their specific peculiarities which could be interesting for the user.

## 3. Absolute magnetometers

The name "absolute" was given to the devices for measuring the magnetic field. The operation principle of these devices is based on fundamental physical constants. There are few known types of absolute magnetometers. The most common types are proton precession magnetometers. They are

Table 13. Principle of absolute magnetometer operation.

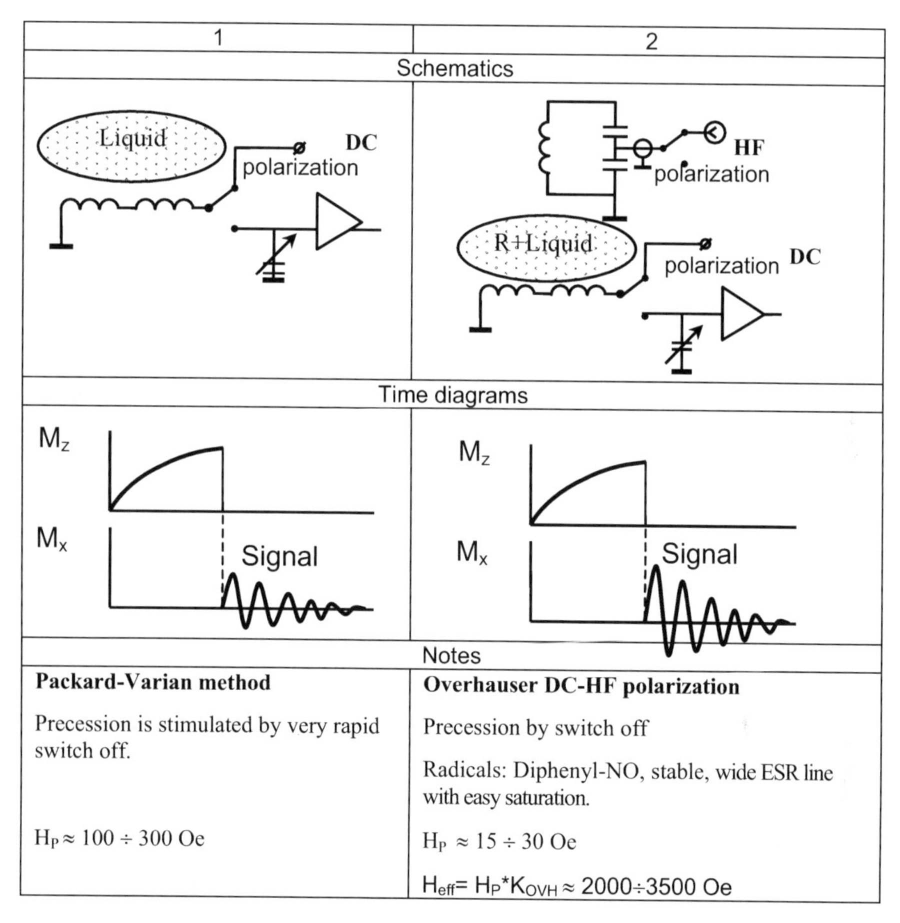


| 1 | 2 |
|---|---|
| Schematics | |
| Time diagrams | |
| Notes | |
| **Packard-Varian method** | **Overhauser DC-HF polarization** |
| Precession is stimulated by very rapid switch off. | Precession by switch off<br>Radicals: Diphenyl-NO, stable, wide ESR line with easy saturation. |
| $H_P \approx 100 \div 300$ Oe | $H_P \approx 15 \div 30$ Oe<br>$H_{eff} = H_P{}^*K_{OVH} \approx 2000 \div 3500$ Oe |

used for measuring the magnetic field strength in the range of 20 μT to 100 μT with accuracy of 0.1nT. They are based on the Packard-Varian method where a proton-rich liquid (in the simplest case - water) is polarized by a strong magnetic field created with the help of a wire winding around this liquid volume, into which a DC current pulse is inserted (Table 13, item 1). After the pulse cessation, the free precession frequency, f, of protons can be observed and the formula:

$$2\pi f = \gamma'_p B$$

allows the measurement of the magnetic field module by way of the measurement of a frequency. The standard for the magnetic induction is thus conveniently converted to a frequency standard, which is widely available. The quantity $\gamma'_p$ is the fundamental physical constant - proton gyromagnetic ratio.

Its recommended value, at low field for a spherical $H_2O$ sample, at 25 degrees Celsius given by CODATA and adopted by IAGA in 1992 is (Cohen and Taylor, 1987):

$$\gamma'_p = 2.67515255 \mathrm{T}^{-1} s^{-1}.$$

The magnetic sensor coil serves both to periodically apply a polarizing field to the liquid and to pick up the signal from the precessing protons after cutting off the polarizing field. An electronic console amplifies the precession signal and performs a frequency measurement with the required accuracy. This measurement is then scaled using $\gamma'_p$ to give the field module intensity in Teslas. Such magnetometers mostly operate once per 5 seconds or less and their precision is often as high as 0.1 nT.

Modern proton magnetometers can be quite compact and have higher resolution and sampling rates. The Overhauser type magnetometer (Sapunov et al., 2001), in particular, is responsible for this progress and has the additional benefit of low power operation.

In addition, it applies an AC field in the radiofrequency band (Table 13, item 2) which allows it to get a much higher precession signal amplitude. One of the best magnetometers of this type is the POS-1 Overhauser magnetometer produced by Ural State Technical University (Russia) (Sapunov et al., 2001) shown in Figure 78.

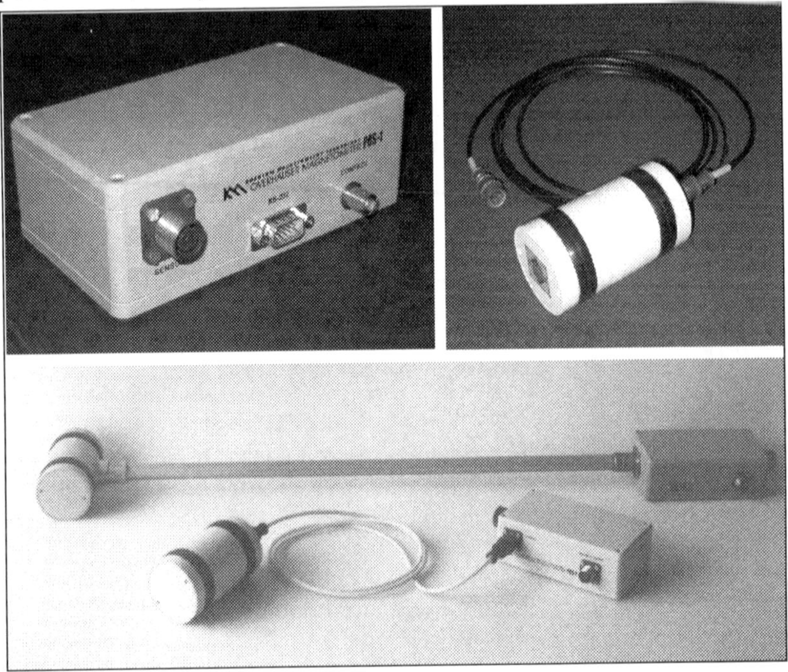

Figure 78. POS-1 Overhauser magnetometer.

The POS operation principle is similar to that of the standard proton magnetometer. Polarization by POS-1 Overhauser occurs in a bias DC-magnetic field (15-30 Oe strength) with an alternating HF-field (frequency about 55 MHz). In this way, it is possible to avoid a sharp decrease in the proton Overhauser signal in the range of 20000-40000 nT and to exclude systematic errors produced by the feedback circuit in other types of Overhauser magnetometers. This design uses a new stable chemical substance with lifetime of up to 10 years.

The POS digital processing of the proton precession signal (Denisov et al., 1999) ensures high sensitivity of measurements (up to 0.01 nT at a 3 sec cycling rate, or 0.1 nT at 1 sec. for the standard Overhauser head). The processing algorithm also provides simultaneous assessment of the measurement error by a quantity named QMC (quality of measuring conditions). QMC is a parameter of sensitivity estimate for the real measurement conditions in field units (nT) available for each single measurement.

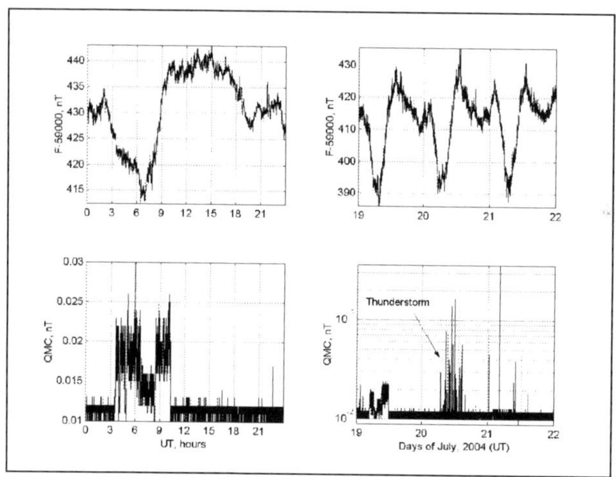


Figure 79. The records of POS-1 with QMC increase: because of technogenic noise (left panel) and thunderstorm (right panel).

The value of QMC is registered together with the total field and can be controlled visually by the magnetologist during processing. The usual QMC for POS-1 is about 10-12 pT. But it is also possible to see significant increases in QMC from time to time. The duration of these intervals vary, but changes correlate mostly with external factors. We assume that the origin of noise is technical. Figure 79 (left panel) shows POS-1 data with an increase of QMC during the daytime. The total field itself is not correlated with changes of QMC. The simultaneous measurements by two POS-1 showed that the increase of QMC is present in both records, and hence is not caused by sensor failure.

Not all noise appearing as changes of QMC is technogenic. It was noticed that QMC varied significantly during the passing of a thunderstorm close to the observatory (up to a few kilometers). The example of such an effect is shown on Figure 79 (right panel) – the thunderstorm moved near the observatory in the evening of July, 20 2004 (about 13UT). This meteorological event did not influence the total field.

An additional application of QMC for repeat station measurements may be the selection of the place for the installation of the POS sensor. The main technical specifications of the POS-1 magnetometer are given in Table 14 and its attributes and advantages are given in Table 15.

Table 14. Main technical specifications of POS-1 magnetometer.

| Range of measurement | 20000-100000nT |
|---|---|
| Resolution | 0.001nT |
| Sensitivity (mean-square error at the optimal sensor orientation) | 0.01 – 0.02nT at 3 sec (the best condition); 0.05 - 0.1nT at 1 sec cycle |
| Absolute accuracy | ± 0.5nT |
| Gradient tolerance | up to 20000nT/meter |
| Operating modes | the user can select single or continued operation by commands via RS232 port |
| Reading intervals | 1.0, 2.0 3.0 ... sec (optional 2, 3, 4, 5 Hz) in cycle mode |
| Data output | three wire RS232 port (binary and/or text format) |
| Power | 10-15 VDC (15 VDC at 0,35 A max and 10 VDC up to 0,5 A max in polarization period) |
| Operating temperature | -10 to +60 C° (at -30 to +75 C° able to work) |

Table 15. The attributes and advantages of POS-1 magnetometers.

| POS-1 attributes | Possibilities |
|---|---|
| Stable sensor fluid with lifetime up to 10 years | Long use of magnetometer without re-filling sensor |
| Short time of sampling up to 1 s during continuous recording | Easy synchronization of variation measurements with POS-1 data |
| Small polarization current | Decrease the magnetic field disturbance from POS-1 (about 0.07 nT at 1.5 m) allowing location of the POS close to another sensor. |
| The digital connection of POS-1 and PC with cable length of up to 100 m | The recording system can be located a long distance from the sensor |
| Wide range of working temperature (down to -30°C) | Keeps recording during a long cut-off of the hut thermostatic system in the winter |

| POS-1 attributes | Possibilities |
|---|---|
| Small size and weight of POS-1 sensor head | Allows use of the POS-1 as the magnetic sensor of the component system |
| Parameter QMC | Allowing controls of the signal quality and noise level during the measurements |

So, the parameters of this magnetometer fully correspond to necessary requirements. Additionally, it has a relatively low price. Therefore it is recommended for repeat survey practice.

## 4. Flux-gate magnetometers

Geomagnetic investigations need high accuracy data. Among various types of magnetometers, the flux-gate magnetometers (FGM) get high quality results at relatively low cost. They are the most widely used magnetometers for both observatory and repeat station observations of the Earth's magnetic field components. Recent developments in the technology, design, and manufacture of flux-gate sensors (FS) have enabled the sensors to operate within acceptable noise levels of a few pT. However, the sensors are especially sensitive to temperature changes and lack in long-term stability. These are important considerations for repeat stations surveys. To satisfy these temperature stability requirements, new theoretical and technological low cost approaches to the design of the FGM, were studied (Berkman et al., 1997).

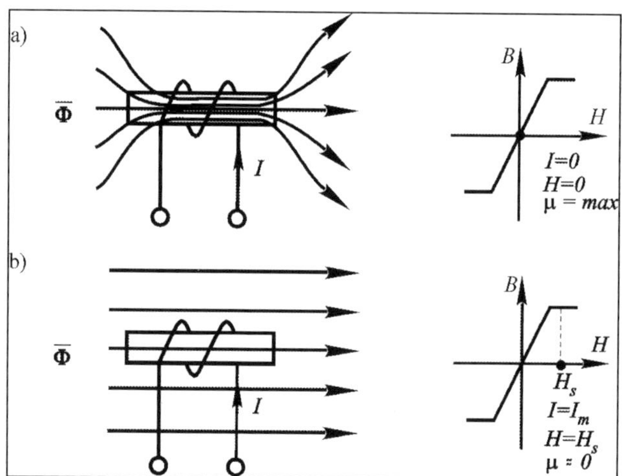


Figure 80. Principle of FGM operation.

In practice, FGMs are used for the measurement of the components of the stationary magnetic field vector. In repeat station survey applications, FGM's are used to measure the Earth's magnetic field vector. Their operation principle is based on Maxwell's law of electromagnetic induction which, in integral form can be written as

$$e = -\frac{d\Phi}{dt},$$

where $e$ is the electromotive force (*emf*), $\Phi$ is the magnetic flux, and $t$ is time.

According to this expression, only an alternating magnetic field can produce *emf* when intercepting a sensor coil. So, to measure a DC field it is necessary to modulate the DC magnetic field component in order to get an output signal. It works in the following way. The FGM sensor (FS) consists of a high-permeability, ferromagnetic core with a winding around it (Figure 80).

In the normal state of the ferromagnetic core, the magnetic field flux $\Phi$ is concentrated in the core (Figure 80, a), due to the cores high relative permeability,. If a strong current (*I*) is applied into the winding, the core becomes saturated and its relative permeability trends to unity. In this state, the core does not concentrate the magnetic flux (Figure 80, b). By introducing alternating current (*I*) into the winding (excitation current), it is possible to gate the magnetic flux $\Phi$ in and out of the core, or in other words, to modulate it, transforming from DC to AC flux. The AC flux, intercepting winding, generates *emf* at its output, according to Maxwell's law.

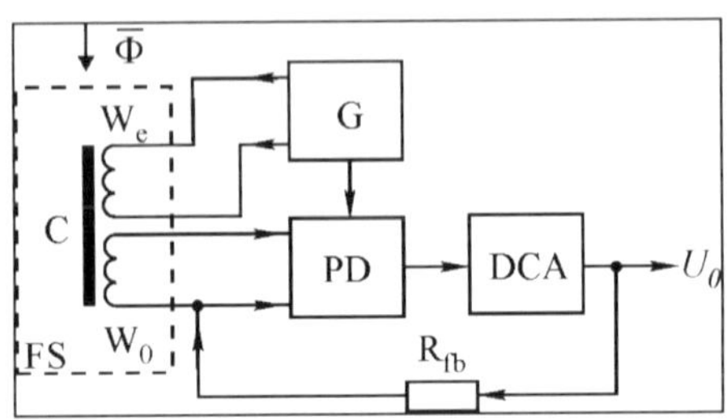


Figure 81. Simplified FGM operation diagram.

The simplified functional diagram of FGM is presented in Figure 81. The bar-type flux-gate sensor (FS) consists of a magnetic core C, an excitation winding $W_e$, and an output/feedback winding $W_0$. The Oscillator/Drive unit (G) provides alternating current through the excitation winding $W_e$, which drives the magnetic core of the sensor in and out of saturation. In the moments of core saturation (two times per one period of excitation current) its relative magnetic permeability falls from a maximum value to 1. This leads to the modulation in the core twice for an excitation period of the total magnetic flux, produced by the external magnetic field. Total magnetic flux changes induce the output signal in the output coil $W_0$ of the sensor at the second and all even harmonics, which is dependent on both magnitude and polarity of the external field. The sensor (FS) output signal goes to the phase synchronous detector PD with a low-pass filter at the output. The PD uses the clock signal of excitation, double frequency signal, from the Oscillator/Drive circuit G and forms the output DC signal proportional to the amplitude and polarity of the measured magnetic field. The output signal of the PD is further amplified by the DC amplifier DCA and forms the FGM output signal $U_0$. To stabilize the transfer factor of the FGM, a feedback resistor $R_{fb}$ is coupled between the FGM output and the $W_0$ winding. The DC current, through resistor $R_{fb,}$ creates C magnetic flux in the FS core. This compensates the measured flux Φ and stabilizes the total FGM transfer function.

The high quality FGMs use a ferroresonance excitation mode (FEM), when the excitation winding $W_e$ of flux-gate core C is shunted by capacitance $C_k$ and they both form a non-linear oscillator with low active losses (Berkman et al., 1997) (Figure 82). Figure 83 shows current and voltages in the circuit.

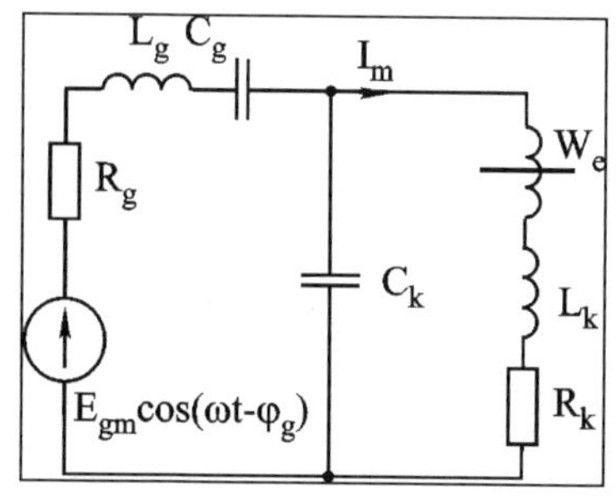


Figure 82. FEM schematic diagram.

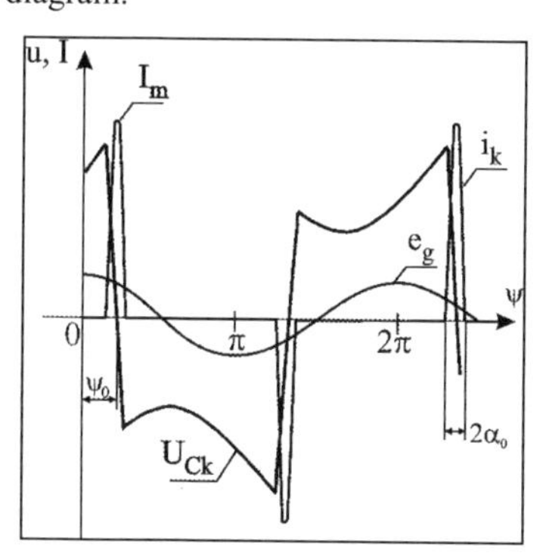


Figure 83. Voltage and current curves of FEM.

Here $R_g$, $L_g$, $C_g$, and $E_g$ are output parameters of the excitation current generator G. $L_k$ and $R_k$ are inductance and resistance of the winding $W_e$, φ – $E_g$ describes the phase and $2\alpha_0$ is the excitation pulse width.

The series circuit $L_gC_gC_k$ in FEM is tuned near the frequency ω of the excitation source $E_g$. The main function of the FEM is to store capacitance

($C_{k)}$ charge which is created at the end of the demagnetization interval for the generation of the current discharge pulse ($I_m$). The pulse is of great amplitude at the saturation interval. Because of the relatively short time of the saturation interval, ($2\alpha_0 << \pi$), the energy input is sharply decreased, especially when the discharge circuit $L_kC_k$ has a high Q- factor.

Using FEM, it is possible to have current pulses $I_m$ (Figure 83) in the excitation winding achieving amperes, when mean consumed current is only tens of milliamperes and active losses in the winding are less than 0,05 watts. A high amplitude of the excitation field (2000 A/m and more) eliminates hysteresis zero drift, and its short duration lowers heat dissipation in the sensor volume. Additional advantages of the FEM are the sensitivity stability and low noise level: typical values are 20 pT rms, and the lowest values about 3 ÷ 5 pT.

A reference, three component flux-gate magnetometer LEMI-018 that uses this excitation mode was developed especially for measurements in difficult field conditions. Its external view is shown in Figure 84 and its technical parameters are given in the Table 16.

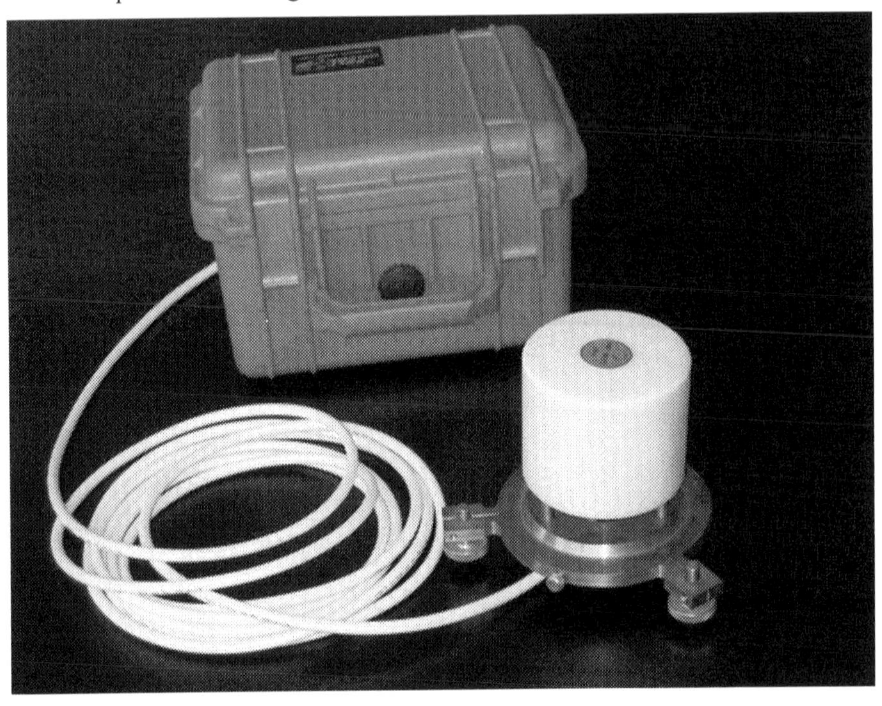

Figure 84. Three component flux-gate magnetometer LEMI-018.

Table 16. LEMI-018 main technical parameters.

| | |
|---|---|
| Measurement range along each component | ±65 000 nT |
| Resolution | 10 pT |

|  |  |
|---|---|
| Noise level at analog output in the band 0.01 –0.5 Hz | < 10 pT |
| Long-term zero drift | < ± 5 nT/year |
| Temperature drift | < 0.2 nT/°C |
| Transformation factor linearity error | < 0.01% |
| Components orthogonality error | < 30 min of arc |
| Time of samples averaging | 1, 2, 5, 10, 60 s |
| Internal FLASH-memory volume | 512 MB |
| Operating temperature range | minus 20 to +40 °C |
| Power supply source | 12 V |
| Power consumption | < 0.6 W |
| Weight: | |
| sensor with rotating basement | 1,7 kg |
| electronic unit with built-in battery | 4 kg |
| Length of connecting cable between sensor and electronic unit | 7 m |
| Optional | memory volume and cable length extension |

As was previously stated, in order to be suitable for repeat station surveys, the FGM has to have long term baseline stability and very low thermal drift. Both these parameters werc investigated and the following results were obtained.

Figure 85 displays FGM thermal test data. Figure 85 a, shows the FGM baselines change with temperature for all three components. Even without compensation, the thermal drift for LEMI magnetometers fits in the limits ±0.2 nT/°C. In some applications thermal drift values can be decreased further with specially designed drift compensation hardware or software. Results of the thermal dependence of baselines for a compensated FGM is shown in Figure 85 b. Thermal drift for a compensated FGM is negligible.

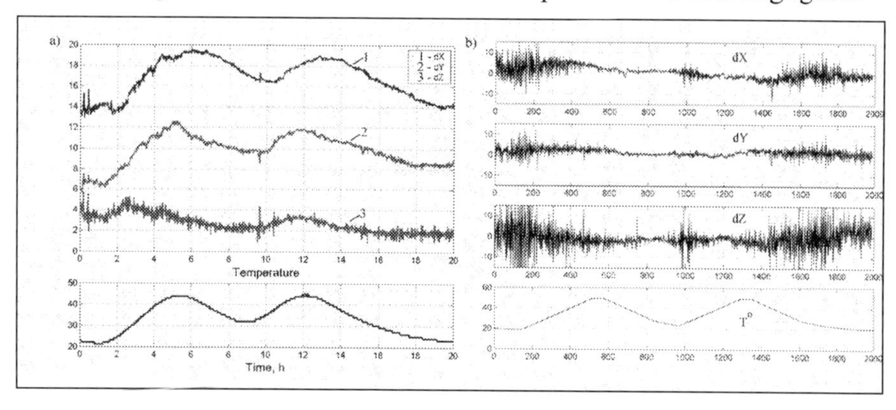


Figure 85. Thermal tests results.

The LEMI FGMs also demonstrate temporal stability. Figure 86 shows the results of absolute measurements of FGM LEMI-008 baselines for almost three years. It can be seen that this component of drift is well below the international standard of ± 5 nT per year.

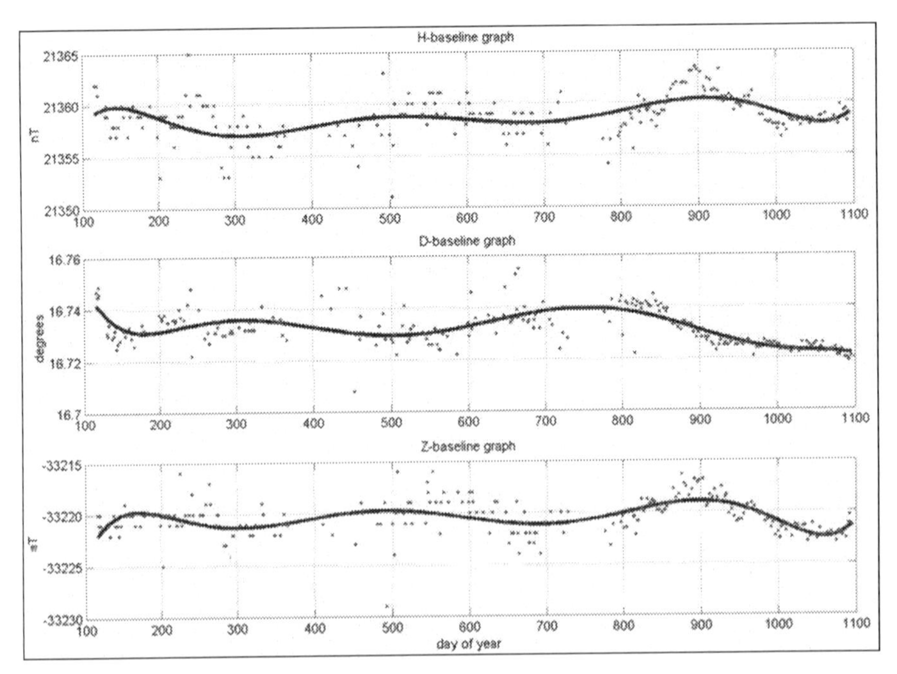


Figure 86. Long-term baseline stability.

In summary, because the FGM is highly accurate it is appropriate to use in field applications, for repeat station surveys, especially in regions with geomagnetic anomalies or sites far from geomagnetic observatories (~ more than 50-70 km).

## 5. Theodolite-mounted declinometer-inclinometer

The ultimate goal of repeat station survey work is to determine the absolute values of the Earth's magnetic field components. In practice, it is made on the basis of exact (absolute) measurement of the scalar value (modulus) F of the magnetic field with the help of an absolute magnetometer (see section 3) and measurements of the Earth's magnetic field, declination angle (D) and inclination angle (I). The declination angle (D) is the angle between the direction to geographic North (True North) and the direction to the geomagnetic North. The inclination angle (I) is the angle between the vector of the Earth's magnetic field and its horizontal component. Having declination (D), inclination (I), and F it is then easy to calculate the X, Y, Z components. Details of this procedure and minimizing measurement error are described in (Newitt, 1996).

The classic instrument used for D and I measurements is a portable one-component, flux-gate magnetometer. Its sensor is mounted on the telescope of a non-magnetic theodolite so that the sensors' magnetic axis and the telescope optical axis are parallel. The sensor is coupled to a battery-powered electronics unit by a long flexible cable. The instrument is commonly called a Declination-Inclination Magnetometer (DIM).

Before providing a detailed description of the DIM, it is necessary to mention that, starting from about 1960 (Alldredge, 1960) new instruments for absolute measurement of Earth's magnetic field were proposed. The majority of these devices measured a given component of the magnetic field using a scalar absolute magnetometer mounted in a three-component coil system where the other two magnetic field components are compensated with current in the windings. There were many attempts to build and commercialize such instruments. The most successful design the dIdD magnetometer (Pankratz et al., 1999). Recently a new device which uses both an absolute sensor with a field compensating coil and a theodolite has been designed. It is known as the DIMOVER (Sapunov et al., 2004). A comparative evaluation of the new devices with traditional methodology shows that they are not the quality of the DIM (Kotzé et al., 2004). Here only the DIM will be described.

The most widespread, accurate instrument in use at the end of the 20th century was the DIM, produced by Bartington in Great Britain. The single component fluxgate sensor is mounted on a Zeiss-Jena 010B non-magnetic optical theodolite, which has 1 second of arc resolution. The production of these theodolites was discontinued because of low market demand and now most theodolites are produced with electronic read-out and have magnetic components. These are more accurate and easier to use than optical theodolites, but they can not easily be demagnetized. Today the only 1-second, steel, optical theodolite on the market is the 3T2KP, produced by the Ural Optical-Mechanical Factory (UOMZ) in Ekaterinburg, Russia. These theodolites have nearly the same construction and specifications as the Zeiss-Jena 010B and attempts to demagnetize them were successful. Lviv Centre of Institute of Space Research (LCISR), Ukraine, and Mingeo Company, Hungary, now produce DIMs based on this demagnetized theodolite.

The 1-second resolution, demagnetized theodolite, 3T2KP-NM has technical specifications similar to those of Zeiss-Jena 010B (see Table 17 below).

Table 17. Main parameters of 3T2KP-NM theodolite.

| | |
|---|---|
| Measuring accuracy face left/face right | 2" |
| The same for zenith angles (mz) | 2.4" |
| Measuring range for zenith angles | 30-145° |
| Telescope | |
| Image position | erect and true-to-side |
| Magnification | 30× |
| Angle of field of vision | 1°35' |
| Shortest sighting distance | 1.5 m |
| Stadia factor | 100±05 |
| Addition constant for stadia factor | 0 |
| Vertical index stabilization | |
| Operating range of the pendulum | ±3′ |
| Mean square setting error | 0.8" |
| Plate levels | |
| Tubular level angular value | 15"/division |
| Circular level angular value | 5'/division |
| Graduated circles | |
| Horizontal/vertical circle dial diameter | 90 mm |
| Coarse finder scale division value | 10° |
| Circle scale division value | 10' |
| Micrometer scale division value | 1" |
| Built-in optical plummet | |
| Image position | erect and true-to-side |
| Magnification | 2.5× |
| Angle of field of vision | 4°30' |
| Focusing range | 0.6 m to infinity |
| Weight (kg) | |
| Instrument | 4.0 |
| Support | 0.7 |
| Instrument in plastic container with accessories | 9.2 |
| Tripod | 5.6 |
| Dimensions (cm) | |
| Height of instrument with support | 34.5 |
| Height of horizontal axis from the lower support | |
| plain with footscrews in middle position | 23.2 |
| Container | 47×24×21 |
| Tripod | d16 x (100-160) |
| Operating temperatures | -10 - +50° C |

The demagnetization of the theodolite is a complicated process because it is impossible to predict which parts of the theodolite will be from magnetic material. The possible influence of remanent magnetization on the D and I measurement precision was studied in depth. The results of Rasson's study (Rasson, 1994) show that the remanent magnetization of telescope parts has little or no influence on the measurements (see Table 18 and Table 19 from (Rasson, 1994)).

Table 18. VFO31 variometer declination baseline.

| $D_0$ | ε (") | δ (") | $S_0$ (nT) | Conditions | Orientation | Location |
|---|---|---|---|---|---|---|
| 2°52'03" | -12 | 53 | -1.9 | no magnet | | |
| 2°52'17" | -42 | 50 | 558 | magnet | horizontal | on telescope |
| 2°51'54" | -8 | 63 | -1940 | magnet | vertical | on telescope |
| 2°51'56" | -7 | 55 | -2.2 | no magnet | | |
| 2°52'01" | -8 | 53 | -2.4 | magnet | horizontal | on alidade |
| 2°52'06" | -2 | 787 | -28 | magnet | vertical | on alidade |
| 2°52'00" | -1 | 56 | -2.2 | no magnet | | |

Table 19. VFO31 variometer horizontal and vertical component baselines.

| $H_0$ | $Z_0$ | $S_0$ (nT) | δ (") | Conditions | Orientation | Location |
|---|---|---|---|---|---|---|
| 20880.4 | 42176.9 | -2.2 | 54 | no magnet | | |
| 20880.7 | 42177.0 | 929 | 55 | magnet | horizontal | on telescope |
| 20880.7 | 42177.0 | -2114 | 65 | magnet | vertical | on telescope |
| 20880.0 | 42177.1 | -2.2 | 51 | no magnet | | |
| 20908.5 | 42163.0 | -4 | -2419 | magnet | horizontal | on alidade |
| 20632.2 | 42299.1 | -0.3 | -271 | magnet | vertical | on alidade |
| 20880.3 | 42176.9 | -1.6 | 53 | no magnet | | |

These results were encouraging and it was concluded that no special attention should be paid to the demagnetization of such a complicated component as the telescope. The influence of the remanent magnetization on the most important part of the theodolite, the theodolite main shaft, which determines the theodolite precision, was tested. The acceptable magnetic cleanliness limits were determined to be below 10 nT at 5 cm distance from the shaft axis (see the results given in Table 20).

Table 20. Shaft magnetism influence.

| | X, nT | Y, nT | Z, nT |
|---|---|---|---|
| Slightly magnetic shaft | 18860,0 | 1465,5 | 46024,3 |
| Non-magnetic shaft | 18862,4 | 1466,1 | 46023,2 |
| Belsk observatory bases | 18862,5 | 1466,5 | 46023,5 |

LCISR continues its production line of LEMI type magnetometers with the LEMI-203 instrument (LCISR, 2005) and their comparative tests with Bartington and DIM-France instruments showed the same levels of precision of D and I measurements (Pajunpaa et al., 2001).

The LEMI-203 magnetometer sensor housing can be mounted on both Zeiss-010 and 3T2KP-NM non-magnetic theodolites (Figure 87).

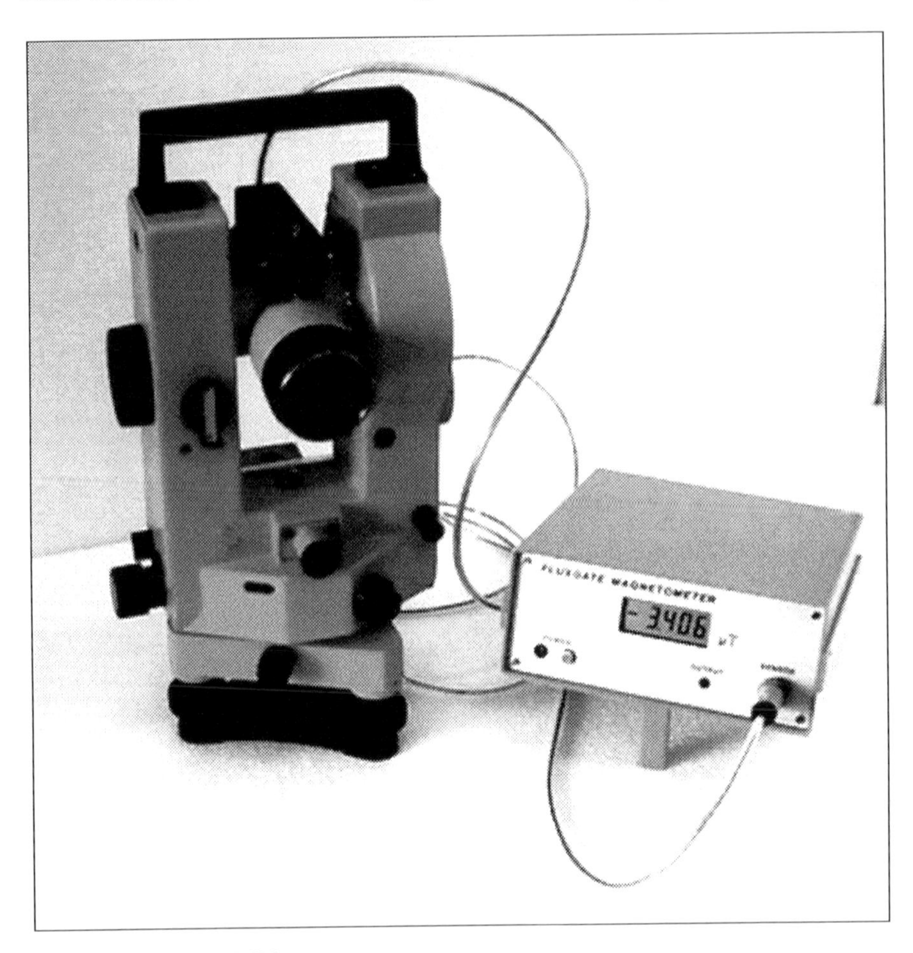


Figure 87. LEMI-203 DIM.

The sensor is coupled with the electronic unit by a flexible cable. The main technical parameters of LEMI-203 DIM are given in Table 21.

Table 21. Main technical parameters of LEMI-203.

| Three measuring ranges, switched automatically: | |
|---|---|
| Range I | ± 70.00 mcT |
| Range II | ± 20.000 mcT |

| Range III | ± 2.0000 mcT |
|---|---|
| Resolution at each range:<br>Range I<br>Range II<br>Range II | <br>10 nT<br>1.0 nT<br>0.1 nT |
| Analog output transformation factor | 0.05 mV/nT |
| Frequency bandwidth of analog output | DC-10 Hz |
| Output resistance | ≤100 Ohm |
| Analog output noise in the frequency band 0.03 - 1 Hz | < 20 pT rms |
| Operating temperature range | minus 5 to +40°C |
| Internal power supply, battery | 12 V, 1,2 Ah |
| Weight:<br>sensor with support<br>electronic unit with battery | <br>0.2 kg<br>2.5 kg |
| Dimensions:<br>sensor with support<br>electronic unit with battery | <br>27x27x75 mm<br>174x78x200 mm |
| Length of connecting cable | 4,5 m |

The simplified functional diagram of the LEMI-203 single-axis magnetometer is shown in Figure 81. The LEMI-203 can be used conveniently both in the observatory and in field conditions. The sensor is fixed to the theodolite's telescope so that sensor's magnetic axis and the telescope's optical axis are parallel. The sensor's magnetic axis alignment at Bartington is made only by them and is very costly. The LEMI-203 sensor has a user friendly design (Figure 88). Alignment can be made by any qualified user following a simple procedure described in the manual (LCISR, 2005).

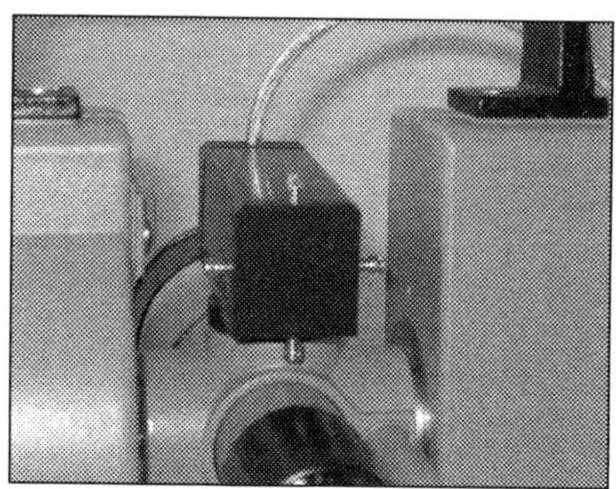

Figure 88. LEMI-203 sensor construction.

At the XIth IAGA Workshop on geomagnetic observatory instruments, data acquisition and processing (Japan, November 9-17, 2004) the comparative tests of 19 DIM instruments took place. In spite of a magnetically disturbed time, all four LEMI-203 instruments taking part in the comparative measurements showed high levels of accuracy (Masami, 2004).

## 6. Conclusion

Geomagnetic instrumentation used in repeat station surveys was described. Recent interest in field surveys has been driven by the need to study secular variation and the practical need to calibrate airport swing bases. This has promoted the development of a new generation of instrumentation. New technology allows a united instrumentation set and user's software for repeat station surveys (Figure 89), instead of the commonly used set of separate devices from different manufacturers. A complete instrument and software set has been developed at LCISR (LCISR, 2005). LCISR has great experience in theoretical, technological, and experimental studies and they propose their instrument set as a cost-effective and user-friendly solution for repeat station and airport swing base surveys.

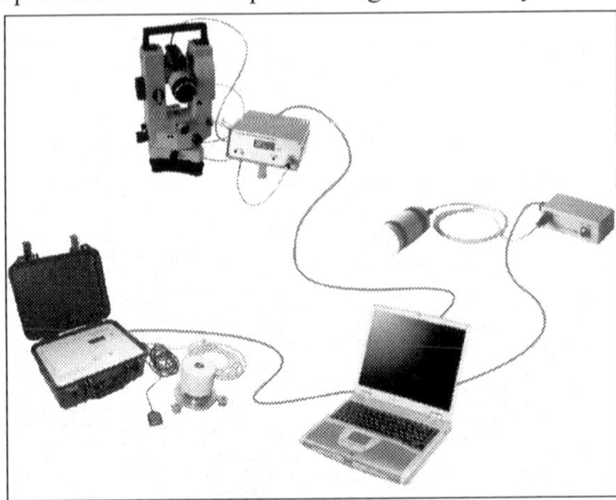

Figure 89. Advanced instrumentation set for repeat station survey.

Further progress in the creation of such instrumentation can be expected in the future, both in the improvement of parameters of the traditional DIM, variometer, absolute magnetometer set and in the creation of new technologies based on absolute magnetometer applications.

## Acknowledgements

The support of Dr. Jean Rasson, the head of Dourbes Geomagnetic Observatory, without which this paper would never appear, is highly appreciated.

## References

Alldredge L. R., 1960, A proposed automatic standard magnetic observatory, Journ. Geophys. Res., 65, 3777-2786.

Berkman R., B. Bondaruk, V. Korepanov, 1997, Advanced flux-gate magnetometer with low drift, *XIV IMEKO Word Congress. New measurements - challenges and visions,* Tampere, Finland, , Vol. IVA, Topic 4, pp. 121-126.

Cohen E. R. and Taylor B. N., 1987. The 1986 CODATA recommended values of the fundamental physical constants. Journal of Research of the National Bureau of Standards (U.S.), 92, 85-95.

Denisov A., Sapunov V. and Dikusar, 1999 Calculation of the error in the measurements of a digital-processor nuclear-precession magnetometer, Geomagnetism and aeronomy, 39, 68-73.

Kotzé P. B., L. Loubser, H. Theron, 2004, Comparative evaluation of a suspended dIdD, an unsuspended dIdD, and a FGE fluxgate system, Proceedings of the XIth IAGA Workshop on geomagnetic observatory instruments, data acquisition and processing (Kakioka and Tsukuba, Japan, November 9-17, 2004), 176-180.

LCISR (2003); http://www.isr.lviv.ua.

Masami O., 2004, Intercomparisons and tests of geomagnetic instruments and measurement training at the Kakioka Magnetic Observatory, Japan, in 2004, Proceedings of the XIth IAGA Workshop on geomagnetic observatory instruments, data acquisition and processing (Kakioka and Tsukuba, Japan, November 9-17, 2004), 6-23.

Newitt L. R., C. E. Barton and J. Bittely, 1996, Guide for magnetic repeat station surveys, Published by IAGA, J. A. Joselyn, Boulder, USA.

Pajunpaa K., J. Bitterly, H.-J. Linthe, V. Korepanov, 2001, Absolute measurements: comparative study of instrumentation, Contributions to Geophysics and Geodesy, Vol. 31/1, 131-136.

Pankratz L. W., Sauter E. A., Körmendi A., Hegymegi L., 1999, The US-Hungarian delta I – delta D (DIDD) quasi-absolute spherical coil system. Its history, evolution and future, Geophysical Transactions, 42, 195-202.

Rasson J., Progress in the design of an automatic DIflux, 1994, Timetable and abstracts for the Lectures Session of Geomagnetic Observatories Instruments, Data Acquisition and Processing, Dourbes, Belgium, September 18-24.

Sapunov V., Denisov A., Denisova O. and Saveliev D., 2001. Proton and Overhauser magnetometers metrology. Contributions to Geophysics & Geodesy, 31, 119-124.

Sapunov V., Denisov A. et al., 2004, Theodolite-borne vector overhauser magnetometer: DIMOVER, Proceedings of the XIth IAGA Workshop on geomagnetic observatory instruments, data acquisition and processing (Kakioka and Tsukuba, Japan, November 9-17, 2004), 159-164.

## DISCUSSION

Question (Spomenko J. Mihajlovic): What does it mean compensation for thermal drift?

Answer (Valery Korepanov): This means that the thermal drift has to be calibrated first, i.e., the zero shift for given temperature change has to be determined during thermal tests and then you have possibility to compensate the drift, because all our magnetometers have the measurement of the temperature and the compensation may be made automatically or during data processing, using thermal factor and value of temperature change.

Question (Angelo De Santis): I believe that the future of magnetic monitoring in seismic and/or volcanic areas will see the use of vector magnetometers.

Do you think that this can be reached in a nearby future with your instruments?

Answer (Valery Korepanov): Yes, because one of the major problems of component magnetometers – too big temporal and thermal drifts – has already been solved. Some of the models of our magnetometers, e.g., LEMI-018 with special calibration, may have resulting temporal drift below +- 3 nT per month and thermal drift below 0.1 nT per centigrade, what is very close to scalar sensors.

Question (J. Miquel Torta): Can the temperature effect be compensated with your observatory variometer even if the variation hut has no temperature control?

Answer (Valery Korepanov): Yes. But the magnetometer has to be specially prepared, i.e., thermal drift has to be calibrated first, and then you have the possibility to compensate it. All our magnetometers also measure the ambient temperature. The temperature effect compensation may be made automatically or later during data processing, using the calibrated thermal factor and value of temperature change.

# PROBLEMS OF SUPPLY IN DI-FLUX INSTRUMENTS

LÁSZLÓ HEGYMEGI[12]
*Eötvös Loránd Geophysical Institute*

**Abstract.** For the most part, the D/I fluxgate magnetometer is the only absolute instrument used in observatories and in the field. At present no company produces a nonmagnetic theodolite and there are only a few organizations converting geodetic theodolites to nonmagnetic instruments. A classic theodolite has many parts which have a magnetic moment. These parts have to be replaced before the instrument can be used as an absolute instrument for magnetic measurements. The most problematic part of these theodolites is the main axis. The precision of this part determines the overall quality of the instrument. Most observatories require 1" resolution. This can be achieved by converting the Zeiss THEO 010 or the Russian 3TK2П theodolites. The Zeiss models have been out of production for more than ten years; the Russian theodolites exhibit various qualities. In the future it will be necessary to introduce new measuring techniques and new instrumentation.

**Keywords:** magnetic absolute measurement, observatory measurements, field measurements, nonmagnetic theodolites, DI-flux instruments, Overhauser dIdD magnetometer

## 1. Introduction

DI-flux instruments are basic instruments in geomagnetism. They are used in observatories and in the field to determine declination and inclination angles of the geomagnetic field vector, **F,** in reference to geographic north and to the horizontal plane. In the past other methods existed to determine these values, but because of the very time consuming observation procedure and sometimes low precision, these methods can be regarded as obsolete. Nowadays, the DI-flux is practically the only instrument having worldwide acceptance.

---

[12] To whom correspondence should be addressed at: Eötvös Loránd Geophysical Institute, Columbus utca 17-23, 1145 Budapest, Hungary. Email: hegymegi@elgi.hu.



*J.L. Rasson and T. Delipetrov (eds.), Geomagnetics for Aeronautical Safety,* 167–175.

This instrument is a steel-free classic theodolite with a single axis fluxgate magnetometer mounted on its telescope. The most commonly used instrument was produced by Carl Zeiss Jena for many years but after the reunification of Germany, the company was closed down, and since that time there is no company producing steel-free theodolites with the same characteristics.

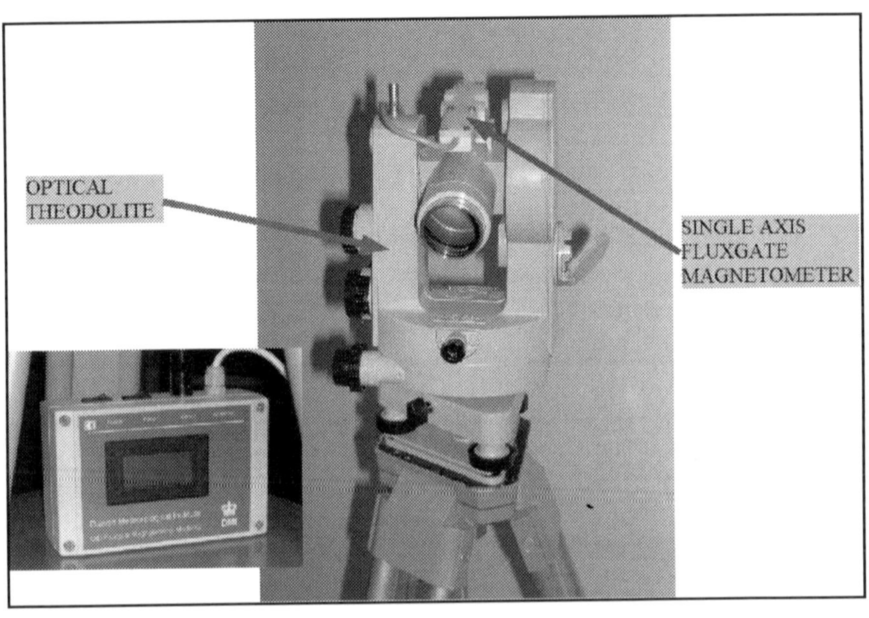


Figure 90. DI-flux magnetometer with theodolite Zeiss THEO 010A and magnetometer DMI model G.

## 2. Production of DI-flux instruments

The easiest way to build a DI-flux instrument is to take a classic, geodetic theodolite and change all the steel parts to brass or other nonmagnetic materials. This is not a simple task and extreme care must be used. All parts of the disassembled instrument have to be magnetically checked as aluminum or plastic parts can have magnetic impurities or internal magnetic particles. The springs are made from steel and are magnetic.

In some cases, the magnetic parts can be replaced easily by copies made of brass or aluminum. This method however cannot be used in all cases. For instance, the main axis of the instrument is constructed with ball bearings. To reproduce the ball bearings with nonmagnetic materials would require costly technology and since very few instruments are required by users such cost would be prohibitive.

For theodolite conversion, a new plain bearing assembly for the axis was constructed. This part is satisfactory for DI-flux magnetometers since the frequency of use is low compared to geodetic instruments. But they must be manufactured with high precision and a careful selection of parts before assembly.

Figure 91. Nonmagnetic vertical axis for a Zeiss THEO 010 theodolite.

The most common instruments used for conversion are the Zeiss THEO 010, 015, and 020 and the Russian YOM 3TK2П. The conversion method for all instruments is nearly the same. There are 40 to 60 parts that must be changed.

There are very few new or second-hand Zeiss instruments on the market and at some time in the future, production of the classic theodolites will end since production costs of graduated glass circles is very high compared to modern angular encoders. Angular encoders also have a convenient direct digital output. Unfortunately, modern instruments are very magnetic and the encoders are encased in magnetic parts.

## 3. Measurement characteristics

The main technical characteristics of converted theodolites have not changed. The angular resolution for a THEO 010 and a 3TK2П is one arc second (and for THEO 020 it is six seconds). The angular resolution and the resolution of the magnetometer determine the precision of the geomagnetic measurement. Using a G type fluxgate magnetometer from the Danish Meteorological Institute, which has 0.1 nT resolution observers can get a 5 to 9 arc second measurement error for declination and a 2 to 4 arc second error for inclination in observatory conditions. In the field, the measurement errors can be in the range of 15 to 25 arc seconds for declination and 6-12 arc seconds for inclination. These statistical values were recorded during the last repeat station campaign in Hungary. At different latitudes or under different external conditions, the errors could be different.

## 4. Increasing measurement precision

Measurement error can occur when the geomagnetic field changes during the measurement procedure. This error depends on the amount of field activity and is usually larger at high magnetic latitudes. If the observations are made faster, the errors are smaller.

Another source of error in field measurements is that the time correction made with observatory variation data can be slightly different at the field site even if the distance to the observatory is small.

To eliminate this problem an on-site recording variometer should be used. Recording variometers however can have high thermal sensitivity and outside temperatures can change significantly in field conditions. A solution to this problem is to use a temperature controller, but this requires more power. The best solution would be to design a magnetic recording system with low temperature sensitivity.

The precision of absolute measurements can be increased by performing the observations during a magnetically quiet period of the day. This is possible if we automatize the measurement procedure. To do so an automatic absolute instrument would have to be constructed using either the present manual method or some new method. Jean Rasson and Sebastien Van Loo from the Institute Royal Meteorologique in Belgium are developing an instrument which uses the present manual procedure. A new instrument, based on a modified version of the dIdD instrument using a different measurement method than the classic DI Flux, is presently being developed in Hungary.

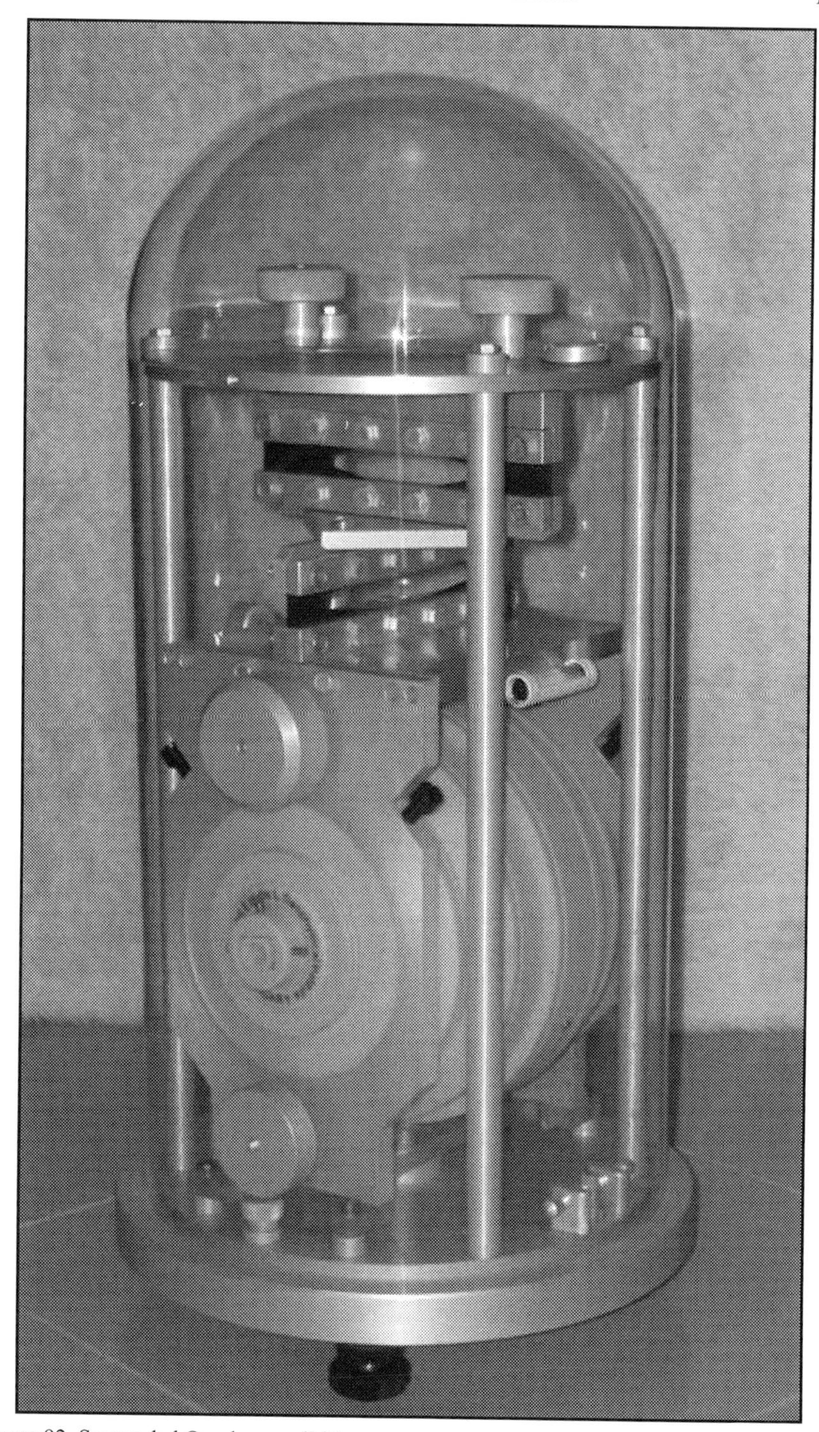

Figure 92. Suspended Overhauser dIdD magnetometer.

## 5. Basic principle of dIdD measurement

The Hungarian dIdD instrument is based on the Overhauser proton precession magnetometer. It measures the absolute value of the geomagnetic field vector, **F**, together with the declination angle, **D,** and the inclination angle, **I,** with reference to the position of the coil system. The instrument is an absolute instrument in its own coordinate system. If position control is employed for the coils absolute values are obtained for **D** and **I** without additional magnetic measurements.

The basic equations of the dIdD method are:

$$\Delta I = \frac{F_{Ip}^2 - F_{Im}^2}{4A_D F \cos I}$$

$$A_I = \frac{1}{\sqrt{2}} \sqrt{F_{Ip}^2 + F_{Im}^2 - 2F^2}$$

Where:

$\Delta I$ is the difference between measured $I$ and initial $I$

$A_I$ is the deflection from the I coil

$F_{Ip}$ and $F_{Im}$ are Overhauser magnetometer readings due to opposite deflection currents in the I coil

$F$ is the undeflected Overhauser magnetometer reading

$$\Delta D = \frac{F_{Dp}^2 - F_{Dm}^2}{4A_D F \cos I}$$

$$A_D = \frac{1}{\sqrt{2}} \sqrt{F_{Dp}^2 + F_{Dm}^2 - 2F^2}$$

Where

$\Delta D$ is the difference between measured $D$ and initial $D$

$A_D$ is the deflection from the D coil

$F_{Dp}$ and $F_{Dm}$ are Overhauser magnetometer readings due to opposite deflection currents in the D coil

$F$ is the undeflected Overhauser magnetometer reading

$I$ is the mean inclination angle

The individual readings are obtained from the Overhauser magnetometer, and the expressions do not contain scale factors or constants which need calibration. The amplitude of the deflection field (if it is within a reasonable range) has no effect on the measurement if it is equal in both directions during one measurement period (e.g. 1-2 seconds).

If the direction of the axis of the coils can be determined with precision, the dIdD is an absolute instrument for determination of the geomagnetic field. The instrument can also work as a recording instrument and can make one complete series of D, I, and F measurements in one second.

A variometer version of the instrument is being tested. To make a complete instrument, devices must be added to measure the direction of one axis in reference to the horizontal line in the vertical plane and the other axis in reference to the geographic north in the horizontal plane. This is possible using angular encoders available today.

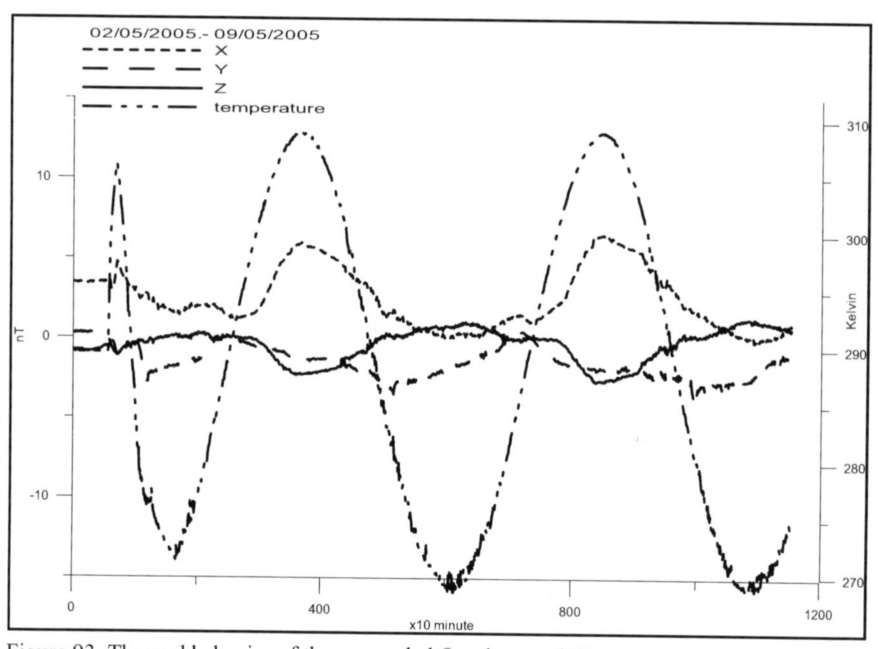


Figure 93. Thermal behavior of the suspended Overhauser dIdD magnetometer.

## 6. Thermal behavior of Overhauser dIdD instrument

External temperature changes have nearly no effect on the Overhauser magnetometer. The baseline change of a suspended dIdD instrument comes from mechanical instabilities. The amplitude of a dIdD's baseline drift is about one third of the best triaxial fluxgate instruments available today. All sources of thermal instability are not yet determined. A wide range of programmed temperature changes have been applied in the test chamber at the Tihany observatory. So far, some of the sources of thermal instability have been detected. There is hope for further improvements in the near future. The present stability of the dIdD makes the instrument suitable for use in the field. It can be used as a recording instrument to obtain variation data for time correction as well.

## 7. Conclusions

Only theodolites with graduated glass circles are suitable for conversion to magnetic absolute instruments, but in the near future there will be no more theodolites available for conversion. In addition there is a demand to increase the precision of absolute measurements. To overcome these challenges it may be advisable to change the measurement method used by most observatories and field survey parties.

We believe that the Overhauser dIdD instrument is stable enough to use in observatories and in the field. One possible application for use in the field is as a local reference instrument to reduce errors caused by the difference in the geomagnetic variation between observatories and field stations. By producing a modified version of this instrument, a recording absolute vector magnetometer can be obtained.

## References


Alldredge, L.R., 1960,A proposed automatic standard magnetic observatory, *J.Geophys. Res., 65, 3777-3786*

Alldredge, L.R. and Salducas, I., 1964, An automatic standard magnetic observatory, *J.Geophys. Res., 69, 1963-1970,*

Hegymegi, L., Heilig, B., Csontos, A., New suspended dIdD magnetometer for observatory (and field?) use, 2004, *Proceedings XIIth IAGA Workshop on geomagnetic observatory instruments, data acquisition and processing, 28-33, Kakioka and Tsukuba*

Kotzé, P., Loubster, L., Theron, H., Comparative evaluation of a suspended dIdD, an unsuspended dIdD, and FGE Fluxgate system, 2004. *Proceedings XIIth IAGA Workshop on geomagnetic observatory instruments, data acquisition and processing, 176-180, Kakioka and Tsukuba*

Pankratz, L.W., Sauter, E.A., Körmendi, A., Hegymegi L., 1999, The US-Hungarian Delta I – delta D (DIDD) Quasi-absolute spherical coil system. Its history, evolution and future, *Geophysical Transaction, 42, 195-202,*

Rasson, J.L., 1996, Report on the progress in the design of an automatic Diflux, *Proceedings of the VIth Workshop on geomagnetic observatory instruments data acquisition and processing,190-194, Bruxelles*


## DISCUSSION

Comment (Jean Rasson): Your device, even with encoders to measure the I and D angles, will not be absolute as long as you don't measure the collimation errors (difference between magnetic axis and optical axis).

We can speak about the absolute D, I measurement when you are able to fully orient magnetic vector with respect to geographic north and vertical.

Question (Angelo de Santis): First, I would like to make several comments and then just 1 question.

The comment is that all the geomagnetic community thanks and appreciates the work you and other people, as Valery Korepanov, do as designers of very good magnetometers. Without your work operators and modelers of the geomagnetic field could not make their own work.

The question is related to the errors you mention for your instrument in Observatory or field conditions. Are they associated to a single measurement or to a series of measurements?

Answer (Laszlo Hegymegi): Those are the minimum and maximum of calculated errors for series of measurements taken in the observatory and field stations but experienced with the same instrument. The difference comes from the influence of external effects which changes from place to place and time to time.

# DEVELOPMENT OF AN AUTOMATIC DECLINATION-INCLINATION MAGNETOMETER

SEBASTIEN A. VAN LOO[13]

JEAN L. RASSON

*Institut Royal Météorologique de Belgique*

**Abstract.** The first results in the design of an automatic DIM are presented. This instrument should be completely operational in 3 years. By associating it with other instruments which are already automatic, like variometers and proton magnetometers, it will be possible to install absolute magnetic observatories, all around the Earth, even in inaccessible places like on islands and on the ocean floor, since there is no need for an operator or maintenance. Automation is difficult because several key components generate considerable magnetic disturbances. Solutions to carry out the operations of rotation of the sensor, precision reading of the angles, and the pointing of an azimuth reference without disturbing the magnetic field, are proposed.

**Keywords:** DIM, absolute magnetic observation, declination, inclination, theodolite, fluxgate, piezoelectric motor, electronic angular encoder, automation

## 1. Introduction

Many automatic instruments are able to provide recordings of the value of the total geomagnetic field as well as its variations. But the declination and the inclination still must be measured manually by an observer, using a DIM (declination-inclination magnetometer). If this instrument could be automated, it would become possible to establish completely autonomous magnetic observatories, working without need of an operator or maintenance (Rasson 1996). The Earth could then be totally and uniformly covered with magnetic observatories, by adding new stations to the current

[13] To whom correspondence should be addressed at: Institut Royal Météorologique de Belgique, Centre de Physique du Globe, B-5670 Dourbes, Belgium. Email: sebvl@oma be



*J.L. Rasson and T. Delipetrov (eds.), Geomagnetics for Aeronautical Safety,* 177–186.

network, with installations far from base observatories, at inaccessible places like the ocean floor (70% of the Earth's total surface), high altitudes, deserted areas, etc.

Since January 2004, we have worked on the development of an absolute, automatic instrument for measurement of the direction of the geomagnetic field. At the beginning of 2006, the first phase of the project will be completed. The objective of this first phase is to obtain a completely functional and automatic instrument, with a total error on the measured angles *D* and *I* smaller than 6 arc-seconds, and using a far target as azimuth reference.

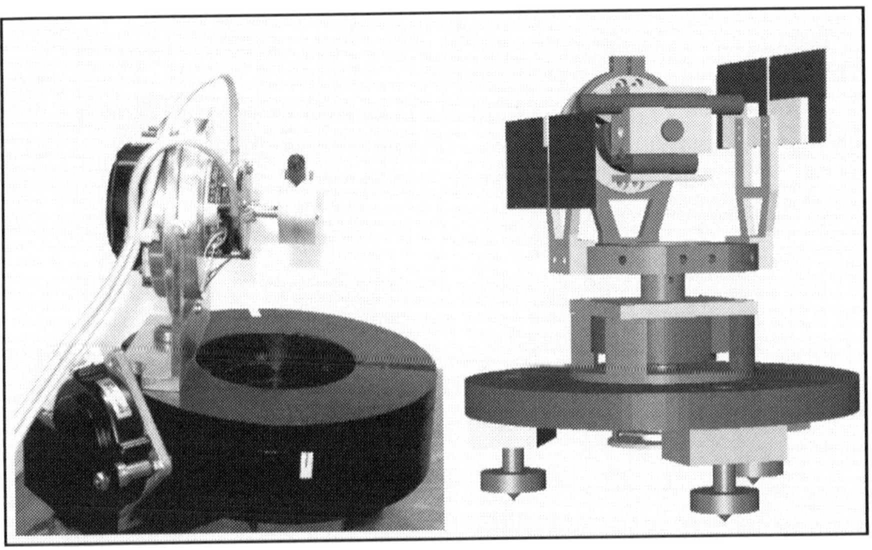

Figure 94. At left, a prototype of the theodolite; At right, a plan of the final theodolite.

The second phase, which ends in January 2008, will be devoted to the development of an automatic gyroscopic North-seeker, which will be used as azimuth reference for the instrument (Chave 1995). The errors on the measured angles will then be kept smaller than 6 arc-seconds for *I*, and smaller than 20 arc-seconds for *D* (Table 22).

Table 22. Specifications of the automatic declination-inclination magnetometer.

| Time | Error on D | Error on I | Azimuth reference |
|---|---|---|---|
| Jan 2006 | < 6 arc-seconds | < 6 arc-seconds | automatic pointing of a far target |
| Jan 2008 | < 20 arc seconds | < 6 arc-seconds | automatic gyroscopic North-seeker |

The instrument will be similar to a robotized DIM system. The fundamental principles leading to the automation of the measurement are first presented. Then technological solutions to minimize error are proposed so that the instrument will meet the high precision and magnetic cleanliness constraints. Last, the electronic system for reading the angles, the use of non-magnetic piezoelectric motors, and the automatic pointing of the target are covered in depth.

## 2. Automation of the measurement

An automatic measurement must have the same metrological qualities as a manual measurement. Thus, the same operations should be reproduced (see Table 23).

In accordance with the traditional method of measurement with a DIM, the declination and the inclination are measured in 4 different positions (Rasson 2005). The instrumental errors should be equivalent to those of a traditional theodolite. The target is also measured in two positions. The execution of this protocol for each measurement ensures the absolute character of the result.

Table 23. Operations to carry out in order to make an absolute measurement of the direction of the geomagnetic field.

| | |
|---|---|
| 1. | Synchronization with universal time. |
| 2. | Leveling of the instrument. |
| 3. | Pointing an azimuth reference (2 positions). |
| 4. | Measurement of the declination (4 positions). |
| 5. | Measurement of the inclination (4 positions). |
| 6. | Pooling the results with those of scalar magnetometer, and variometer. |

It was necessary to design and use a non-magnetic theodolite. Instead of a telescope, the theodolite is equipped with a directional magnetic sensor (fluxgate), and with a laser to point at the target. To make 4 positions of measurement for declination and inclination, the sensor must be able to make a complete rotation around the horizontal and vertical axes. Finally, the angular position of the sensor must be measured very precisely.

## 3. Technological solutions

The two principal problems to overcome are avoiding magnetic parts or parts which cause a magnetic disturbance, and designing a precision device (from the mechanical and electronic points of view). Ferromagnetic materials cannot be used in construction, nor can electric lines conveying detectable DC current. Electronic circuits must be kept far away from the magnetic sensor. Figure 94 shows the present status of the theodolite. Its final version has not yet been realized. A device for controlling and correcting the leveling is also under development.

The signals produced by the electronic acquisition system (readings of the angle, fluxgate, level, and pointing at the target) are collected by a microcontroller which uses analog to digital converters. Signals are then processed, and instructions are sent to the motor drivers in order to carry out the desired operation (Figure 95). The data storage, time control, and user interface are ensured by a computer, connected to the microcontroller via a USB bus.

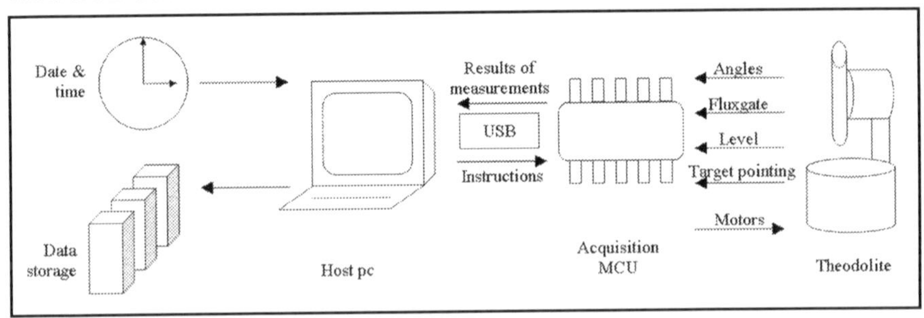


Figure 95. Interactions between the different subsystems.

### 3.1. THE ANGULAR ENCODERS

In order to electronically evaluate angles, optical angular encoders are used. One system is used for each of the two orthogonal axes of the theodolite.

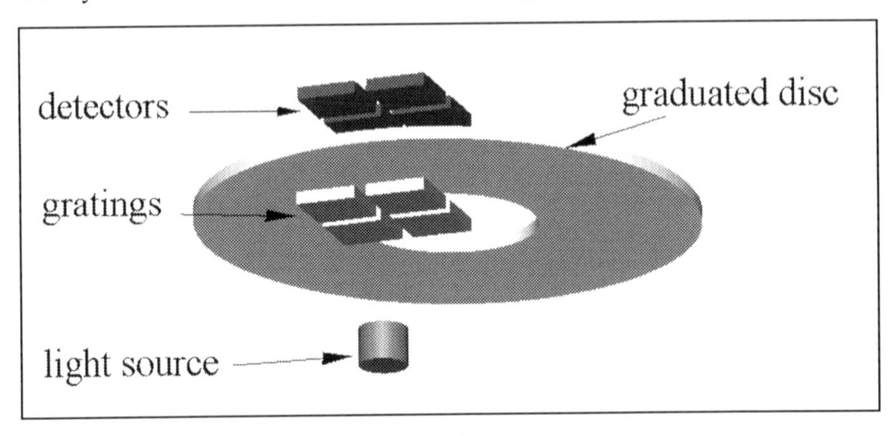


Figure 96. General diagram of an optical encoder.

A graduated disc, fixed on one axis of the theodolite, rotates between a light source and a detection system (Figure 96). Gratings, with the same period as the graduated disc, are placed behind the light source in order to amplify the signal by the optical moiré effect. There are four gratings and one photodiode for each graduated disc. The gratings are shifted by a quarter of a period (Figure 97). By subtracting the light signals *c* from *a*, and *d* from *b*, we obtain two sinusoidal signals free from the common mode. The disc is also equipped with a third track which produces only one reference pulse per rotation.

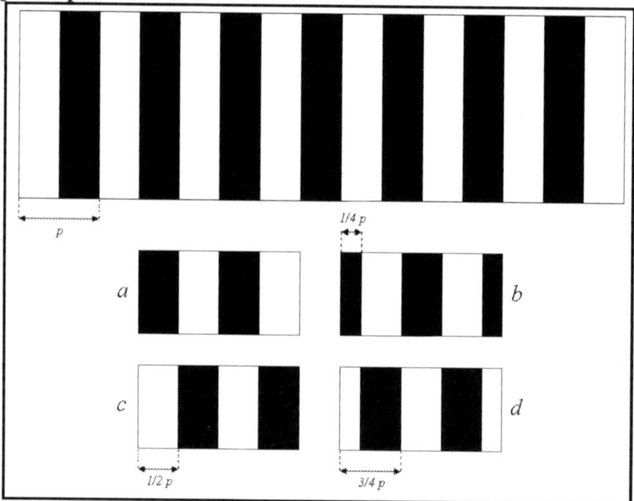


Figure 97. Graduated disc and gratings.

Since discs with 2500 graduations are used, a resolution of 0.144° is obtained (simply by counting the graduations). Then, because the two sinusoidal signals are in phase quadrature (Figure 98), calculating the arctangents of the signals sine/cosine leads to an analog signal having a linear dependence on the angle. Depending upon the quality of the encoders, the electronic disturbances, and the mechanical alignment of the system, a precision of up to 1 arc-second can be achieved. The reference pulse is used to make this incremental encoder absolute.

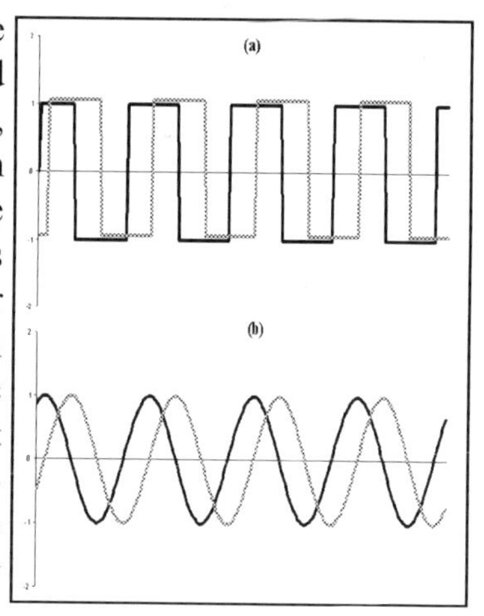


Figure 98. Electric signals allowing (a) the period count and (b) the continuous evaluation of the angle by interpolation between the graduation period increments.

Good signals lead to good precision. So the errors related to encoder and electronics quality, like amplitude modulation and undesired offset, are corrected in real-time by a digital processing algorithm (Figure 99). Errors, related to mechanical misalignment of the encoder compared to the rotation axis, are corrected by placing two encoders around the same disc 180° apart (Figure 100). Taking the average of the two measured angles provides a result free from eccentricity errors.

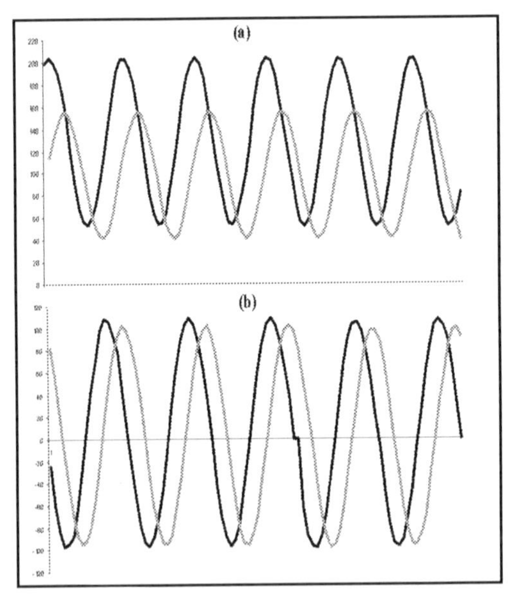


Figure 99. Signals before (a) and after shaping (b).

Figure 100. Two encoders placed around the horizontal axis.

Available encoders are generally not magnetically clean, and cannot be placed symmetrically in pairs on the same disc. Some parts (the detector board and others) have to be replaced by specially designed circuits (Figure 101).

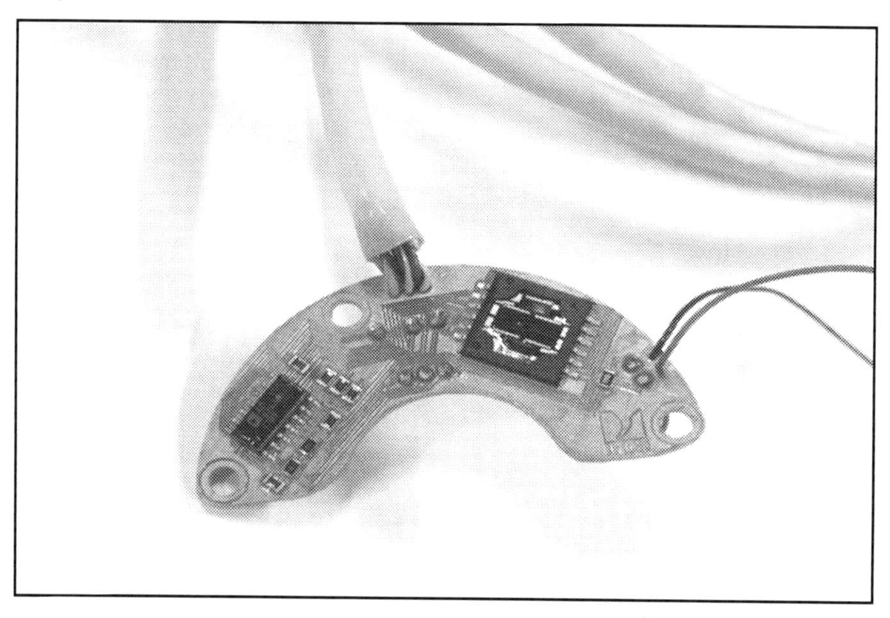

Figure 101. Example of a detector board for angular encoders where a ready-made IC is used as a detector, and linear amplifiers are included on the board.

Preliminary tests show that the error can easily be made lower than 3.6 arc-seconds. More rigorous tests are presently under development.

## 3.2. THE PIEZOELECTRIC MOTORS

The movement around the axis of the theodolite is driven by piezoelectric motors, which can be bought in totally non-magnetic versions.

The rotational movement of the shaft is obtained by pressing its base against an annular piezoelectric crystal, on the surface of which a revolving traveling wave is maintained (Figure 102). This traveling wave is obtained by stimulating the crystal with two high voltage signals (300Vpp), one cosine and one sine, at a frequency of about 40 kHz. In this way, power is produced as a small, non-disturbing AC current.

Sometimes, a slow speed is necessary, primarily because of the computing and reaction times of the electronic circuits (for example when the angle has to be calculated precisely, or when a position has to be reached very finely). Other times, in order to save time, large displacements can be carried out at high speed. Smooth accelerations and decelerations are also necessary to avoid vibrations at start and stop. For these reasons it is very important to have total control of the rotation speed. The motor shafts can be used directly as axes for the theodolite, with no need for a transmission or reduction system.

Three parameters of the motor drive sine waveforms can be varied to control the motor rotation speed: amplitude, phase, and frequency. Changes in amplitude led to a loss of torque at slow speed. Tests varying the phase demonstrated that speed variation was strongly non linear, and repeatability was too low for effective control. Adjusting the frequency of the excitation signals allowed us to obtain satisfactory motor speed control with adequate torque, linearity, and repeatability.

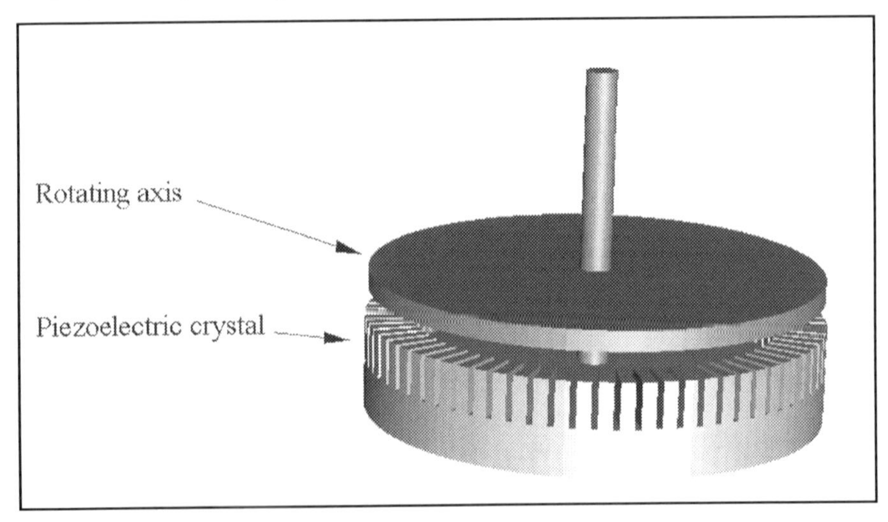


Figure 102. General diagram of a rotary piezoelectric motor.

### 3.3. THE AZIMUTH REFERENCE

In order to reference the horizontal angle measurements to True North the theodolite must acquire a known azimuth reference. This process is traditionally performed by an observer who points the telescope at a far target. To automate the process, the following method is presented.

A laser diode module is installed in place of the telescope. It points toward a corner cube reflector which is centered at the point whose azimuth is precisely known (actually the visual target). According to the properties of the corner cube reflector, an incident light ray is reflected along the incoming beam, but offset by a distance, $e$, (Figure 103) depending on the angle, $\alpha$, between the incident ray and the line which connects the center of the corner cube to the vertical axis of the theodolite. Two solar cells are positioned around the laser in order to evaluate the offset of the reflected ray. The difference of light touching the two solar cells is directly related to the pointing of the center of this electronic target: when the reflected ray returns precisely in the center, the laser exactly points to the center of the target.

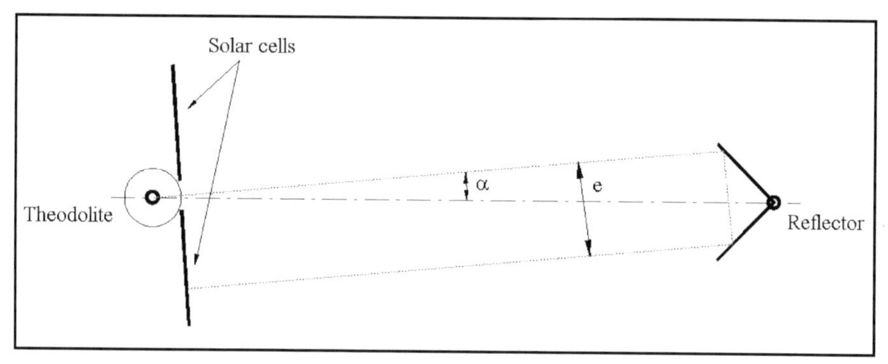


Figure 103. The corner cube reflector.

The goal of the second phase of the project is to replace this azimuth reference system with an automatic gyroscope.. This would allow the instrumentation to work in a closed system with no need to connect to external references.

## References

Chave, A.D., Green, A.W., Evans, R.L., Filloux, J.H., Law, L.K., Petitt, R.A., Rasson J.L., Schultz, A., Spiess, F.N., Tarits, P., Tivey, N. and Webb, S.P. (1995). Report of a Workshop on Technical Appoaches to Construction of a Seafloor Geomagnetic Observatory, Technical Report WHOI-95-12, Woods Hole Oceanographic Institution, Woods Hole, USA.

Rasson J.L. (1996). Progress in the design of an automatic DIflux, in Proceedings of the VIth Workshop on Geomagnetic Observatory Instruments, Data Acquisition and Processing (JL Rasson Ed.), Publ. Sci. et Techn. No 003, Institut Royal Meteorologique de Belgique, Brussels p190-194.

Rasson J.L., (2005). About Absolute Geomagnetic Measurements in the Observatory and in the Field, Publication Scientifique et Technique No 040, Institut Royal Meteorologique de Belgique, Brussels, 43 p

## DISCUSSION

Question (Jordan Zivanovic): Is the microcontroller with 8 gates or more?
Answer (Sebastien van Loo): I currently use a microcontroller with a 16 bit digital port, having 8 analog inputs (ADCs), and a USB interface
Question (Jürgen Matzka): How to find the zero-position of the fluxgate sensor (slow movement or stepwise moving)?

Answer (Sebastien van Loo): Piezoelectric motors offer the possibility to rotate so slowly that the zero-position of the sensor can be found by moving continuously.

Question (Spomenko J. Mihajlovic): What about magnetic influence of electronic parts. Can you use photo-resistors?

Answer (Sebastien van Loo): The majority of the electronic systems are kept far away from the sensor. For the circuits which must be closer, like the angular encoders, I take many precautions to minimize the disturbances, like avoiding current loops, and choosing SMD-packaged parts. Actually, I use photodiodes rather that photo-resistors (angular encoders, target pointing). But if the use of photo-resistors appeared essential later, I think that it would be possible to find some models which are magnetically clean enough.

Question (Valery Korepanov): How do you find true azimuth in small closed volume?

Answer (Sebastien van Loo): Initially, the azimuth reference will be obtained, by the automatic pointing of a far target.

The second phase of the project is devoted to the replacement of this system by an automatic gyroscope. It would then be possible to obtain true azimuth in a small volume.

Question (Angelo de Santis): In your automatic system have you considered the possibility to make an absolute measurement of D and I practically simultaneously by placing the fluxgate element at a given nonzero inclination with respect to horizontal plane and rotating it at the usual four positions of zero-current findings?

Answer (Sebastien van Loo): The measurement algorithm that I chose consists in measuring the declination while the fluxgate is placed horizontally and the inclination while the fluxgate is in the magnetic meridian.

But the instrument can be programmed to execute any other algorithm, without need of hardware adaptations.

# NEW MAGNETIC MATERIALS

STANOJA STOIMENOV[14]
*Institute of Physics*
*Faculty of Natural Sciences and Mathematics*
*University "St Cyril and Methodius"*

**Abstract.** New, sophisticated magnetic materials can be found as essential components in computers, sensors, and actuators, and in a variety of telecommunications devices ranging from telephones to satellites. Some of these materials exhibit unique structure and magnetic properties. Nano, aerogel, superconducting, and liquid magnets belong to this group of materials.

**Keywords:** hysteresis, maximum energy products, squareness factor, nanomagnets, Ferro fluids, spintronics, Ferro gels, magnetic multilayers, Giant Magnetoresistance (GMR).

## 1. Overview

Since the discovery of loadstones in the early days of our civilization, magnetic phenomena have been observed. The use of magnetic needles in compass construction has made navigation one of the most profitable areas of application for geomagnetism.. A compass needle works the way it does because it reacts to the Earth's magnetic field. All magnetic objects produce invisible lines of force that extend between the poles of the object.

With relevant historic contributions such as the experiments of M. Faraday and the theoretical compendium of J. C. Maxwell magnetism nowadays is one of the scientific pillars of human knowledge. The theoretical advances and the strong involvement of magnetism in many branches of technological development make this discipline one of the most relevant areas of current research.

[14] Address for correspondence: Institute of Physics, Faculty of Natural Sciences and Mathematics, University "St Cyril and Methodius", Skopje, R Macedonia. Email: stanojs@iunona.pmf.ukim.edu.mk



*J.L. Rasson and T. Delipetrov (eds.), Geomagnetics for Aeronautical Safety,* 187–199.

The impact of magnetism in diagnosis and therapy constitutes a rapidly developing field. With regard to the development of experimental techniques, the emergence of imaging possibilities based on the use of magnetic force microscopy opens a new area that will allow a better understanding of magnetism at the microscopic and nanoscopic levels. Several contributions are devoted to one of the expanding areas of magnetism: "spintronics". This field embraces the interface between magnetism and electronics. Considerable interest in spintronics is raised by the large demand for new electronic devices that use a magnetic field to act on the spin of the electrons in order to control the electrical current.

Bacteria, sharks, dolphins, honey bees, salamanders, and homing pigeons, as well as other organisms, seem to detect the direction of the Earth's magnetic field. Indirect but reproducible evidence suggests that bees and birds can also respond swiftly to changes in its intensity. The mechanisms behind this sensitivity are not completely known. For humans the Earth's magnetic field is invisible, but in certain conditions man can be sensitive to it.

The Earth's magnetic field can be registered by a compass needle on the Earth's surface. This magnetic field reaches thousands of miles out into space. Studies in magnetism explain how compasses work.

## 2. The origin of magnetism

Magnetism is a phenomenon that occurs when a moving charge exerts a force. Moving charges set up magnetic fields. The direction of the magnetic field is perpendicular to the electrical current direction.

The repulsion or attraction between two parallel wires carrying current is of particular importance especially to magnetic levitation. If the currents flow in the same direction, the wires attract. If the currents flow in opposite directions, the wires repel.

In a static magnet, the electrical current moves in terms of electrons orbiting around atomic nuclei. Electrons circling atoms set up small magnetic fields. In most materials, these fields are aligned in a fairly random manner so that they cancel each other. In a non-static magnet, however, these fields line up to create a net magnetic dipole, with a magnetic field extending into the surrounding space. The circulating electron produces its own orbital magnetic moment, measured in *Bohr magnetons (*$\mu$B*)*, and there is also an associated spin magnetic moment due to the electron itself spinning, like the earth, on its own axis.

When we look at the smallest length scale accessible to us, molecules and atoms become giant structures, and we even regard protons and neutrons "from inside". The nuclear constituents belong to the class of particles called ***hadrons.*** These constituents are subject to very strong interaction with a range of 1 femtometer. Currently we imagine that this femto-world is described by quantum chromodynamics (QCD) within the framework of the standard model of elementary particles. QCD is the fundamental theory of the strong interaction of matter. The fundamental constituents of hadronic matter are the so-called ***quarks.*** They carry a new form of charge (color charge). In contrast to atoms or atomic nuclei, they are not directly detectable as free objects since they are always bound in "color-neutral" hadronic states.

## 3. Magnetic parameters

Ferromagnetic materials show ***hysteresis.*** Their magnetization (*B*) does not return to zero after the application of a magnetic field (*H*). Figure 104 shows typical hysteresis loops. The hysteresis loop is a means of characterizing magnetic materials, and various parameters can be determined from it. The field produced by the magnet after the magnetizing field has been removed is called the remanence ($\boldsymbol{B_r}$). The reverse field required to bring the induction to zero is called the coercitivity ($H_c$). The maximum value of the product of B and H (Figure 104 *a*) is called the ***maximum energy product,*** $\boldsymbol{(BH)_{max}}$ and is a measure of the maximum amount of useful work that can be performed by the magnet. *(BH)+* is used as a figure of merit for permanent magnetic materials. The development of permanent magnets in the 20th century shows that the ***maximum energy product*** $\boldsymbol{(BH)_{max}}$ has improved exponentially, doubling every 12 years.

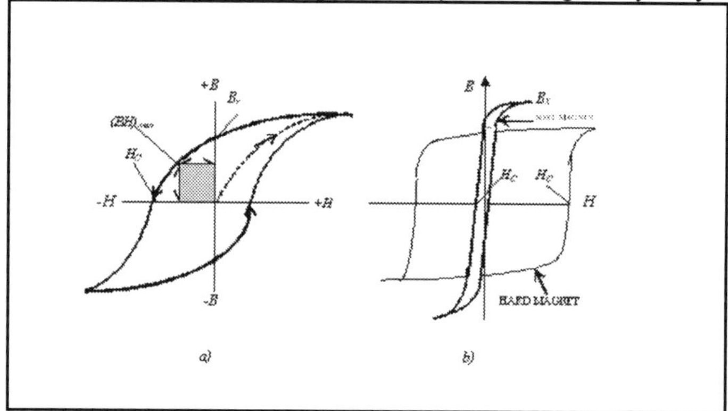


Figure 104. Typical hysteresis loops for a magnetic material.

The two loops (*b*) represent *hard* and *soft* magnetic materials. In addition, the shape of the initial magnetization curve and the hysteresis loop can provide information about the magnetic domain behavior within the material. The ***squareness factor*** is a measure of how the square loop is and is a dimensionless quantity between 0 and 1.

***Magnetic domains*** exist in order to reduce the energy of a system. A uniformly magnetized specimen has a large magnetostatic energy associated with it. This is the result of the presence of magnetic free poles at the surface of the specimen generating a demagnetizing field. The break up of the magnetization into two domains reduces the magnetostatic energy by half. In fact, if the magnet breaks down into N domains, then the magnetostatic energy is reduced by a factor of 1/N.

There are various methods of increasing or decreasing the ***coercitivity*** of magnetic materials, all of which involve controlling the magnetic domains within the material. For a hard magnetic material it is desirable to prevent both the rotation of the direction of magnetization of the domain and the moving of the domain walls or nucleation of domains is difficult. To prevent easy rotation of domains, the material should have strong uniaxial magnetocrystalline anisotropy.

## 4. Nano magnetism

One of the most important trends in contemporary physics research has been the constant drive towards smaller and smaller structures. Each step in the direction of finer scale and lower dimensionality has been accompanied by fascinating new scientific discoveries that have resulted in novel applications. Modern technology, which enables fabrication of structures on

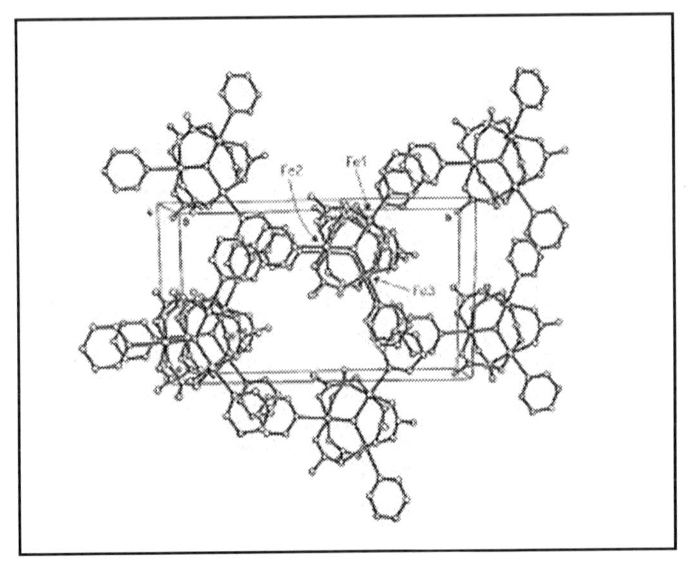


Figure 105. Magnetic nanomaterial.

the nanometer scale, has opened up exciting new research opportunities that previously were considered unreachable and has also led the way to a new generation of technology. The dynamic fields of *nanoscience* and *nanotechnology* promise to revolutionize our world. The field of nanotechnology has caused excitement within many of the traditional scientific disciplines. By manipulating structure on atomic and nanometer length scales, (Figure 105, *Wilson et al,* 2000), new properties may emerge. *Nanoscale* building blocks may then be used to construct larger structures that perform multiple functions with increased speed. The latest generation of magnetic materials provides an example of this development process.

*Nanomagnetism* involves studying how ferromagnetic materials behave when they are geometrically restricted in at least one dimension. Apart from 2D *thin films,* such objects can be 1D "***nanowires***", or zero-dimensional "***magnetic islands***". Thanks to new high-resolution fabrication techniques, these objects are now relatively easy to make. Indeed, physicists have been able to create nanomagnets with structures that range from relatively large micron-sized domains to individual atomic chains.

The magnetostatic interaction between neighboring dipoles of ferromagnets is caused by the overlap between the wave functions describing the spins of neighboring electrons. The exchange interaction creates an effective torque on neighboring magnetic dipoles that causes them to line up (*Koltsov and Perry,* 2004).

Therefore to determine the magnetic distribution in a nanomagnet of a particular shape and size, one has to take all three contributions into account. But attempting to calculate the precise electron wave functions at each atomic site in a material would take vast amounts of computer time. Instead, scientists ignore the atomic nature of the material and introduce an approximate phenomenological expression for the exchange of interaction. This approximation lets scientists simulate the distributions of magnetization in complex lithographically fabricated structures ranging in size from 10 nm to 10 μm.

Nanomagnets may behave very differently from their bulk counterparts. This may be because the new crystallographic structures can be stabilized in a nanomagnet, or because the atoms at the crystal surfaces and edges represent a larger fraction of the total volume. These atoms have fewer nearest neighbors and experience electric and magnetic fields of different symmetry to those within the interior of the crystal. They may possess a different magnetic moment and magneto-crystalline anisotropy that significantly modifies the overall behavior of the nanoparticle. Finally, as the size of the particle becomes comparable to the domain wall width, it is no longer possible to accommodate a domain structure. The response of the magnetization to an external field is then qualitatively different. The magnetization remains nearly uniform and rotates towards an applied field as its strength is increased.

A computer disk, for example, contains a 2D ferromagnetic thin film on which information is stored in sub-micron-sized "bits" made of hundreds of domains. The magnetic moments within different domains are forced to align with the magnetic field produced by the read head, which consists of a current-carrying coil wound round a magnetic yoke. Since the moments in the magnetic domains remain stable, the material "remembers" whatever information has been recorded.

## 5. Spintronics

Spintronics, at the interface between magnetism and electronics, is an expanding new field of research (*S. A. Wolf et al,* 2001). Spintronics (short for spin-based electronics), sometimes called *magnetoelectronics,* is the term given to microelectronic devices that function by exploiting the spin of electrons. The basic concept of spintronics is the manipulation of spin currents in contrast to mainstream electronics in which the spin of the electron is ignored. Adding the spin degree of freedom provides new effects, new capabilities, and new functionalities.

The most common use of spintronics today is in computer hard drives. Here, memory storage is based on *giant magnetoresistance* (*GMR*), a spintronic effect. Current research is focused on bringing magnetic random-access memory (MRAM) to market. Spintronic-based MRAMs should rival the speed and rewritability of the conventional RAM and retain their states (and thus memory) even when the power is turned off. Beyond today's applications to hard disc and memories, the potential of spintronics is very promising for new advances and will have important impacts on science and technology in the 21$^{st}$ century.

The first step on the road to the utilization of the spin degree of freedom was the discovery of the ***Giant Magnetoresistance of magnetic multilayers*** in 1988 by *Baibich et al.*

The resistance of such a multilayer is lowest when the magnetic moments of ferromagnetic layers are aligned and highest when they are antiparallel. As the relative change of resistance can be as high as 200%, this effect has been called ***Giant Magnetoresistance (GMR).*** In the antiparallel configuration each electron is alternately a majority and minority spin electron. The short circuit effect does not exist so resistance is much higher. In specially designed multilayers, known as spin valves, the magnetic configuration can be switched between parallel and antiparallel configuration, by a field of only a few Oersted, so that a large change of resistance can be induced by a very small field. The first spintronic devices were based on spin valves. Today most people have spintronic devices on their desktops, since all modern computers use spin valves for the read heads of hard discs. Because they can detect very small fields and very small magnetic bits, spin valve-based read heads have led to an increase in the density of stored information by almost two orders of magnitude.

The ***magnetic tunnel junction (MTJ)*** (Julliere M. 1995) is the type of spintronic device that will soon have important applications. An *MTJ* is a structure in which two ferromagnetic layers (electrodes) are separated by a very thin insulating layer, commonly aluminum oxide. The electrons can tunnel through the insulating layer and, because the probability of tunneling from a ferromagnetic electrode depends on the spin direction, the resistance of the MTJ is different for the parallel and antiparallel orientations of the magnetic moments of the electrodes. For electrodes of conventional ferromagnetic alloys, the relative change of resistance (Tunnel Magnetoresistance or TMR) can reach 70% at room temperature. The MTJ is of very small size, below the micron range. An important application of these small sized MTJ's will be for a new type of computer memory, the MRAM (Magnetic Random Access Memory). As illustrated in the upper part of Figure 106 (Albert Fert et al., 1995), each junction can store one bit of data, say "0" for the parallel configuration of the magnetic moments of the electrodes and "1" for the antiparallel configuration.

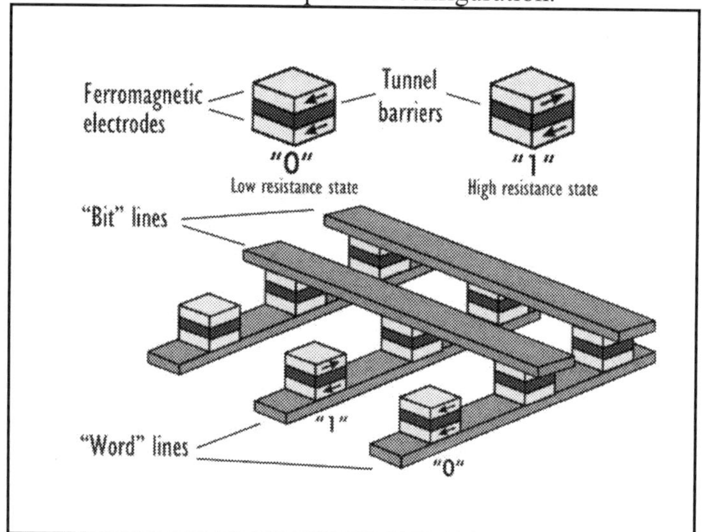


Figure 106. Top: Memory cells of an MRAM (Magnetic Random Access Memory).

Spintronics focuses on two types of materials. Ferromagnetic metallic alloys are currently used for magneto electronic devices. Ferromagnetic semiconductors, however, are attracting greater attention. If the manufacture of ferromagnetic semiconductors becomes practical, the current microchip industry could switch to this type of spintronic device with relatively little change in its infrastructure. The primary barrier to the synthesis of ferromagnetic semiconductors is finding a way to inject spin-polarized currents (spin currents) into a semiconductor.

As new and better techniques for synthesizing ferromagnetic are developed, their prospect for revolutionizing the microelectronic industry increases. Spintronics will surely play a major role in the next generation of information storage devices.

## 6. Aerogel magnets

The researchers from the Physics Department at the Universitat Autònoma de Barcelona (UAB) have created a new, ultra-light, transparent magnetic material by combining silica aerogels. Aerogels are extremely light, solid, very porous materials. They are made up of 99% air and extremely fine magnetic particles composed of neodymium, iron, and boron ($Nd_2Fe_{14}B$). Thanks to its properties, this new magnetic material could have interesting technological applications. It could be used to create a new type of flat screen and a magneto-optical memory device for computers. These new materials were orientated through a magnetic field during synthesis. They retain the transparent, light properties of the aerogel as well as the magnetic properties of the chemical composition.

Until now, all aerogels created with magnetic properties were too "soft", from a magnetic point of view, for storing information. (Weak external magnetic fields easily erase stored information). This prevented its use in many technological applications.

The new material created by UAB researchers firmly retains the orientation of its magnetic field, just like with a traditional magnet, making it very attractive for using in permanent magnetic memories. Because this new material allows light to travel through, its properties could simplify the design of magneto-optical memory devices, which could eventually be read by a laser beam. Furthermore, the material can be transparent or opaque according to the direction in which it is observed, making it potentially useful for creating a flat screen similar to an LCD. With 99% air content, aerogel is the lightest material made to date. It is so light that some have called it "frozen fog". Due to its extremely porous composition, it has the lowest levels of thermal, electrical, and sound conductivity, making it the material with the best isolation properties.

Aerogels are produced by an extremely complex process. First, a chemical solution containing water – the "gel" – is dried in special conditions in order to eliminate water molecules and to substitute them for air, so that an extremely porous solid material is obtained.

## 7. Magnetic fluids

Magnetic fluids are stable colloidal suspensions of small magnetic particles such as magnetite ($Fe_3O_4$). The particles, about 10 nm in diameter, are dispersed in nonmagnetic carrier liquids that include water, hydrocarbons, fluorocarbons, esters, diesters, and polyphenyl ethers. Stabilizing dispersing agents (surfactants), such as oleic acid or polymers, coat the particle surfaces to keep them separated and evenly dispersed within the carrier liquid. Surfactants overcome the attractive Van der Waals and magnetic forces between particles and prevent agglomeration and sedimentation. Ferrofluids generally are weak magnetic materials - they have a low "saturation magnetization".

In the absence of a magnetic field, the particles' magnetic moments are randomly distributed. Applying a magnetic field orients the particles along field lines almost instantly. Ferrofluids respond immediately to changes in the applied magnetic field, and removing the field quickly randomizes the moments. In a gradient field, the fluid responds as a homogeneous liquid and moves to the region of highest flux. This permits precise positioning and control of the ferrofluid by an external magnetic field. Forces holding the magnetic fluid in place are proportional to the gradient and strength of the magnetic field. Changing the fluid magnetization properties or magnetic field intensity lets users adjust the ferrofluid retention force.

Operating seal-life depends on ferrofluid volatility. Products needing long lives must have low evaporation rates. Also, seals operating in a high vacuum require low-vapor-pressure ferrofluids. Simple magnetic-fluid seals consist of an annular, axially polarized permanent magnet in contact with two stationary pole pieces, the magnetic fluid, and a magnetically permeable shaft. A single ring-shaped permanent magnet, or several magnets, are spaced along the bore of a nonmagnetic retainer and generate magnetic flux. Standard applications use AlNiCo permanent magnets. Special applications may require more-powerful, rare-earth magnetic materials such as samarium-cobalt and neodymium-iron-boron.

Most applications use multistage magnetic-fluid seals. Each stage supports a pressure differential proportional to the magnetic-field strength below the projection and the ferrofluid's magnetization saturation value. Typically, a single stage handles a pressure differential of 10 to 25 kPa. The entire seal's pressure capacity is approximately the sum of the individual stages' pressure capacities.

Unconventional Ferrofluid materials are Ferrogels. Ferrogels (Jakova E. et al., 2003) have been obtained by dissolving the magnetic particles in polymer solutions with subsequent cross linking. They can be superparamagnetic and isotropic as well as ferromagnetic and anisotropic.

Here, shape and volume changes induced by magnetic fields are of major interest for applications (artificial muscles). In ferrogels the elastic degree of freedom takes over a role similar to the nematic one in ferronematics. The magnetoelasticity comes in the form of $\Delta E$ effects (Stoimenov S. et al, 1995) and magnetostriction, and through the magnetic part of the Maxwell stress, and makes the system anisotropic in an external magnetic field. This gives rise to a field contribution in the sound spectrum at low frequencies that depends on the angle between the field and wave vectors. Various dynamic couplings of the elastic degree of freedom with the magnetization and flow are found. In the high frequency limit (above the magnetic relaxation frequency) the sound velocities are shifted due to those couplings. Uniaxial magnetic gels are obtained by freezing-in a finite magnetization during cross linking in the presence of an external field. The combination of a preferred direction, the magnetic degree of freedom, and the elasticity makes this material unique and very peculiar (Bohlius S., 2003).

The macroscopic description of ferronematics differs from that of ordinary nematics in several ways. First, the magnetic susceptibility anisotropy is dramatically enhanced, thus allowing for a convenient orientation of ferronematics in external magnetic fields. In addition there are several dynamic cross couplings, which are linear in the field. These effects are present in ordinary nematics, but generally neglected there, since they are assumed to be very small. In ferronematics, however, the response to external fields is very much enhanced (Jarkova E., 2002).

In many applications, ferrofluid seals operate for several years without maintenance. Seal life depends on the application, but many ferrofluid seals have operated for over ten years without maintenance. Ferrofluid seals are also suitable for chemically reactive and radioactive environments.

## 8. Superconducting magnets

At first glance, superconducting (SC) magnets seem more complicated than electromagnets, especially because they require low temperatures to keep magnet solenoids in a superconducting state. However, many of the technologies involved are the same in practice, and SC magnets have significant advantages over their electromagnetic and permanent counterparts. SC technology allows users to produce extremely high magnetic fields without the kW or MW power supplies needed for electromagnets. Once SC magnets are energized, or brought to field, users can disconnect them from their power sources and they will remain energized, which significantly reduces electricity costs. SC magnets can also generate a far higher field than permanent magnets, which are limited to 2T. SC magnets are used worldwide for many applications. In health care, MRI, (the medical term for NMR) is often used for clinical diagnosis. This technology depends on high quality SC magnets. MRI systems hold the largest share of the SC magnet market.

Most superconducting magnets are wound using conductors which are comprised of many fine filaments of a niobium-titanium (NbTi) alloy embedded in a copper matrix. These conductors have largely replaced the single filament conductors since their magnetic field more readily penetrates the fine filaments, resulting in greater stability and less diamagnetism. Consequently, the linearity of the magnetic field and the magnet current is greatly improved. Another advantage of these conductors is the more rapid rate at which a magnet can be charged and discharged, typically a few minutes for most laboratory size magnets.

Single filament NbTi magnets are preferred where the stability of the magnetic field over a long period of time is essential — usually in nuclear magnetic resonance measurements. Better persistent mode operation can be obtained with this material, and since the field is held constant for long periods of time, the extra time required to charge the magnet is inconsequential.

High temperature superconductors (HTS, Bednorz and Müller, 1986) have played two roles in the development of superconducting magnets. HTS leads have made possible new classes of LTS magnets, and magnets employing HTS material in the windings have come on the market offering unique advantages.

Magnetic-levitation is an application where superconductors perform extremely well. Transport vehicles such as trains can be made to "float" on strong superconducting magnets, virtually eliminating friction between the train and its tracks. The only friction that exists is between the carriages and the air. Consequently, maglev trains can travel at very high speeds with reasonable energy consumption and low noise levels (systems have been proposed that operate at up to 650 km/h, which is far faster than is practical with conventional rail transport).

Electric generators made with superconducting wire are far more efficient than conventional generators wound with copper wire. In fact, their efficiency is above 99% and their size about half that of conventional generators. These facts make them very lucrative ventures for power utilities.

In a Tokamak, for controlled nuclear fusion, two superimposed magnetic fields enclose the plasma: this is the toroidal field generated by external coils on the one hand and the field of a flow in the plasma on the other hand. In the combined field, the field lines run helicoidally around the torus centre. In this way, the necessary twisting of the field lines and the structure of the magnetic areas are achieved. Apart from the toroidal field generated by the external field coils and the field generated by the flow in the plasma, the Tokamak requires a third vertical field (poloidal field), fixing the position of the flow in the plasma container. The flow in the plasma is mainly used to generate the enclosing magnetic field. In addition, it provides effective initial heating of the plasma. The flow in the plasma is normally induced by a transformer coil. Because of the transformer, the Tokamak does not work continuously, but in pulse mode.

Low-temperature superconductors are expected to continue to play a dominant role in well-established fields such as in MRI and scientific research, with high-temperature superconductors enabling the newer industries. This is, of course, contingent upon a linear growth rate. Should new superconductors with higher transition temperatures be discovered, growth and development in this exciting field could explode virtually overnight .Superconducting magnets could get bigger and better.

## 9. Summary

During the last 50 years, an impressive development has occurred in the field of magnetism stimulated by the better understanding of the characteristic behaviour of different types of magnetic materials. More recently, the discovery of behaviours which are specific to nanomagnets, spintronics, SC magnets, etc. open a new focus for research. These few examples represent a vast field of investigation for the discovery of new extraordinary applicable magnetic properties.

## References


1. Albert Fert *et al.*, 2003, *Europhysics News*, 34/6, 227
2. Baibich M., *et al.*, 1988, *Phys. Rev. Lett.***61**, 2472,
3. Bednorz, J.G, Műller A.K., 1986, *Z. Phys.* B64, 189;
4. Binasch *et al.*, 1989, *Phys. Rev.* B39, 4828,
5. Bohlius S., Brand R.H., and Pleiner H., 2003, "Macroscopic dynamics of uniaxial magnetic gels" *Phys. Rev.* E68, 041706
6. Jarkova E., Pleiner H., Müller H.-W, and H.R. Brand, 2003,"Hydrodynamics of isotropic ferrogels" *Proceedings Arbeitstagung Flüssigkristalle (Mainz)* 31, P26
7. Jullière M., 1975, *Phys. Lett.* 54A, 225,
8. Koltsov D., and Perry M., 2004., *Physics World*, July 2004.
9. Stoimenov S, Milosevski M, Ristic M., 1995, *Sintering and Materials (International Academic Publishers)*, 459-464
10. Wilson *et al.*, 2000, *JACS* 122, 11370.

**DISCUSSION**

Question (Jürgen Matzka): Can magnetically levitated trains affect close-by magnetic observatories?
Answer (Stanoja Stojmenov): Generally it depends on distance.
Question (Spomenko J. Mihajlovic): What does ferromagnetic superconductivity mean?
Answer (Stanoja Stojmenov): Ferromagnetic superconductivity may be a term from technological procedure. Most superconducting magnets are wound using conductors which are comprised of many fine filaments of NbTi alloy embedded in a copper matrix. These conductors have largely replaced the single filament conductors since their magnetic field more readily penetrates the fine filaments, resulting in greater stability and less diamagnetism. Consequently, the linearity of the magnetic field and the magnet current is greatly improved.
Question (Valery Korepanov): Can you tell something about Barkhausen noise minimization in new magnetic materials?
Answer (Stanoja Stojmenov): One reason for Barkhausen noise minimization it possible to be the grain structure. The small grain size results in many atoms being placed in grain boundary positions, which are not part of the crystalline lattice. This leads to new material properties, and it is possible to use this phenomenon, as an inhibitor of noise in new magnetic materials.

# PROGRESS AND LIMITATIONS IN MAGNETIC FIELD MEASUREMENTS

NENAD NOVKOVSKI[15]
*Institute of Physics,*
*Faculty of Natural Sciences and Mathematics*

**Abstract.** Magnetic field measurement techniques have evolved signify-cantly through the centuries. Nowadays, there are new challenges in magnetics in micro-miniaturization and in calibration toward natural standards. Recent development in the metrology provides a new basis for the definition of the standards used in the magnetic field measurements. Technological developments offer solutions for further miniaturization of the next generation of magnetic sensors. Possibilities for development of both the standards and the magnetic sensors are discussed in connection with the development of the metrology and nanotechnology.

**Keywords:** magnetic field sources, magnetic field sensors, magnetic fields standards

## 1. Introduction

Magnetic field measurements have been performed for centuries, for various purposes. Determination of the inclination and declination of the magnetic field at any place on the earth's surface is needed for navigation. The magnetic field is also important for biological beings, since it can have strong influences on them. Many biological processes are connected with magnetic fields, and hence the external fields can strongly modify their regular behavior. For example, the magnetosomes in some bacteria show various responses to their surroundings, such as oriented motion in a geomagnetic field, and intracellular storage of iron etc. (Vainshtein, 1998).

[15] Address for correspondence: Institute of Physics, Faculty of Natural Sciences and Mathematics, Gazibaba b.b., 1000 Skopje, Macedonia. Email: nenad@iunona.pmf.ukim.-edu.mk



*J.L. Rasson and T. Delipetrov (eds.), Geomagnetics for Aeronautical Safety,* 201–212.

It was also shown that European robins (Erithacius rubecula), in the presence of 1.315 MHz, 480 nT fields oriented at an angle of 24 degrees to the geomagnetic field lines, become disoriented (Thalau, 2005). Strong static magnetic fields (about 4 T) can cause acute responses of various natures (Saunders, 2005). Power-frequency magnetic fields in homes, coming from a variety of sources (appliances and domestic wiring, or electricity distribution and transmission circuits), influence the occupants. The strongest influence was identified as the presence or absence of overhead power lines of 132 kV or more within 100 m of the home (208 nT near lines) (Merchant, 1994). The geomagnetic field has an effect on cardiovascular regulation (Gmitrov, 2004), and human lymphocyte activation (Capri, 2004). The field also has a connection with the blood leukocyte radio sensitivity on gender determination in humans (Ivanov, 2003). Magnetic field measurements are also important for the development of neuronal electrical activity imaging, because the existing methods, such as electrical impedance tomography (EIT), are close to their threshold of detectability. A method using superconducting, quantum interference devices (SQUIDs) used in magneto encephalography (MEG) was proposed (Ahadzi, 2004).

## 2. Origin of the magnetic field

In all cases, the magnetic field results from currents flowing through a conductor or, the equivalent, charges moving in space. The magnetic field strength ($\vec{H}$) at some point in space ($\vec{r}$) of a current with the density distribution $\vec{j}$ flowing in some domain $V$, in general terms, can be expressed by the generalized Biot-Savart law

$$\vec{H}(\vec{r}) = \frac{1}{4\pi} curl \int_V \frac{\vec{j}(\vec{r}\,')}{|\vec{r} - \vec{r}\,'|} \mathrm{d}V', \quad (1)$$

The simplest case is that of a current, $I$, flowing through an infinite straight conductor, where the value of the field at distance, $r$, from the conductor is

$$H(r) = \frac{I}{2\pi r} \quad (2)$$

The field measured at any point is the result of both the macroscopic currents and the internal motion of the charges in the atoms of the materials the magnetic field is passing through. This involves the orbital motion of the electrons and the spin of the elemental particles. There is a mathematical expression for the field inside a material, described by the magnetic flux density (*B)* that is equal to

$$B = \mu_{\mathrm{r}} \mu_0 H \ , \tag{3}$$

with the permeability in a vacuum

$$\mu_0 = 4\pi \cdot 10^{-7} \frac{\mathrm{T}}{\mathrm{A/m}} \ , \tag{4}$$

where $\mu_{\mathrm{r}}$ is the relative permeability of the material.

Expression (3) though often used, has many limitations. First, the magnetic materials exhibit hysteresis to various degrees: from small to large. So, the magnetic flux density depends not only on the material properties and the external sources of the magnetic field, but also on the direction of variation of the field, and in many cases also on the whole past evolution of the external fields. In the case when the material properties are studied, such difficulties are removed by a special preparation of the initial state of the material and a convenient variation of the fields applied on it. But, in the case when the sources of the field are not controlled, as is the case with the geomagnetic field, such an approach cannot be applied. Second, the dependence of $B$ on the magnetic field strength is not linear even for materials with less pronounced hysteresis. Nevertheless, when considering small variations of the magnetic field, the exact expressions can be linearized and the expression (3) used with very high precision, especially where specific well defined conditions can be reproduced.

## 3. Development of the instrumentation

For more that 150 years the main instrument for measuring the magnetic field used a magnetic needle suspended at its middle from a fine fiber, (or some variation of this set-up). It was a rather delicate type of instrument with limited accuracy. The working principle is based on the determination of the moment of force ($M$) acting on a magnetic dipole ($\mu$) in a magnetic field with a given flux density ($B$)

$$\vec{M} = \vec{B} \times \vec{\mu} \ . \tag{5}$$

The expression (5) provides a basis for highly accurate measurements, but the limitations arise from the inertia and fragility of the mechanical construction.

In the middle of the past century, the fluxgate magnetometer, based on the saturation of magnetic materials was devised. After a very fast development, it reached outstanding performance, with very high sensitivity and precision. Resolutions as low as 10 pT combined with outstanding linearity were obtained. The output signal depended only on the component of the field parallel to the coil axis, and thus allowed measurement of the three perpendicular components of the field. The fluxgate magnetometer is based on the hysteresis of some ferromagnetic materials, usually ferrites, and hence it is material based and requires calibration with a standard. Long term stability is also astonishingly good – some nT per year, thus not requiring periodic calibration if higher precisions are not required.

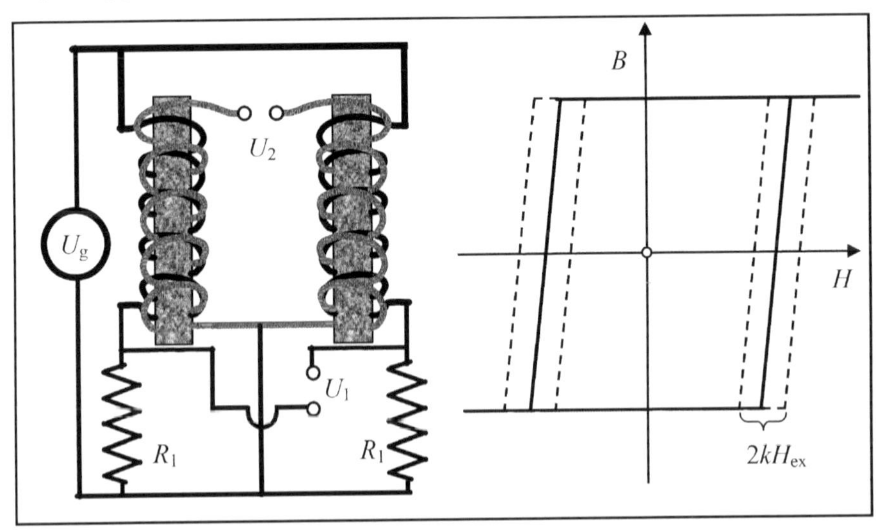


Figure 107. Fluxgate Magnetometer Function.

As the properties of the magnetic materials are dependent upon the signal frequency and temperature, and the magnetic flux density in general is a rather complicated function of the magnetic field strength, it is surprising that exceptionally good measurements are obtained with the fluxgate instrument. In order to explain this issue, an illustration of how a fluxgate magnetometer functions is given in Figure 107. First, the current driven by the source ($U_g$) in the primary coils has a defined frequency and enough amplitude to provide saturation in a part of the cycle. Therefore, a well defined hysteresis loop and a repeatable magnetization curve are obtained. As the instrument works at a defined frequency, the accuracy is further improved. In addition, the differential principle of functioning of the fluxgate magnetometer, results in a compensation of the variations due to external factors. The measurement is affected by the values in the parts where the magnetic flux density is close to zero and hence only the part of the linear dependence between the magnetic field strength ($H$) and the magnetic flux density ($B$) has to be considered, thus justifying the use of expression (3) with high accuracy. The sampling rate of fluxgate magnetometers can attain the value of the magnetization frequency of the core, typically on the order of 100 kHz (Son, 1989).

A standard for absolute calibration of the fluxgate magnetometers is provided by the proton precession magnetometer. For calibration of the magnetic field measurements and sensors at higher fields (on the order of mT and T), calibration with a magnetometer based on the nuclear magnetic resonance (NMR) can be achieved. The proton precession magnetometer is appropriate for land based geomagnetic field measurements if the total amplitude of the Earth's magnetic field is required (total field measurements). The resolution of the proton magnetometer is comparable to that of the fluxgate magnetometer (about 100 pT). The readings of the proton precession magnetometer are a function only of the field and the fundamental properties of the particles (protons). They are independent of the material used in the sensor, and do not vary with time. The accuracy of the measurement is determined by the accuracy of the measurement of the frequency of the electrical signal produced by the proton precession. As frequency measurements done by electronic counters are extremely precise, the measurement error, in principle, can be as low as $10^{-8}$.

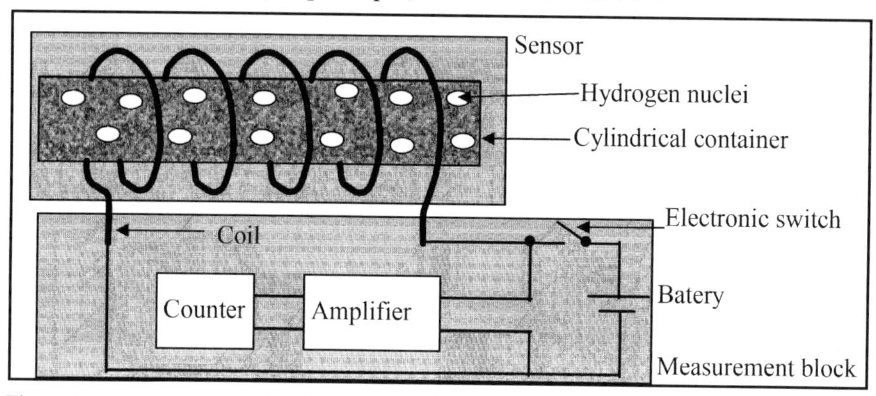


Figure 108. Proton Precession Magnetometer Function.

The resolution of the proton precession magnetometer is limited by noise generated in the coil of the instrument itself. Measurements are based on the signals generated by the proton precession in the coil. These signals are very weak and need important amplification. For very low fields, the noise induced in the same coil and in the circuits can largely cover the useful signals. This results in a lower resolution than that of the fluxgate magnetometer. Miniaturization of the sensor is not yet successful, because for smaller sized instruments the useful signal becomes smaller. Additionally, the sampling rate of the proton precession magnetometer is limited by the instrument function itself, and hence this instrument can only be used for measuring slow variations and for calibration of fluxgate magnetometers.

Further progress in resolution is achieved by the SQUID (Superconducting Quantum Interference Device) magnetometer. It combines both a measurement independent of materials' properties and extremely high resolution. Based on macroscopic quantification of the magnetic field, it gives a measure of the variation of the magnetic flux through a frame formed by two superconductors (Figure 109).

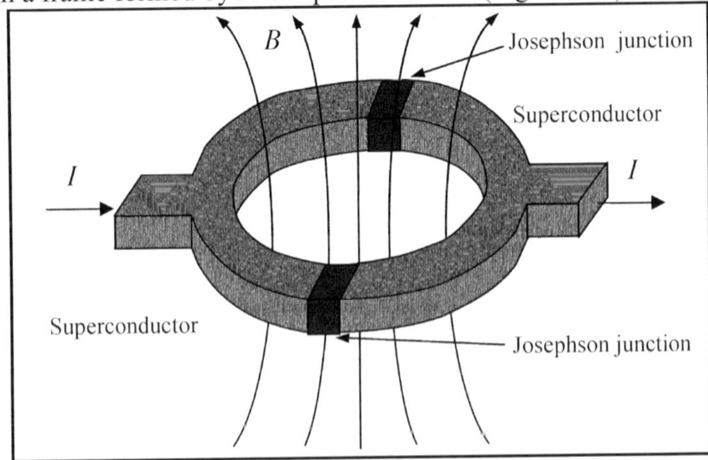


Figure 109. SQUID Magnetometer Function.

The magnetic field is quantified in units

$$\Phi_0 = \frac{h}{2e} \approx 2{,}0678 \cdot 10^{-15}\ \mathrm{T \cdot m^2}\ . \tag{6}$$

where $h$ is Plank's constant and $e$ is the electron charge.

When constant current is maintained through the device, voltage oscillations appear when the flux varies. Magnetic flux changes are obtained by multiplying the number of oscillations with the value $\Phi_0$. The resolution threshold of the SQUID magnetometer is approximately 10 fT. It is therefore useful for measuring extremely low fields such as that of the human heart ($10^{-10}$ T) and brain ($10^{-13}$ T). The measurement accuracy, in principle, is similar to that of the proton precession magnetometer, because counting the periods of an electrical signal is done. Because the signal amplitude is not as low as it is in the proton precession magnetometer, the SQUID magnetometer is less noise sensitive. SQUID magnetometers measure only the changes of the magnetic field and can be used only with a given baseline. The method of differential counting is advantageous for recording fast variations of the field. As far as the electronic circuits can count the oscillations, the measuring instrument functions as a counting

A/D converter and can give the correct value of the field. For very fast variations of the magnetic field, the instrument is not able to count each quantum. This produces a baseline shift. So, plots of magnetic field measurements will show sharp peaks and fluctuations of various natures. The instrument must also function at low temperatures, to maintain a superconductive state. The temperature requirements present additional problems for maintaining the equipment. The use of high-temperature superconductors, which are nowadays being investigated for this use, could alleviate some of the problems.

The Hall device (Figure 110) is quite often used in magnetic field measurements, mainly in the mT range. Its main advantage is that it can be easily miniaturized and integrated within electronic circuits. The instrument is based on specific materials' properties and requires calibration.

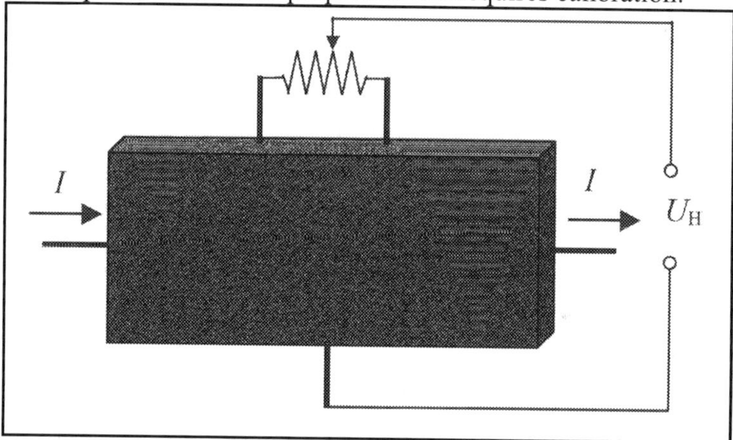


Figure 110. Hall Sensor Function.

The magnetoresistive effect is also used for construction of magnetic field sensors. The main advantage is high sensitivity and accuracy. The resistance measurements can be easily obtained and small variations of the resistance easily measured. The main disadvantage is a high temperature coefficient.

## 4. The challenge of miniaturization

Many different sensors, based on various principles, are used for measurement of the magnetic field. The miniaturization and microelectronic circuits impose some requirements on these sensors. The magnetic sensors with the most potential for future applications include: Hall devices, magnetoresistors, inductive coils, and fluxgates (Popovic, 1996). The Hall device, while very compatible with microelectronics, suffers from a limited sensitivity to silicon, a high level of 1/F noise, and a relatively large offset. Ferromagnetic magnetoresistors generally have a high sensitivity in a low field; associated problems are the flipping effect and hysteresis. Inductive coils have many applications in proximity and distance sensors, but the miniaturization of the coils is difficult. The fluxgate is a highly sensitive magnetic sensor. In principle, it could be integrated, but the main challenges are the three-dimensional structure of the coils and the low magnetic permeability of integrated ferromagnetic cores.

Recently, a solution combining the advantages of the magnetoresistors and the fluxgate magnetometers was proposed and tested (Malinowski, 2005). The pick-up coil of the fluxgate was replaced by a magnetoresistive device. A single magnetic tunnel junction was used in order to make a two-dimensional (2D) magnetometer based on the principle of fluxgate sensors. Both components of the magnetic field are extracted from the pulse stream by using the pulse position method. A magnetic tunnel junction is based on spin dependent tunneling effect (Slonczewski, 1989).

A contemporary field of investigation is nanomagnetics, comprising nanostrucures such as: nanowires, multilayers and nanojunctions. Future applications are envisaged as permanent magnets, soft magnets, magnetic recording media, sensors, structures, and materials for spin electronics (Skomski, 2003). For example, the magnetoelectronic device called the spin-valve transistor (Anil Kumar, 2000) can be used as a magnetic field sensor. It has a ferromagnet–semiconductor hybrid structure. Using a vacuum metal bonding technique, a spin-valve transistor structure Si/Pt/NiFe/Au/Co/Au/Si has been obtained. It employs hot electron transport across the spin valve (NiFe/Au/Co). The hot electrons are injected into the spin valve across the Si/Pt Schottky diode. After traversing the spin valve, these hot electrons are collected across the Au–Si Schottky diode with energy and momentum selection. The output current is found to be extremely sensitive to the spin-dependent scattering of hot electrons in the spin valve (Anil Kumar, 2000). Nanostrucutred ferrites are investigated as promising candidates for spintronic devices. It was found that $Zn_{1-x}Co_x$ $Fe_2O_4$, $Mn_{0.5}Zn_{0.5}Fe_2O_4$ and $CoFe_2O_4$ exhibited ferrimagnetic and superparamagnetic behavior, respectively, at room temperature (Tomar, 2005).

## 5. Reconstruction of the currents producing the magnetic field

Each magnetic field is the result of distributed currents, as given by (1) and the distribution of ferromagnetic materials around the sensor. While measuring the field on a grid of points, it is possible, in principle, to reconstruct the currents. There are several applications for this reconstruction. These include non-destructive testing and monitoring of fuel cells. Even in the simplest cases, where there are no ferromagnetic materials, the situation is rather complex, because an inverse problem has to be solved. The question of uniqueness and non-uniqueness of the reconstruction was raised and discussed (Hauer, 2005). The situation with the geomagnetic field is even more complex, because of the presence of ferromagnetic materials. As the magnetic state of the ferromagnetic constituents is not a unique function of the surrounding field sources, but depends on the previous state, it is practically impossible to determine the currents. The approach where magnetic dipole distribution is reconstructed is possible, but the dipoles do not give the entire picture of the sources of the magnetic field. As the hysteresis behavior can be connected to self-organizational patterns, some synergetic approaches can be fruitful (Jones, 2005).

Similarly, the determination of the history of the geomagnetic field can suffer from related limitations. Nevertheless, the global features that are usually determined do not require precise solutions (Doell, 1969; Dunlop, 1990; Aitken, 1999; Korte, 2003; Gratton, 2005).

## 6. On the standard of the magnetic field

The issue of calibration of magnetic field measurements in connection with the proton precession magnetometer was previously discussed. A natural standard, independent of the peculiar materials' properties was provided by fluxgate magnetometers. Alternatively, the SQUID magnetometer, (where the flux is quantified in units of $\Phi_0$ ) can provide a magnetic flux standard. To obtain a standard for the magnetic field strength, it is possible to develop a new standard based on the current flowing through a straight, infinite conductor (2). Nowadays it is possible, because new standards for voltage (Josephson Junction Arrays, JJA) (Kohlmann, 2005) and resistance (Quantum Hall Effect) (Jeckelmann, 2005) were adopted in 1991. Based on these standards, the current, I, can be determined with accuracy as high as $10^{-8}$, and measuring the normal distance from the conductor with similar accuracy by interferometry, the magnetic field strength, *H,* can be calculated with comparable accuracy (Figure 111). Thus existing metrology can be used for the resolution of the metrological triangle *H-I-L* (magnetic field strength-current-length, where the length is the normal distance from the conductor to the measurement point) given by the expression (2), i.e. by comparison of independent high accuracy standards to confirm the basic physical principles. A similar situation exists for the resolution of the *U-I-R* metrological triangle, with the introduction of a new standard for current based on the counting of elementary charges.

In the case of the magnetic field, the quantum, $\Phi_0$, or the precession constant factor has to be compared to the result obtained by measuring distance and current (the voltage over resistance ratio).

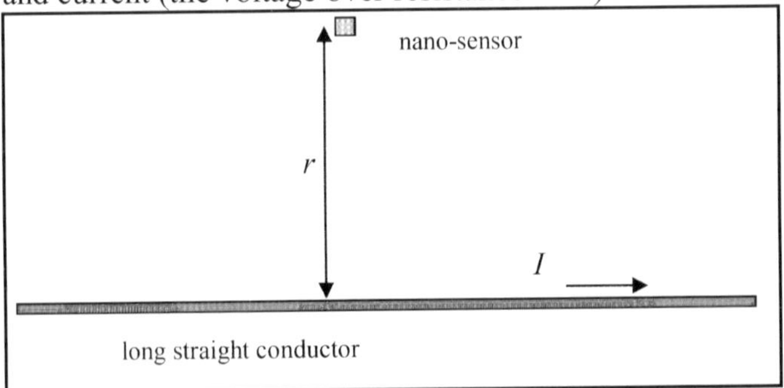


Figure 111. A possible principle for the standard of the magnetic field strength.

## 7. Conclusion

Measurements of magnetic fields are important for navigation, understanding Earth's magnetism, paleogeology, and biology. Important developments have been made in the past 60 years in magnetics. New challenges imposed by the technological development of microelectronic circuits and miniaturization require further development of magnetic sensors. Nanodevices are promising solutions that are presently extensively studied. The recent development of the electrical standards and emerging solutions provide a basis for development of new standards for magnetic field measurements.

## References

Ahadzi, G. M., Liston, A. D., Bayford, R. H. and Holder, D. S., 2004, Neuromagnetic field strrength outside the human head due to impedance changes from neuronal depolarization, *Physiol. Meas.* **25**(1): 365–378

Aitken, M. J., 1999, Archaeological dating using physical phenomena, *Rep. Prog. Phys.* **62**(9):1333–1376

Anil Kumar, P. S. and Lodder, J. C., 2000, The spin-valve transistor, *J. Phys. D: Appl. Phys.* **33**(22): 2911–2920

Capri, M., Mesirca, P., Remondini, D., Carosella, S., Pasi, S., Castellani, G., Franceschi, C. and Bersani F., 2004, 50 Hz sinusoidal magnetic fields do not affect human lymphocyte activation and proliferation in vitro, *Phys. Biol.* **1**(4): 211–219

Doell, R. R., 1969, History of the Geomagnetic Field, *J. Appl Phys.* **40**(3): 945–954

Dunlop, D. J., 1990, Developments in rock magnetism, *Rep. Prog. Phys.* **53**(6), 707-792

Gmitrov, J. and Gmitrova, A., 2004, Geomagnetic field effect on cardiovascular regulation, *Bioelectromagnetics* **25**(2): 92–101

Gratton, M. N., Show, J. and Herrero-Bervera, E., 2005, An absolute palaeointensity record from SOH1 lava core, Hawaii using the microwave technique, *Phys. Earth Planet. Inter.* **148**(2-4): 193–214

Hauer, K.-H., Kühn, L. and Potthast, R., 2005, On uniqueness and non-uniqueness for current reconstruction from magnetic fields, *Inverse Problems* **21**(3): 955–967

Ivanov, S. D., Iamshanov, V. A., Koshelevskii, V. K., Ivanova, A. S., Ivanova, T. M., Glushkov, R. K., Semenov, E. V. and Petrov, A. N., 2003, The influence of geomagnetic field on blood leukocyte radiosensitivity and gender determination in humans, *Radiats Biol. Radioecol.* **43**(2): 245–8

Jeckelmann, B. and Jeanert, B., 2005, The quantum Hall effect as an electrical resistance standard, Meas. Sci. Technol. **14**(8): 1229-1236

Jones, B. A. and O'Grady, K., 2005, Magnetically induced self-organization, *J. Appl. Phys.* **97**(10): 10J312 (3 pages)

Kohlmann, J., Behr, R. and Funck, T., 2005, Josephosn voltage standards, *Meas. Sci Technol.* **14**(8): 1216–1228

Korte M. and Constable C., 2003, Continuous global geomagnetic field models for the past 3000 years, *Phys. Earth Planet. Inter.***140**(1-3): 73–89

Malinowski, G., Hehn, M., Kammerer, J.-B., Sajieddine, M., Jouguelet, E., Hébrard, L., Alnot, P., Braun, F. and Schuhl, A., 2005, Flux-gate like 2D magnetometer based on a single magnetic tunnel junction, *Eur. Phys. J. Appl. Phys.* **30**(2): 113–116

Merchant, C. J., Renew, D. C. and Swanson, 1994, J., Exposures to power-frequency magnetic fields in thc home, *J. Radiol. Prot.* **14**(1): 77–87

Popovic, R. S., J. A. Flanagan, and P. Besse, 1996, The future of magnetic sensors, *Sensors and Actuators A: Physical* **56**(1-2): 39–55

Saunders, R., 2005, Static magnetic fields: animal studies, *Prog. Biophys. Mol. Biol.* **87**(2-3): 225–239

Skomski, R., 2003, Nanomagnetics, *J. Phys.: Condens. Matter.* **15**(20): R841-R896

Slonczewski, J. C., 1989, Conductance and exchange coupling of two ferromagnets separated by a tunneling barrier, *Phys. Rev. B* **39**(10): 6995–7002

Son, D., A new type of fluxgate magnetometer using appaernt coercitive field strength measurement, 1989, *IEEE Transactions on Magnetics* **25**(5): 3240–3242

Thalau, P., Ritz, T., Stapput, K., Wiltschko and R., Wiltschko, W., 2005, Magnetic compass orientation of migratory birds in the presence of a 1.315 MHz oscillating field, *Naturwissenschaften* **92**(2): 86–90

Tomar, M. S., Singh, S. P., Perales-Perez, O., Guzman, R. P., Calderon, E. and Rinaldi-Ramos, C., 2005, Synthesis and magnetic behavior of nanostructured ferrites for spintronics, *Microelectron. J.* **36**(3-6): 475–479

Vainshtein., M, Kudryashova, E., Suzina, N., Ariskina, E. and Sorokin, V., Functions of non-crystal magnetosomes in bacteria, Proc. SPIE: Instruments, Methods, and Missions for Astrobiology, Richard B. Hoover, Editor, July 1998, pp. 280-288

## DISCUSSION

Question (Jean Rasson): Can you explain if giant magnetoresistance can be useful to us?

Answer (Nenad Novkovski): In the case of giant magneto resistive (GMR) sensors it is expected to obtain higher sensitivities that in the case of fluxgate magnetometers (0.1 nT), according to some authors down to 0.01 nT.

Orientation sensitivity like in the case of fluxgate and the zero stability when perpendicular to the field can be obtained, as one with the possibility to make the measurements in the proximity of magnetic field sources and under high gradients. The spatial resolution is nowadays about 1 mm and hence it is substantially lower than that of the fluxgate magnetometers (15 nm). In parallel, the high bandwidth (> 300 kHz) will allow following fast changes.

Question (Jürgen Matzka): Magnetoresistive sensors are directional?

Answer (Nenad Novkovski): The anisotropic magneto restive sensors (AMR) are highly directional. There are well suited for precise determination of the Earth's magnetic field direction. In addition, they can be integrated on silicon wafers together with the integrated circuits used in the measurements.

# GEOMAGNETIC MEASUREMENTS FOR AERONAUTICS

JEAN L. RASSON[16]
*Institut Royal Météorologique de Belgique*

## 1. Introduction

Anybody who has been on the open sea without visual clues for orientation or, worse, trapped in the dark by a sudden failure of the lighting system, knows how easy it is to feel lost. If a navigation system is available, then there is a means to calming down and finding your way.

The geomagnetic declination allows one to do just that: navigate and find a way to destination with the aid of a specialized instrument called a "magnetic compass".

The principle function of the compass is to indicate the North direction on a graduated horizontal disk. This disk is free to rotate around a vertical axis and is actually moved by the torque exerted by the horizontal component of the geomagnetic field on a magnet inside it. The compass indicates the direction of Magnetic North, which is different from True North. The difference between the two directions is the magnetic declination.

For the compass to work accurately as a navigation device, various conditions must be met:

- The compass must work properly.
- The compass must not be perturbed by artificial magnetic fields or it must be compensated for them.
- The horizontal component of the geomagnetic field must be strong enough to drive the compass needle (a condition not met at and near the poles).
- Magnetic declination must be known at the location of the compass in order to determine True North from Magnetic North.

Aeronautics uses the magnetic compass extensively as a navigation tool. Of course, aircraft have other navigation devices, but the compass is still

[16] Address for correspondence: IRM/CPG, Rue de Fagnolle, 2 Dourbes, B-5670 Viroinval, Belgium. Email : jr@oma.be



*J.L. Rasson and T. Delipetrov (eds.), Geomagnetics for Aeronautical Safety,* 213–230.

the primary direction indicator on small aircraft and is a very important back-up device on larger planes. Airport infrastructure must include the elements required to perform an aircraft check, and to calibrate or compensate onboard compasses. Airport infrastructure quality is not only nice buildings and runways, it also relates to more sophisticated facilities like the knowledge of the correct and up to date value of the magnetic declination.

This paper will focus on how the geomagnetic community can help aircraft operators and airport authorities with the proper operation and certification of magnetic compasses on board aircraft they own or those that pass through their facilities.

The list of services and products provided by the geomagnetic community follows:

1. Compass rose certification.
2. Runway azimuth determination.
3. Supply of isogonal information and maps.
4. Supply of magnetic declination data.

Compass rose certification is discussed extensively here, because the procedure is less well known among observatory personnel and if not done properly, becomes a very lengthy process as large numbers of measurements need to be made. Runway azimuth determination is more straightforward. They will be reviewed briefly.

Before starting the discussion of geomagnetic services and products, definitions of often used scientific and technical terms related to the subject of geomagnetism and magnetic navigation are defined:

- ***Azimuth***: The angle a direction makes from true North.
- ***Magnetic meridian***: The vertical plane containing the geomagnetic vector.
- ***Magnetic declination***: The azimuth of a horizontal direction in the magnetic meridian.
- ***Magnetic variation***: This expression is used instead of "magnetic declination" in maritime and aeronautical sectors.
- ***Secular variation***: The change of the magnetic declination over time at one location; usually expressed in arc minutes/year.
- ***Compass rose***: The graduated circle of the compass; by extension, the pattern painted on the compass calibration pad at an airport, or the pad itself.
- ***VOR***: Acronym for "VHF Omni directional Ranging"; an electronic aid located at various spots in the country for assisting in the navigation of aircraft.

- ***Heading***: The azimuth of the trajectory (speed vector) of a moving vehicle
- ***Isogonal map***: Map displaying the spatial distribution of the value of the magnetic declination as contour lines.
- ***Hard and soft magnetism***: "hard" refers to a magnet-like durable magnetization, which will remain after any external field has been removed. "Soft" refers to a magnetization existing only when an external field is applied.

## 2. Motives – Geomagnetism and the Commercial Sector

Surely the main motive for a magnetologist is the scientific curiosity. The investigation and the discovery of the internal and external magnetic processes going-on in the Earth and its physical manifestations are what push us all forward. However, the collaboration with the aeronautical community has its rewards also and we give below some additional reasons for going into this activity.

The benefits for an observatory of commercially providing “services and products” are many and the experience and past history show that combining scientific and socio-economic activities really leads to a win-win situation. Here is why:

- Customers need and are ready to pay for what is delivered. The demand for products and services justifies the existence, ventures, and expenses of the geomagnetic community. This justification is regularly required by the political world, which often provides observatories with funding. They regularly assess the usefulness of observatories and hence their return on their investment weighing the money spent in maintaining an observatory against the service provided.
- The commercial relationship may actually result in the delivery of better services or products to the customer. Since there is a financial transaction, the observatory personnel may feel a stronger need to pamper the customers by providing highly accurate data, detailed customer information, and post-delivery services.
- An observatory delivering data and services free of charge may be unhappy if it later finds that it has been sold to a third party. Maybe you remember this advertisement for magnetic field sensors from Honeywell: "Buy this sensor, get the magnetic field for free". The magnetic field is free for anyone to observe, but the use of this information by a Honeywell sensor implies the coordinated and continuous effort of the whole geomagnetic observatory community, which comes at a cost.

- The commercial delivery of data and/or services allows the observatory to earn money. This is important since funding from governments is generally on the decline.

## 3. Inventory of products and services

### 3.1. COMPASS ROSE CERTIFICATION

#### 3.1.1. *Compass Rose*

A compass rose or compass calibration pad is a spot on the airport grounds suitable for performing aircraft compass swings. Swings involve rotating the whole aircraft to known magnetic azimuths and, for each orientation, observing the compass deviations.

The compass needle directional indication is affected by ferrous, metallic components within the aircraft. To reduce the deviation effect, the aircraft compass must be checked and possibly compensated periodically by adjusting compensating magnets. This procedure is called "swinging the compass". During compensation, the compass is checked at, say, 30° increments. Adjustments are made at each of these points, and the difference between the magnetic heading and the compass heading is shown on a compass correction card (see Figure 112). When flying compass headings, the pilot must refer to this card and make the appropriate adjustment for the desired heading.

| FOR | 0° | 30° | 60° | 90° | 120° | 150° | 180° | 210° | 240° | 270° | 300° | 330° |
|---|---|---|---|---|---|---|---|---|---|---|---|---|
| STEER | 359° | 30° | 60° | 88° | 120° | 152° | 183° | 212° | 240° | 268° | 300° | 329° |

COMPASS CORRECTION CARD

Figure 112. Example of a compass correction card.

For an airport compass rose to be useful, the following elements are required:

- A large circular area devoid of magnetic perturbations and accessible to aircraft must be maintained. The horizontal dimensions of the largest aircraft that will be swung will determine the diameter of the area.
- Magnetic azimuth markings must be applied to this area so that the aircraft can be precisely oriented along them (Figure 112). Due to secular variation; the magnetic azimuth markings must be periodically updated.

- The true North direction should be marked.
- The compass rose must be certified for use.

The compass rose can be certified in a specified class if the magnetic perturbations, which affect magnetic declination, produce deviations (from a spatially averaged mean) that remain below a given value.

### 3.1.2. *Compass Rose Construction*

If there is no compass rose at the airport already, a location should be selected for one. A spot that is likely to be non-magnetic, that can accommodate the dimensions of the largest aircraft should be chosen with the aid of the airport staff. A special pad may be constructed for the compass rose, or the end of an infrequently used runway can be reserved (Figure 113).

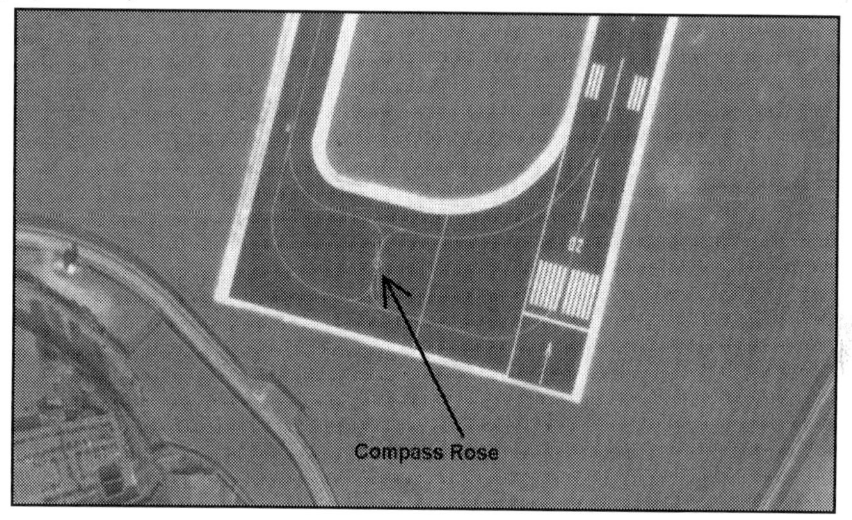


Figure 113. The compass rose at this international airport has been implemented near the end of the less frequently used runway.

When looking for a suitable site for a compass rose, before doing any measurements, an investigation of the site should be made to check for the presence of hard and/or soft magnetized bodies likely to give magnetic perturbations. Iron rebar, present in reinforced concrete, is often used for building runways and produces significant magnetic perturbations. Iron conduits for cables or fluids, located underground are also common sources of magnetic perturbation. Stepper motors inside guided cameras and other robotics often contain hard magnetized elements like magnets, and should be avoided.

When selecting a new compass rose site, and before starting expensive construction work, it is necessary to run a proton magnetometer survey, in order to determine the magnetic hygiene of the underlying terrain. A proton magnetometer is used first, because it is much faster to operate. Its output however is the modulus of the magnetic field and not magnetic declination. (The Tomlinson equation of section 3.1.7 on page 222 is useful to link both quantities.) The results of this first survey indicate if construction and certification can proceed or if another site must be found.

The choice of the spatial sampling interval is a delicate point in measuring the spatial features of the compass rose. Most authors (Civil Aviation Authority UK, 1993; Loubser, 2005) recommend a spatial sampling of 6 m, but some worry (Crosthwaite, 2005) that magnetic anomalies may "hide" between the 6 m separations of the measurement points. A solution exists if a closer spacing is chosen for the proton magnetometer measurements than for the declination measurements, especially if there is reason to believe that sharp, short wavelength magnetic perturbations exist in the underlying terrain. An example of spatial sampling is given in Figure 114.

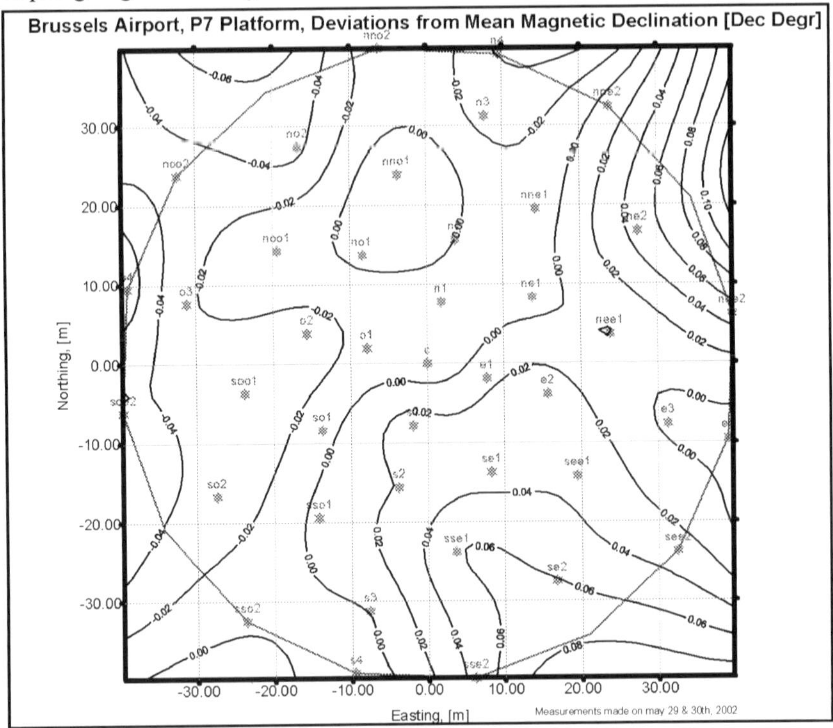


Figure 114. The compass rose and its different measurement stations as implemented at the Brussels International Airport. The isogonal lines show the deviations $dD_i$ from the average value. According to this $dD_i$ distribution, this rose qualifies for a Class 1 certification.

Some electronic navigation and landing aids produce strong electromagnetic radiation likely to perturb sensitive magnetic instrumentation, notably the proton magnetometer. The Instrument Landing System (ILS), for instance, sends out a powerful beam making valid proton precession measurements impossible in front of it. Therefore, it is not a good idea to establish a compass rose in close proximity to those facilities.

### 3.1.3. *Compass Rose Certification: Principles and Classes*

In an airport, a compass rose certification must be performed regularly in order to evaluate the magnetic cleanliness of the site and for keeping track of the changing magnetic declination.

Therefore the magnetic declination $D_i$ is measured at the N points of a grid covering the compass rose with the spatial separation/sampling agreed upon (generally 6m).

Those N measurements are corrected for the diurnal variation $\delta D_i$, thus providing N values $D^*_i$ defined by:

$$D_i = D^*_i + \delta D_i ,$$

and the arithmetical mean is taken:

$$\overline{D} = \frac{1}{N}\sum_{i=1}^{N} D_i^* .$$

The deviations, with respect to the spatial mean $dD_i$, are then computed:

$$dD_i = D_i^* - \overline{D} .$$

The extremes $dD_{max}$ and $dD_{min}$ in the $dD_i$ series are then easily found and the maximum deviation with respect to the mean value is defined:

$$MaxDev = \frac{(dD_{max} - dD_{min})}{2}$$

The quantity *MaxDev,* expressed in units of degrees, is used in aeronautical quarters for evaluating the Compass Rose magnetic cleanliness. According to the *MaxDev* values, compass roses are divided into different classes with corresponding certification standards:

- Class 1 certification: MaxDev $< 0.1°$
- Class 2 certification: MaxDev $< 0.25°$

It should be emphasized that the natural, daily variation of the magnetic declination is about 10 arc-minutes (= 0.17°) on a quiet day in the mid-latitudes, and is even more elsewhere. This is why the daily variation of the magnetic declination should be removed from the $D_i$ series, lest it introduce an unacceptable bias in the magnetic cleanliness data.

There are different approaches for measuring and certifying compass roses. The differences in methodology stem from the way the actual declination measurements are carried out (what instrumentation is used) and the way the diurnal variations are removed from the data. From section 3.1.4 to section 3.1.7 we review the different procedures known to us for performing the compass rose certification task.

### 3.1.4. *Compass Rose survey with the DIflux magnetometer and observatory data reduction*

The DIflux magnetometer is used almost universally in magnetic observatories and for repeat station work to make magnetic declination measurements. This instrument, essentially a non-magnetic theodolite equipped with a fluxgate sensor mounted on the telescope, allows error free measurements of the geomagnetic angles like declination and inclination. Additionally, it can easily be used for geodetic azimuth determination using astronomical sightings on the sun or stars (Rasson, 2005).

To reduce the diurnal variation of the compass rose data, using observatory data, a magnetic observatory must be close by. The observatory must be close enough so that the diurnal variation measured there is similar (within a few arc-minutes) to the one experienced at the airport. Each $D_i$ measured on the compass rose is tagged by the time stamp $t_i$. Then $D_{obs}(t_i)$ is the declination measured at the observatory synchronously with the measurement $D_i$, and $\delta D_i$ is the diurnal variation (supposed to be similar at both the observatory and the airport). We define the constant $D^*_{obs}$ as the observatory data corrected for daily variation by:

$$D_{obs}(t_i) = D^*_{obs} + \delta D_i.$$

If we define:

$$\Delta_i = D_i - D_{obs}(t_i) = D^*_i + \delta D_i - [D^*_{obs} + \delta D_i] = D^*_i - D^*_{obs},$$

we see that this quantity is free from time variations and contains only the spatial information about the magnetic cleanliness of the compass rose. $\Delta_i$ can be used in lieu of $D^*_i$ in the relationships of section 3.1.3 for computing the quantity *MaxDev* since the removal of the mean will eliminate the constant $D^*_{obs}$.

A distant azimuth mark should be chosen so as to be visible from all stations on the compass rose. This azimuth will be measured from the center station of the rose. Since the geometry and the orientation of the other station points is precisely known, the azimuth of the distant mark can be computed from the center station azimuth. This will allow the full measurement of the declination on all stations with only one azimuth measurement task to be performed.

The final product delivered to the airport authorities will be a list of the measured stations, the value of their $dD_i$'s, and an isogonal map similar to the one in Figure 114. According to the results of the survey and the value of the MaxDev parameter, the report may also contain a certificate awarding a class 1 or class 2 status to the compass rose.

### 3.1.5. *Reciprocal sighting with two declinometers*

This procedure is described in aircraft instrument calibration manuals such as the CAP562 leaflet (Civil Aviation Authority UK, 1993). It has been accurately described in a to-be-published IAGA guide (Loubser, 2005) where a detailed analysis and error budget can be found. The procedure is based on the principle of taking reciprocal bearings with two declinometers (or datum compasses).

One declinometer is set up in a fixed position throughout the survey, while the second one is moved from point to point on the grid of survey points on the compass rose. At each point the two declinometers are simultaneously aligned with magnetic North and readings are taken. The two declinometers are then aligned with each other and readings are again taken of the bearings (reciprocal bearings). The two readings, taken in opposite directions, will differ from each other by 180°. Therefore, if 180° is subtracted from one reading, and the two readings are then subtracted from each other, this difference should be a measure of the effect of local magnetic disturbances of geological or other origin. If the readings are taken at a time when the magnetic field is disturbed (magnetic storm or daily variation), the readings on the two declinometers will be affected identically if they are taken within one minute of each other. Thus, when the difference between the two readings is taken, this disturbance effect is cancelled.

This procedure obviates the need for a neighboring magnetic observatory and the subsequent correction of the survey data for the daily variation. It also obviates the need for a distant azimuth reference mark during the survey. Its use is therefore to be advocated in regions devoid of magnetic observatories. However, it imposes the availability of 2 dedicated observers and 2 dedicated and expensive declinometers, not part of the standard equipment at a magnetic observatory.

### 3.1.6. *USGS method*

This procedure is also suitable when nearby magnetic observatory data is not available. It is normally performed with a datum compass, but could also be done with a DIflux magnetometer. The diurnal variation is estimated by performing a sequence of measurements over the compass rose points including frequent re-measurements over the same center point. These closure measurements at the center point allow the extraction of daily field variations during the whole measurement task and its subsequent correction. This method requires one observer and one helper. The helper is employed for several hours and is supplied by the airport.

The procedure is fully explained in the paper by Berarducci in this volume.

### 3.1.7. *Tomlinson method*

This procedure, invented by L. Tomlinson of Eyrewell Observatory in New-Zealand, relies on the measurement of the sole field modulus. Therefore the use of the quick proton magnetometer survey allows a rapid estimation of the magnetic cleanliness of the compass rose.

By using the assumption that the magnetic anomalies underlying the compass rose have the simple structure of a magnetic dipole, Tomlinson (Tomlinson, 2000) arrives at a relationship linking the field modulus $dF_i$ and the magnetic declination $dD_i$ deviations required for the MaxDev parameter computation as explained in section 3.1.3 on page 219:

$$dF_i = \left|0.858 \times H \times \sin I \times \sin(dD_i)\right|,$$

where H is the horizontal field component and I is the magnetic inclination. Therefore, the MaxDev parameter can be replaced by a $\mathrm{MaxDev_F}$ parameter valid for a single airport, according to the values of H and I there (Table 24).

Table 24. Values of the parameter $\mathrm{MaxDev_F}$ in nanoteslas for different locations (year 2004). The Class 1 certification corresponds to magnetic declination deviations smaller than 0.1° and the Class 2 for deviations smaller than 0.25°.

| Location | I [°] | H [nT] | $\mathrm{MaxDev_F}$ Class 1 [nT] | $\mathrm{MaxDev_F}$ Class 2 [nT] |
|---|---|---|---|---|
| North magnetic Pole | 90 | 0 | 0 | 0 |
| Resolute Bay CA | 88.2 | 1790 | 3 | 7 |
| Dourbes BE | 65.5 | 19970 | 27 | 68 |
| Skopje MK | 58.7 | 24290 | 31 | 78 |
| Ohrid MK | 57.7 | 24740 | 31 | 78 |
| Kanoya JP | 45.0 | 32780 | 35 | 87 |
| Huancayo PE | 1.1 | 25950 | 1 | 2 |
| Kakadu AU | -40.3 | 35440 | 34 | 85 |
| Eyrewell NZ | -68.6 | 21110 | 28 | 70 |
| Terra Nova Bay Ant. | -83.0 | 7830 | 11 | 29 |
| South magnetic Pole | -90 | 0 | 0 | 0 |

Obviously, a problem exists for equatorial sites where $\sin I \sim 0$, (which puts the parameter $MaxDev_F$ at an unrealistically low level). The Tomlinson method is not valid in regions near the magnetic equator (Tomlinson, personal communication). At polar sites where $H \sim 0$, the same low value is obtained, but a compass should not be used there anyway.

It should be mentioned that the Tomlinson procedure has been accepted by the New Zealand Civil Aviation Authority as the basis for an alternative method of testing aircraft compass roses (Tomlinson, 2000).

In any case, the Tomlinson method can be used over a well delimited region as a way to:

- get alternative information on the magnetic cleanliness of the compass rose,
- get higher spatial resolution than other methods (Crosthwaite, 2005) thanks to its quicker measurement protocol,
- obtain guidance when surveying using a different procedure.

keeping in mind that the Tomlinson procedure is not valid in regions situated on the magnetic equator.

### 3.2. RUNWAY MAGNETIC AZIMUTH DETERMINATION

At the ends of aircraft runways, large 2 digit number markings are painted in white. These markings represent the magnetic azimuths of the runways in units of 10 degrees. They equal the number a pilot should read on his magnetic compass when his aircraft is correctly aligned with the runway during the landing procedure. Figure 115 illustrates such runway markings.

Figure 115. The beginning of runway 01 at Tete Airport (Mozambique). The number 01 is painted in enormous digits so that a pilot can clearly read it during his landing procedure. The painted number equals the magnetic azimuth of the runway in units of 10 degrees, i.e. the magnetic azimuth of this runway is 10°.

Table 25 gives the markings of several airports' runways in Mozambique, where the magnetic declination is high and varies considerably from place to place. Geodetic and magnetic azimuths are quite different and require regular updates.

Table 25. Listing of the main airports in Mozambique. The geodetic azimuth, magnetic azimuth, and the runway markings are valid for December 2004.

| Airport | Geodetic Azimuth [dec. deg.] | Magnetic Azimuth [dec. deg.] | Magnetic Azimuth [runway marking] |
|---|---|---|---|
| Maputo int. | 28.890 | 47.1 | 5 |
| Pemba int. | 340.741 | 347.0 | 35 |
| Nampula | 46.064 | 53.9 | 5 |
| Lichinga | 70.626 | 75.2 | 8 |
| Tete | 1.370 | 8.7 | 1 |
| Beira int. | 104.940 | 116.5 | 12 |
| Quelimane | 354.191 | 364.1 | 36 |

### 3.2.1. *Measurement Procedure of the Magnetic Azimuth*

The magnetic azimuth is the angle between Magnetic North and the runway axis. Figure 116 shows how to measure the magnetic azimuth of an airport runway with a DIflux magnetometer.

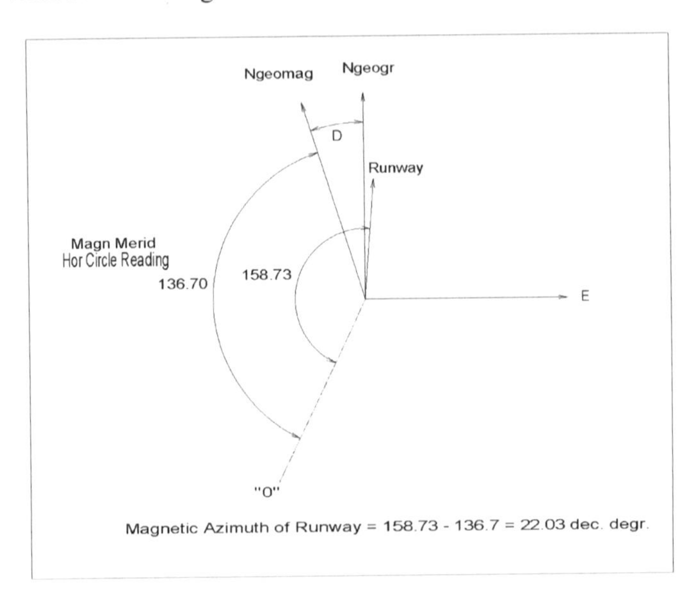


Figure 116. Magnetic azimuth measurement with a DIflux magnetometer. The indicated angles are the readings of the horizontal circle ("0" marks the direction of the starting graduation).

The tasks to be performed for runway azimuth determination are:

1. Perform a proton magnetometer survey of the runway location where the magnetic azimuth will be determined. If the runway is too magnetic, refer to section 3.2.2.
2. Determine trace of the direction of the magnetic meridian, $T_{magmer}$, by using the DIflux horizontal circle. The traditional 4 position DIflux D-measurement protocol is used with the instrument telescope horizontal and the fluxgate normal to the magnetic meridian. The trace is the same as would be used for positioning the DIflux for an I-measurement.
   - Measure the trace of the runway direction, $T_{runw}$, by sighting the runway center-line markings. An example of how to do such a sighting through the telescope of the DIflux is shown in Figure 117.
   - Compute the magnetic azimuth, $A_{mag}$, of the runway:

$$A_{mag} = T_{runw} - T_{magmer}.$$

This procedure does not require the knowledge of the True North direction. Hence there is no need to perform astronomical, gyro, or GPS orientation measurements.

Figure 117. How to sight the runway center line markings through the DIflux telescope. A VOR antenna may be seen aligned with the runway at its terminus.

3.2.2. *Problem solving*

The situation is not always straight forward. Two complications can arise:

1. The runway is too magnetic to perform a valid determination of the magnetic meridian.
2. The runway is not straight or it is not completely flat. This happens often at smaller airports. Sometimes one end of a runway is not visible from the other because the central section is elevated.

To solve the first problem, it is necessary to determine the geodetic azimuth of the runway $A_{runw.}$ This can be done by performing a sun shot. Then, the magnetic declination must be measured on a magnetically clean spot close to the runway (this spot is often a repeat station point as well). The magnetic azimuth of the runway $A_{mag}$ is then computed as:

$$A_{mag} = A_{runw} - D_{runw}$$

For solving the problem of item 2, one should measure the magnetic azimuth on both extremities of the runway, and report this particularity to the airport authorities.

Figure 118. Aircraft landing at the Pemba Airport (Mozambique) while observers are in the process of doing runway azimuth measurements. While the observers could take cover, there was no time to dismantle and evacuate the DIflux and its tripod, visible at the lower left.

### 3.2.3. *Observer's security*

Some airports, usually small ones, will allow normal airplane traffic while magnetic measurements are being made on the runways. Working on the runways can therefore be dangerous. Even if the airport closes its runway for a period of time while measurements are being made, unscheduled or unauthorised aircraft movement may take place as a result of an emergency. Therefore, while on the runway, observers should keep contact with the control tower of the airport via a portable, short wave radio or cellular phone. In the absence of a reliable warning system, the firing of runway lights is a good indicator of an imminent landing. In any case, when on a runway, it is good to regularly scan the surrounding skies to detect incoming aircraft.

When confronted with an imminent landing, while doing runway geodetic or magnetic measurements, it is best to dismantle the instrumentation, including the tripod, and move everything well off of the runway. This requires a second set-up with precise location of the tripod and levelling the DIflux magnetometer after the aircraft traffic has ceased. If there is no time to dismantle equipment before the landing occurs, the observer may decide to leave all equipment on the runway, and take cover. The Figure 118 shows such a situation.

## 3.3. SPATIAL MAGNETIC DECLINATION DATA FOR PREPARING MAPS AND HEADING LISTS

To create detailed isogonal maps, full magnetic declination surveys are required. The detail of the maps will of course relate to the spatial sampling of the survey measurements. Using recent, and some past magnetic repeat station data and by extrapolating normal field maps, it is possible to "refresh" obsolete surveys and to produce detailed, up-to-date, isogonal maps (see more on the computation of normal fields in this volume).

It is also possible to compute magnetic field components, and hence declination, from total field aeromagnetic surveys using Fourier techniques. A few magnetic repeat stations covering the total field survey must be observed or extrapolated at both the total field survey epoch and at the planned map edition epoch (Le Mouël, 1970; Schmidt and Clark, 1998).

With up-to-date maps, aeronautical and airport authorities can compute the magnetic headings which are necessary to fly from one airport to another. This information can then be printed on aeronautical maps. See Figure 119.

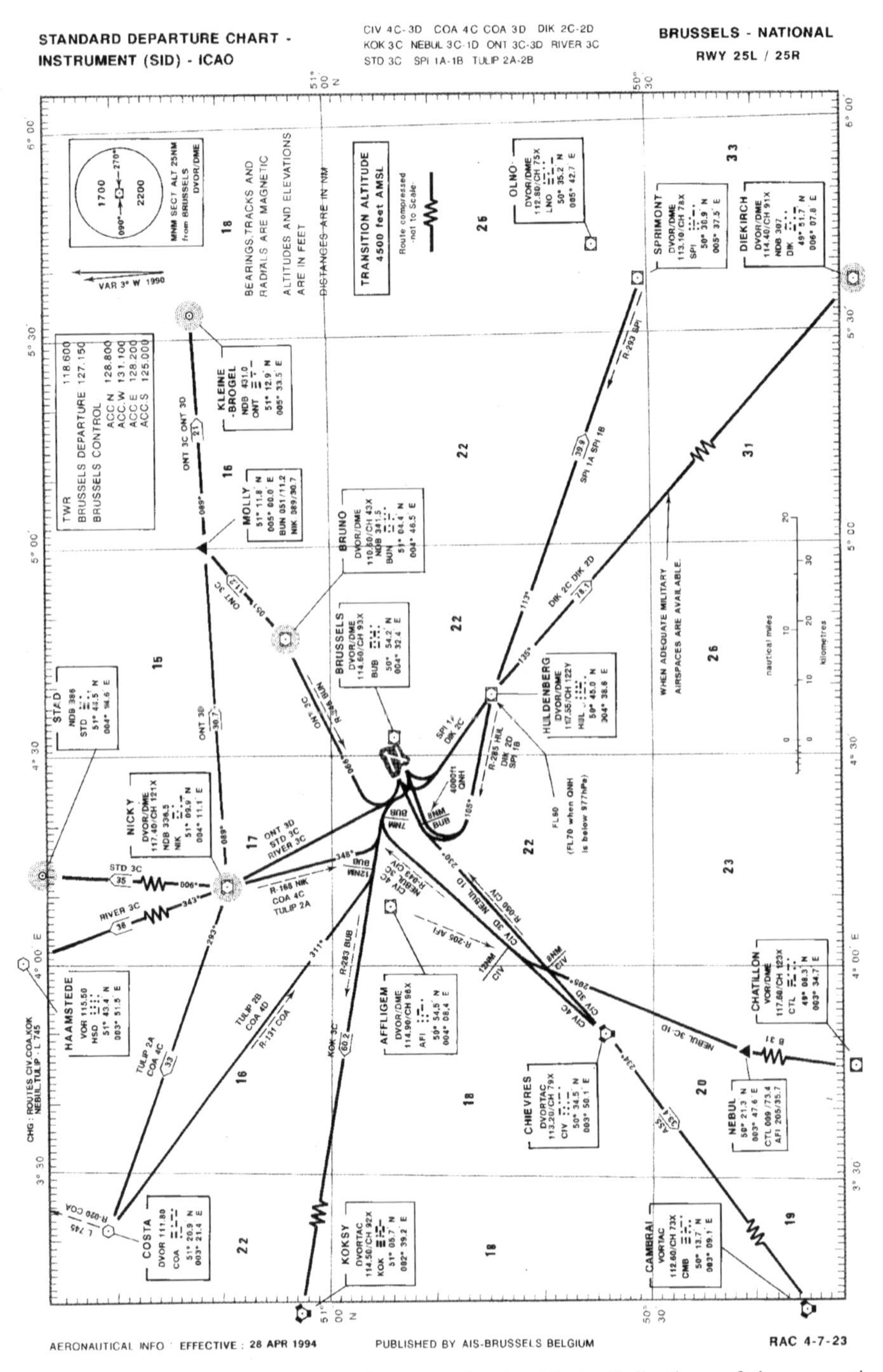


Figure 119. An aeronautical map (departure chart) with the indications of the magnetic headings to follow on the aircraft compass in order to connect from one airport to another.

### 3.4. SUPPLY OF DECLINATION VALUE (VOR STATIONS)

Extrapolated repeat station data is adequate to provide declination values at VOR stations. It can be extracted from an isogonal map. If no adequate isogonal map is available, or if the customer wants better than 0.1° accuracy of the magnetic declination at the VOR site, then an *in-situ* measurement session with the DIflux and Proton magnetometers should be made. The procedure is similar to that used at a repeat station. For the benefit of the geomagnetism program, the VOR declination measurement site could then become a part of a repeat station network.

## References


Civil Aviation Authority (UK), 1993, CAP562 Civil Aircraft Airworthiness Information and Procedures, Part **8**, Aircraft Instruments, Leaflet **8-1**: Compass base surveying, 1-8; Leaflet **8-2**: Compasses, 1-19; available at http://www.caa.co.uk/docs/33/CAP562.PDF

Crosthwaite, P.G., Lewis, A.M., Wang, L. and Hopgood, P.A., 2005, Errors Using Gridded Measurements for Compass Calibration Pad Certification, Proceedings of the XIth IAGA Workshop on Geomagnetic Observatory Instruments, Data Acquisition and Processing, Kakioka, Japan, November 9 – 17, 2004, 111-114.

Le Mouël J-L., 1970, Le levé aéromagnétique de la France. Calcul des composantes du champ à partir des mesures de l'intensité, Annales de Géophysique, **26**(2), 229-258

Loubser L., 2005, Guide for calibrating a compass swing base, IAGA publication, Hermanus, in press

Rasson J.L., 2005, About Absolute Geomagnetic Measurements in the Observatory and in the Field, Publication Scientifique et Technique No **040**, Institut Royal Meteorologique de Belgique, Brussels, 1-43

Schmidt, P.W. and Clark, D.A., 1998, The calculation of magnetic components and moments from TMI: A case study from the Tuckers igneous complex, Queensland. Exploration Geophysics, 29, 609-614

Tomlinson L.A., 2000, Magnetic intensity surveys to determine the suitability of aircraft compass test bases, unpublished; see http://www.caa.govt.nz/fulltext/acs/ac43-7_1.pdf


## DISCUSSION

Question (J. Miquel Torta): Are these services compulsory for the airports? Are these services always done by geomagnetic observatory people?
How expensive can these kind of services be for the airport or in other words, how much can we charge them?
Answer (Jean Rasson): I am not sure if those services are compulsory in Spain, but I guess that in some countries they are by law. Anyway, an airport should be able to provide a compass rose calibration pad to its customers (the visiting aircrafts) as a basic service. Note that the recent increase in aircraft security awareness leads airport managers to a more rigorous approach in those matters.

I have heard that private companies are providing those services, but I never came across an advertisement for them. The armed forces of some countries have specialists in their ranks able to perform some of those services. But the vast majority of those services are provided by magnetic observatory staff.

About the charging policy for those services, many parameters come into play:

- Number of man/hour necessary to perform the task, including data reduction calculations back at the observatory
- Distance of airport to observatory
- Is the airport abroad?
- Is the airport a repeat station serving for your other researches?

In our case, for tasks performed at national airports, we charge the standard hourly rate according to the time spent and the rank of the staff involved, in addition to travel costs. When measuring abroad, we apply a flat rate (up to 8000 EURO) depending on the country where the work is to be done.

# MEASUREMENTS OF MAGNETIC DECLINATION AT THE AIRPORTS IN BULGARIA

ILIYA CHOLAKOV[17]
*Geophysical Institute*
*Bulgarian Academy of Science*

**Abstract.** Such measurements were carried out in Bulgaria for the first time in 2001. The geomagnetic meridian is defined by using the geomagnetic theodolite "Shulze" and the geographic meridian - by GPS and geodetic way. In the beginning the declination was measured at the threshold of the runway. Very big anomalies were detected in these places. For this reason the measurements were made outside of the runway for three of the airports and on the secular stations near to another three of them. The measured values coincide very well with the values obtained by reduction of the declination according to the Panagjuriste geomagnetic observatory from 1990 to 2001.

**Keywords:** meridian; theodolite; azimuth; threshold

Because of the increased requirements for flight safety in Bulgaria, in 2001 a team from Geomagnetic Observatory Panagjurishte (PAG) performed measurements of magnetic declination at the airports in Sofia, Plovdiv, Varna, Bourgas, Rousse, and Gorna Orjahovitza (Figure 120).

Field geomagnetic measurements are usually performed from May – September. However, due to restricted deadlines for data delivery, it was necessary to make the measurements in February, March and April when it is cold and snowy in Bulgaria. The team who carried out the task consisted of two specialists from the Geomagnetic Observatory Panagjurishte at the Geophysiscal Institute (Bulgarian Academy of Sciences) and two specialists from the Central Laboratory of High Geodesy (Bulgarian Academy of Sciences). The direction of the magnetic meridian was determined using a "Schulze"theodolite magnetometer. It was used for all geomagnetic

[17] Address for correspondence: Geophysical Institute, Bulgarian Academy of Science, Acad. G.Bonchev Str. Bl. 3, Sofia 1113, Bulgaria



*J.L. Rasson and T. Delipetrov (eds.), Geomagnetics for Aeronautical Safety,* 231–234.

measurements (magnetic maps and secular variation) carried out between 1937 and 1990, as well as for comparative inter-observatory measurements (Kostov and Nozharov, 1987). The theodolite magnetometer is well studied and its instrumental correction has been reliably determined through numerous measurements in PAG. The horizontal circle of the theodolite has an accuracy of 0.2', and one unit on the scale of the optical tube is equal to 1' from the horizontal circle. Two magnets with magnetic moments $M_1 = 5M_2$ and a brass fiber with a gauge of 20 μm were used. The accuracy of a single measurement of the declination is 0.5'. The geomagnetic meridian measurement was made by two observers.

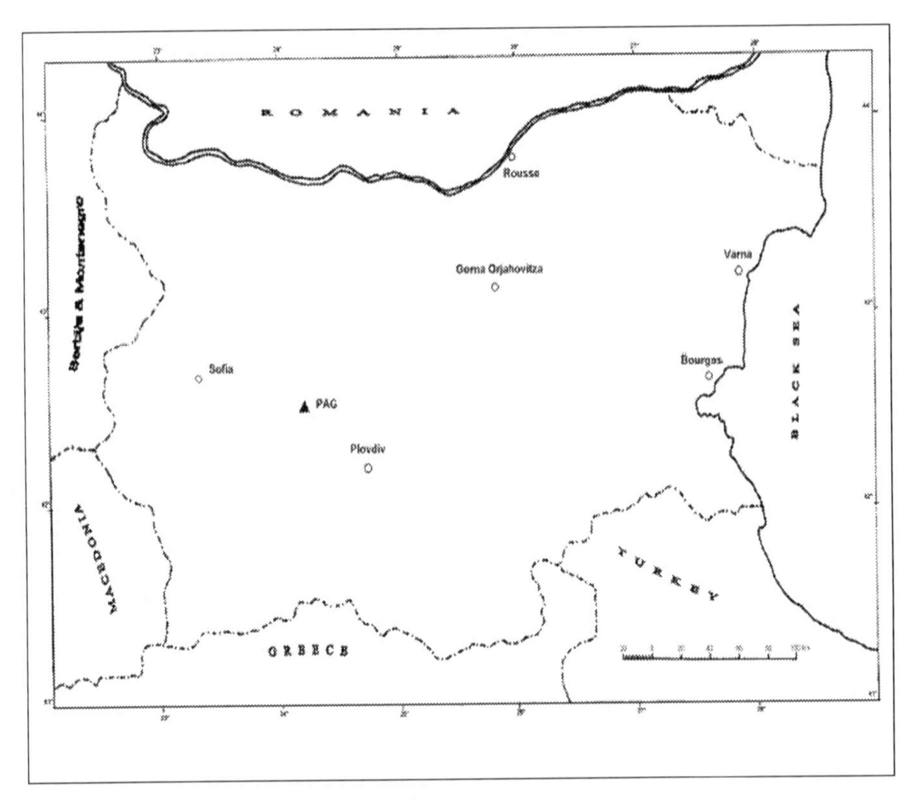


Figure 120. Map of the airports in Bulgaria.

The geographic azimuths at Plovdiv, Rousse and Gorna Orjahovitza airports were determined using a GPS, and at the other airports by using geodetic means. The measurements were reduced from the diurnal variations using data from the Panagjuriste Observatory.

Table 26. Values of the declination, measured at the airports.

| Declination | | | | | | |
|---|---|---|---|---|---|---|
| Airport | THR[1] E[2] | THR W[3] | RS[4] | Airport | P[5] E | P W |
| Sofia | 6°19' | 1°54' | 3°10' | Plovdiv | 3°47' | 3°09' |
| Bourgas | 5°47' | 5°21' | 3°47' | Rouse | 3°36' | 3°36' |
| Varna | 6°37' | 5°04' | 3°49' | Gorna Orjah. | 3°35' | 3°32' |

[1] THR - Threshold

[2] E – East

[3] W – West

[4] RS - Repeat Station

[5] P - Point

Figure 121. Threshold of the run-way.

First the declination was measured at the threshold of the runways (these are the points for take off and landing of the airplanes), Figure 121. Large anomalies were detected at the runway thresholds and it was difficult to find a time interval free of aircraft traffic. For example, the difference between the measured declination on the eastern and on the western part of Sofia airport was 4°25' (Table 26). Our opinion is that the large difference is due to concrete thickness of about 80 cm thickness with substantial steel reinforcement. Knowing that the airplanes use the declination in the air far off the runway we decided to maкe the measurements away from the runway. At the Plovdiv, Burgass and Gorna Oriahovitzathe airports, measurements were made near the airport runways and on the Sofia, Bourgass and Varna airports the measurements were made at the nearest repeat stations of the Bulgarian network.

The repeat stations of the secular network have been occupied many times and they have constant azimuth marks. The last measurements were performed in 1990. The declinations of epoch 1990 were reduced to 2001 using data from the Panagjuriste Observatory. The differences between the reduced and the measured values in epoch 2001 are as follows: for Varna airport - 1.8'; for Bourgas airport - 3.8' and for Sofia airport - 0.9'. These differences reveal a good correlation between the measured and reduced values of the declination.

We recommend that the measurements of declination be performed outside of the runway because of the presence of perturbations.

## References

Butchvarov, I., Cholakov, I., 1986, Investigation of the Ratio between the Magnetic Theodolite Magnets Moments and the Error at the Determination of the Earth Magnetic Field Declination, *Bulg. Geoph. J.* **XII(2):** x-x, (in Bulg.).

Cholakov, I., 1986, Investigation of the Instrumental Correction of the "Schulze" Magnetic Theodolite. *Bulg. Geoph.* J. **XII(3):** x-x (in Bulg.).

Kostov, K., Nozharov, P., 1987, Absolute Magnetic Measurements in Bulgaria 1787-1987, Geoph. Inst. - BAS, Sofiap (in Bulg.).

# GEOMAGNETIC MEASUREMENTS AND MAPPING FOR AERONAUTICS IN GERMANY

JÜRGEN MATZKA[18]
*University Munich (LMU) and*
*Geophysical Observatory Fürstenfeldbruck*

**Abstract.** This report is about activities of the Geophysical Observatory Fürstenfeldbruck in providing aeronautics in Germany with the infrastructure to use magnetic compass navigation accurately in aircrafts. Three prerequisites are important to accurately use the geomagnetic field for aircraft navigation purposes. First, the compass or magnetic field sensor employed on an aircraft has to work properly. Second, the direction of the horizontal component of the magnetic field and its temporal and spatial changes have to be known for all points of a given area. Third, the magnetic influence of the aircraft on the field sensor has to be known for all headings. This report deals mainly with the second and the third requirements described above. It specifically addresses the practical and technical details of magnetic measurements for aeronautics.

**Keywords:** geomagnetism; aeronautics; calibration pad; compass navigation; Fürstenfeldbruck; FUR

## 1. Magnetic declination in Germany

In Germany, maps with magnetic declination, isogonic charts, are a scientific product (e.g. Korte and Fredow, 2001) that can be used by air safety authorities. These maps are not derived from declination measurements at airports or along major air traffic routes but from declination measurements made at German geomagnetic repeat station sites that were set up for scientific research. After the reunification of Germany, a magnetic survey was conducted in 1992.5 (e.g. Beblo et al., 1995). Previous surveys were carried out separately for the German Democratic

[18] Address for correspondence: Geophysical Observatory Fürstenfeldbruck, Ludwigshöhe 8, DE-2556 Fürstenfeldbruck, Germany. Email: matzka@lmu.de



*J.L. Rasson and T. Delipetrov (eds.), Geomagnetics for Aeronautical Safety,* 235–246.

Republic (e.g. Bolz et al., 1969) and the Federal Republic of Germany (e.g. Schulz et al., 1997). Today, repeat station surveys are carried out by the Observatory Niemegk which is run by the GeoForschungsZentrum Potsdam. Approximately 40 repeat stations are occupied every 2 years. In general, the quality of repeat station data is more limited by the ability to separate external and internal field components than by the accuracy of the actual measurements. For the best possible separation of the internal and external field, the repeat station data is reduced to either one of the observatories (Fürstenfeldbruck (IAGA code: FUR), Niemegk (NGK), Wingst (WNG)) or to a temporary variometer station near the repeat station. The details of the data reduction can be found in Korte and Fredow (2001). The distribution of repeat stations, geomagnetic observatories, and temporary variometer stations is shown in Figure 122.

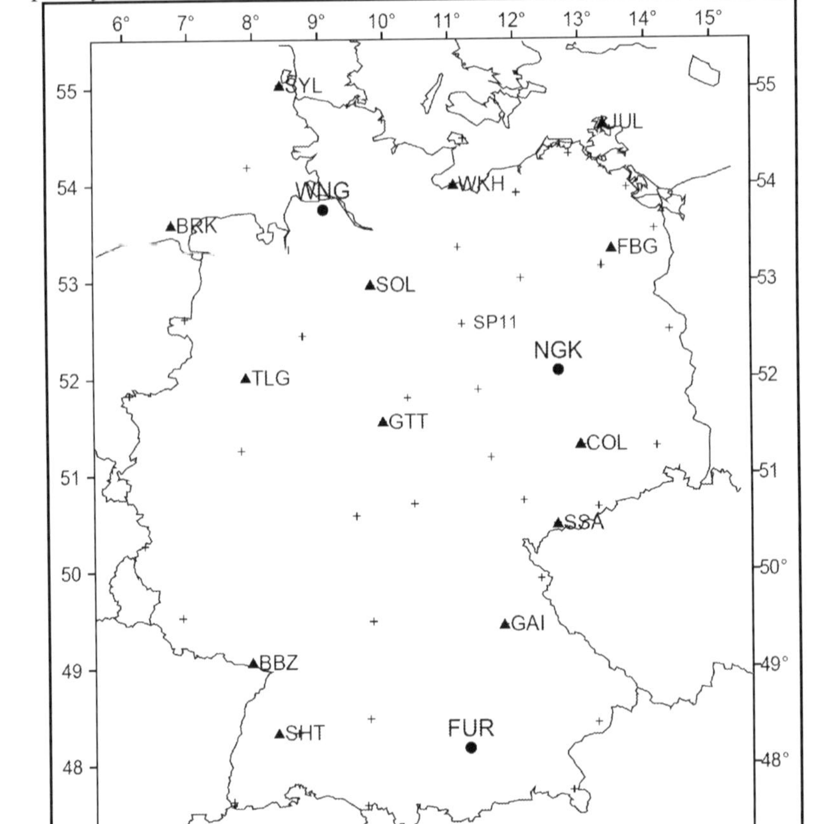


Figure 122. Repeat stations (crosses), temporary variometer stations (triangles), and geomagnetic observatories (dots) in Germany (from Korte and Fredow, 2001), with permission of Monika Korte, GeoForschungsZentrum Postdam.

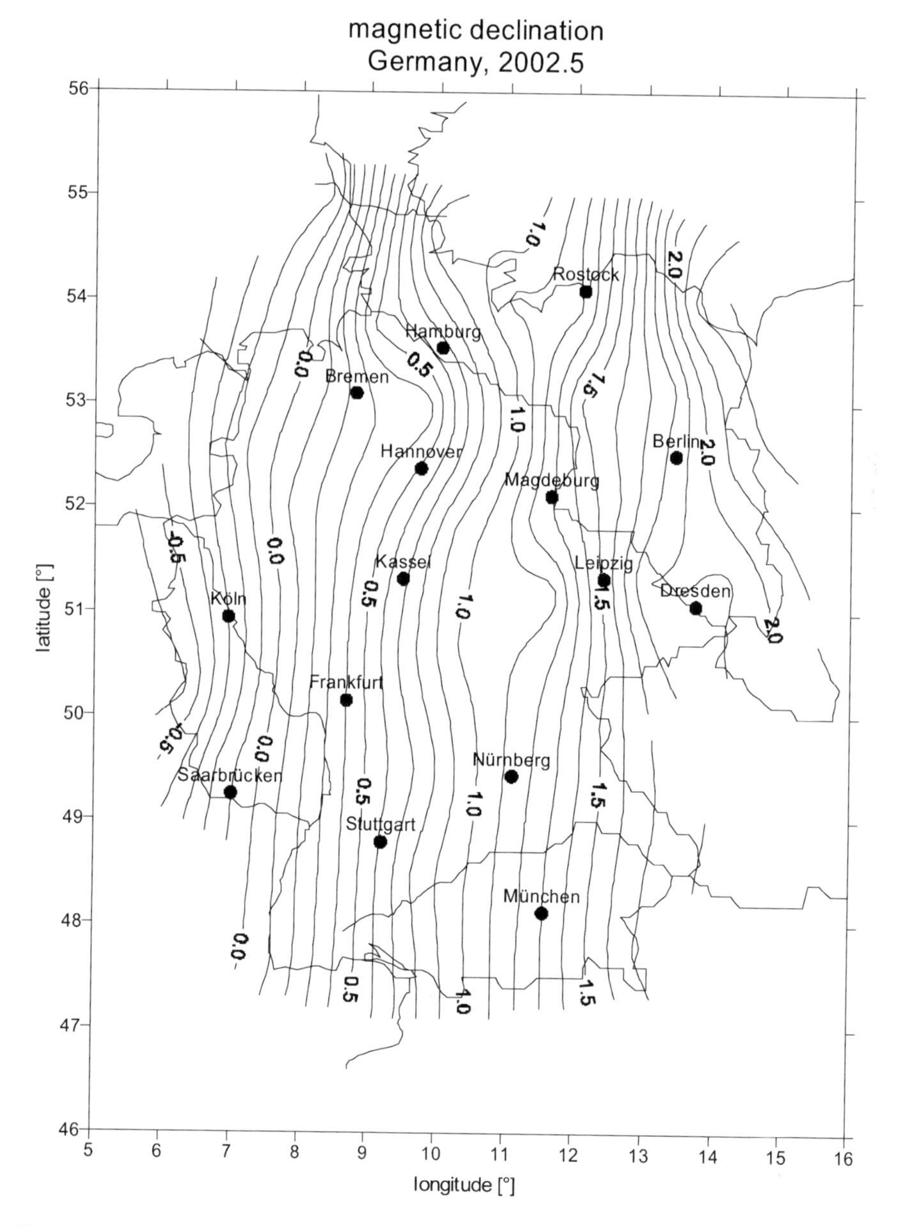


Figure 123. Magnetic declination map for Germany for the year 2002 (reduced to 2002.5) by M. Beblo. For details see text.

Two different approaches to calculating declination maps for Germany are discussed in the following. A declination map for the year 2002.5, established by Martin Beblo (see Figure 123, published on http://obsfur.-geophysik.uni-muenchen.de/images/2002d.gif), was obtained by combining the results of earlier, 3-component ground surveys of high spatial resolution (to quantify regional magnetic anomalies) with repeat station and observatory data of less spatial resolution (to quantify the change by secular variation). The ground surveys are described by Weingärtner (1991). A software package (Erhardt, 1991) is used to combine the ground survey data with the latest German repeat station data and a global magnetic field model. The magnetic declination is calculated on a 0.1° latitude / 0.1° longitude grid for the territory of Germany by Kriging.

The second approach to creating a declination map is to determine the difference between repeat station data, for one year and global, magnetic field model data. Then, using the secular variation information of that magnetic field model, the declination value expected at the repeat station for a subsequent year is derived. The magnetic declination map in Figure 124 is an example of a map hat was derived using this method. The magnetic field model used was the comprehensive model CM4 (Sabaka et al., 2004). The map was calculated during the year 2004, at a time when no annual observatory mean values for the year were available yet. Again, the magnetic declination was calculated on a 0.1° latitude / 0.1° longitude grid by Kriging.

The declination map for 2004.5 is smoother than the map for 2002.5. The method of combining repeat station data with a global magnetic field model has the advantage of extrapolating to a future epoch. Neither method described takes into account local magnetic anomalies that are potentially important at individual airports. Nevertheless, both methods give good approximations of the geomagnetic declination at altitude relevant for air traffic, since the increased distance to magnetic anomalies acts like a spatial low-pass filter.

## 2. Correcting for the magnetic influence of the aircraft on the onboard magnetic field sensor

The magnetic field sensor on board can be either a compass in the cockpit or a magnetic field sensor mounted somewhere in the aircraft. The sensor should be mounted on the wingtips or on the tail fin as far as possible from the most magnetic parts of the aircraft. The sensor consists of at least two magnetic field probes mounted perpendicular to each other in the horizontal plane. The aircraft's magnetic influence on the compass or the sensor (in the following, both the compass and the magnetic field sensor will be referred to as 'sensor') is tested on the ground on a compass calibration pad. The calibration pad must be in an area with a homogenous magnetic field and large enough to accommodate the entire aircraft at all possible headings. As a general rule, the homogeneity of the magnetic North direction at the

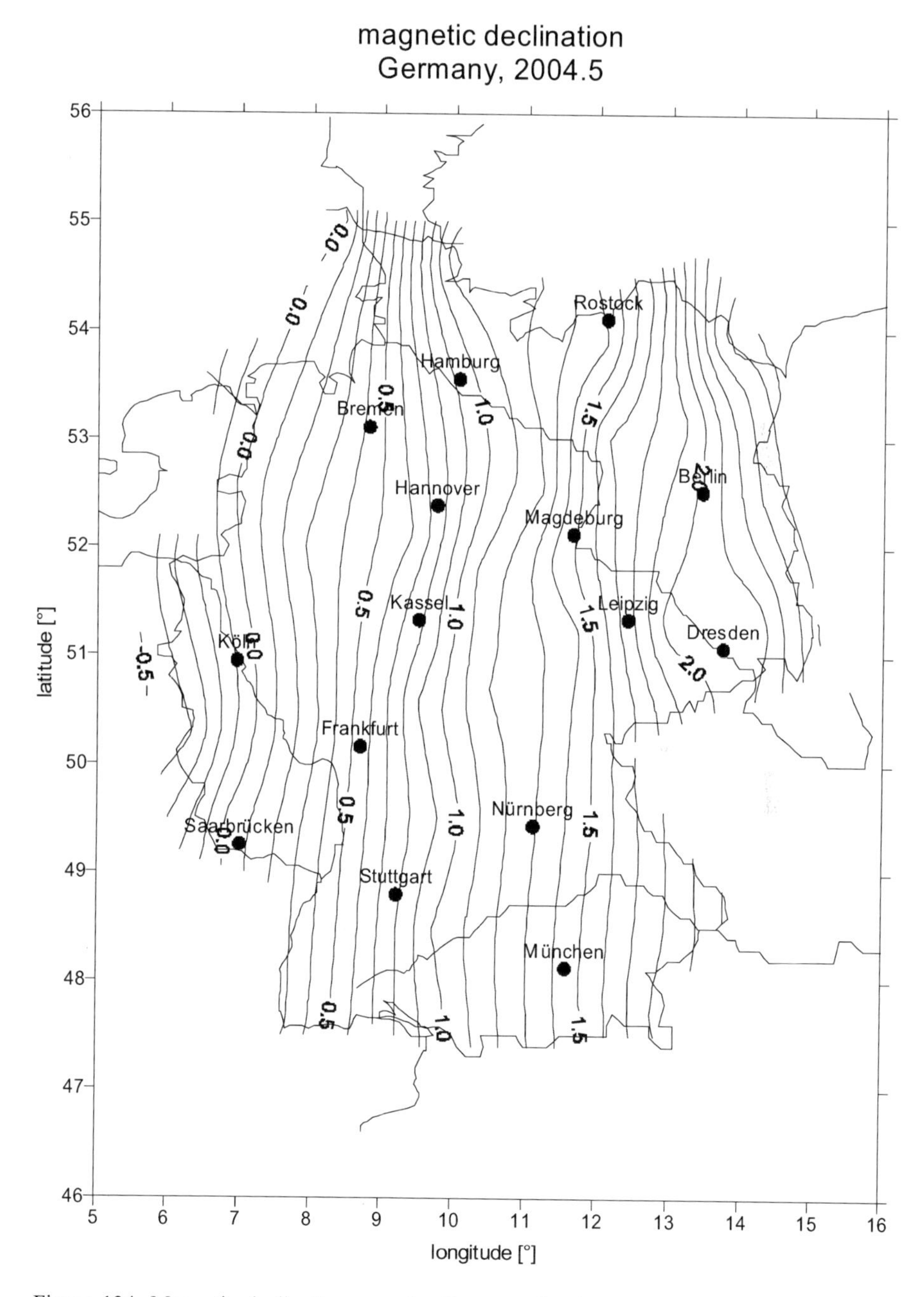


Figure 124. Magnetic declination map for Germany for the year 2004. Calculated by M. Korte. For details see text.

calibration pad should be within 1° for civil aviation and within 0.1° for military aviation. The aircraft is first aligned with magnetic North, which is indicated on the calibration pad by painted marks. Then, the offset between the sensor reading and magnetic North is determined. Next, the aircraft is rotated by a certain angle and the procedure is repeated. For example, if the angle of rotation is 15° , the aircraft is rotated by 15° from magnetic North. This subsequent direction is usually indicated by a painted mark on the calibration pad as well. Then, the difference between the sensor reading and the expected 15° is determined. For a full calibration, this procedure is done for each 15° heading for which an individual offset is determined. The individual offsets will be used to correct the sensor for each of the respective 15° headings.

## 3. Updating magnetic North at a calibration pad

Depending on the secular variation of the declination and the aimed for accuracy of the calibration procedure, the magnetic North direction has to be updated for each calibration pad at regular intervals. In Germany, the expected secular variation of the declination is +0.1° per year. Typically, the calibration pads are marked with a new magnetic North direction every year or two. The measurement of the North direction is performed with a DI-flux instrument, a non-magnetic theodolite with a fluxgate probe mounted on the telescope. The measurements are performed in the four positions (Kring Lauridsen, 1985; Jankowski and Sucksdorf, 1996) with at least 2 readings in each position. The theodolite can measure horizontal angles with an accuracy of 0.2 minutes of arc. The time of each measurement is noted using a GPS or radio controlled clock accurate to within 1 minute. The theodolite has to be leveled and the optical plummet centered above the mark. A non-magnetic umbrella provides shadow for the instrument during the leveling, since direct sunlight makes leveling difficult (due to the thermal expansion of both the level liquid and the instrument). Moreover, the fluxgate's sensor offset depends on temperature, so it is critical that temperature remain constant during measurement. The danger involved in using an umbrella close to the theodolite under windy conditions should not be underestimated.

Magnetic North is determined in the center of the calibration pad, at an external point, or at both points. The advantage of measuring magnetic North at the center of the calibration pad is that its direction can be indicated directly with the theodolite. If there is a small magnetic anomaly in the center of the calibration pad, due to iron parts in its structure, it is preferable to measure the magnetic North direction at an external point off the pad. This point has to be accurately defined and marked on the ground.

To transfer magnetic North from the external point to the center of the calibration pad, the angle between magnetic North and the center must be measured with the theodolite from the external point. Then, the angle, X, at the center of the pad between magnetic North and the external point, is calculated. The theodolite is set up at the center of the pad and magnetic North can be determined by measuring the angle towards the external point and adding this to the angle X. Let us assume that the distance between the center and the external point is approximately 10 meters. To achieve an accuracy of 0.01° when transferring the magnetic North direction from the external point to the center, both points have to be known with an accuracy of 1 mm. To mark the points, screws are fixed in the concrete of the calibration pad. These screws are made from brass and have a thin cross carved into the head.

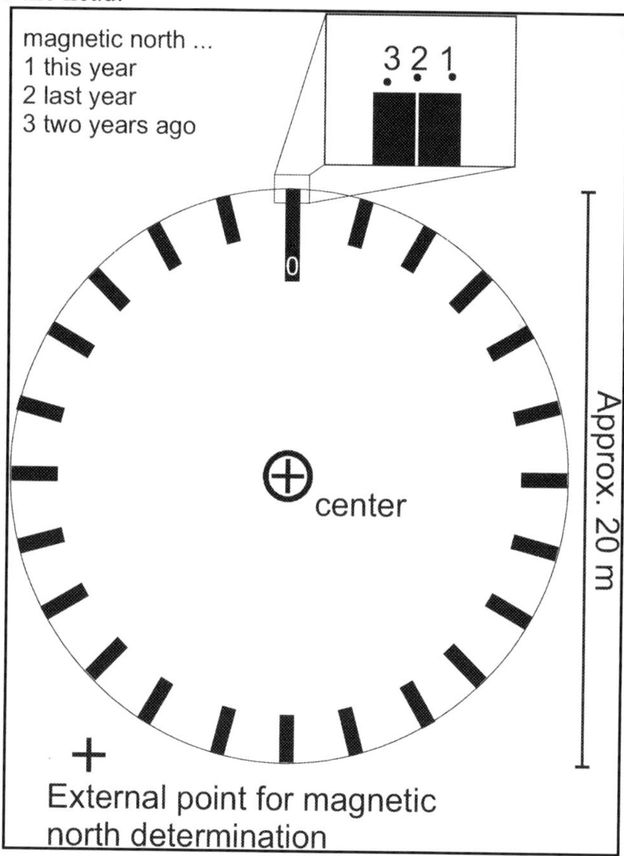


Figure 125. Sketch of a typical compass calibration pad. Two positions for magnetic measurements are indicated: the center and an external point. The inset shows an enlarged view of the painted mark (black with white central stripe) for magnetic North. Drill holes are used to mark magnetic North before the marks are repainted. The numbers 1, 2 and 3 indicate the magnetic North direction (with respect to the center) of this year, last year and two years ago, respectively. Paint marks are not to scale.

In the ideal case magnetic measurements are performed both in the center of the pad and at the external point. Then, the magnetic North directions at both points can be compared, after reduction of the temporal variation of the declination, with data from the nearest geomagnetic observatory. The two magnetic North determinations are usually the same, within 0.01°. The magnetic North direction has to be properly marked on the calibration pad. The magnetic North direction is usually marked by a drill hole of 6 mm diameter at a distance of approximately 10 meters from the center of the pad. The drill is positioned so that it is seen exactly in the cross hairs of the theodolite telescope bearing towards magnetic North. Usually, the drill holes are accurate to within 0.02°. Each 15° heading is marked by a drill hole. Figure 125 shows a calibration pad with the features described above.

## 4. Accounting for temporal variation of the magnetic field

Knowing the variation of the magnetic declination from the nearest observatory during the magnetic North measurements is advantageous for several reasons. First, if multiple measurements were performed, they can be compared with each other after reducing the temporal variation. Second, if the declination measurement was anomalous due to external fields, (e.g. during disturbed days, magnetic storms), it can be reduced to a value corresponding to the mean value of the nearest magnetically quiet day before or after the measurement. Figure 126 shows the quiet daily variation, Sq, of the declination in Fürstenfeldbruck for each month, calculated (for the year 2002) using the model of Campbell (2003). Note that this quiet daily variation of the declination is qualitatively quite similar throughout the year and that the amplitude of the quiet variation is highest in the summer months. During the usual working hours, even during the quietest possible magnetic conditions, declination changes can be on the order of plus or minus 0.1° (or 6'). The daily mean value of the declination is close to the mean value of the minimum and the maximum of the declination variations. It is therefore advisable to reduce the magnetic North direction measured at the calibration pad to a daily mean value at the calibration pad by using the recordings of the nearest geomagnetic observatory and assuming that the variations are similar in both places. The difference between the marked magnetic North direction and the daily mean value are reported to the airport staff who can take this difference into account when painting the marks. To keep this difference low waiting for a magnetically quiet day to perform the measurements and carrying them out close to local midday, when the declination has a value close to its daily mean value is recommended (Figure 126).

Before calibrating an aircraft's sensor, it is important that airport staff checks with geomagnetic observatory personnel to make sure that the geomagnetic conditions are not too disturbed for accurate calibration.

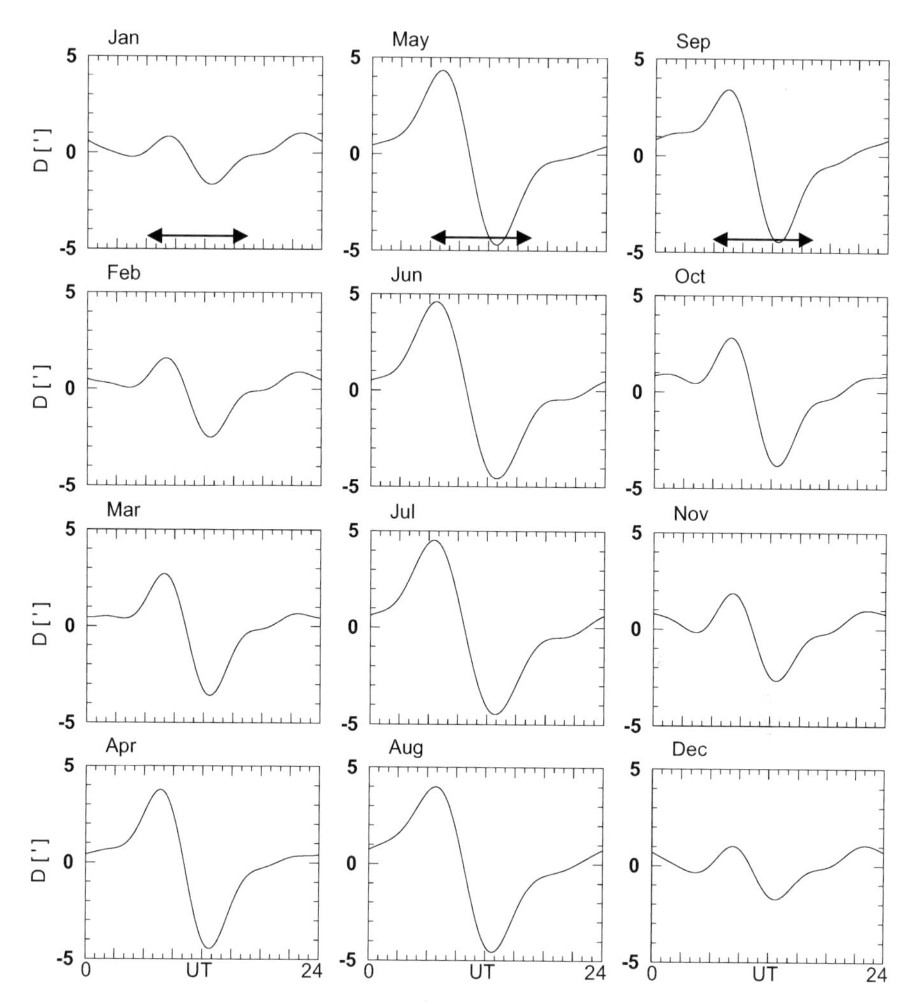


Figure 126. Daily variation of the magnetic declination in minutes of arc at Fürstenfeldbruck for magnetically quiet days. The variation is calculated for each month from the Sq model distributed with Campbell (2003). The double headed arrows indicate working hours, corresponding to the most likely time during the day when magnetic North measurements or compass calibrations are performed.

## 5. Calibration pad report

The calibration pad report should include a description of the measurement procedure (e.g. 4 position DI-flux), the exact location of the measurement points, the height of the instrument, the names and affiliation of the persons performing the measurements, and an estimate of the accuracy of the results. For the magnetic North determination and marking, typically an accuracy of 0.03° can be attained for a reasonably sized calibration pad.

If two or more magnetic North directions were measured, they should be compared to each other in order to check for measurement errors and spatial inhomogeneities of the magnetic North direction. The angles between the magnetic North direction and fixed objects like church towers or airport towers should be reported and compared to the angles measured in previous years. The results should also be compared with the expected secular variation for the region. A visual inspection of the calibration pad and the surrounding area should be made to be sure that no magnetic objects were introduced.

As noted above, the report should also give the difference in declination (observatory data) for the time of the measurements at the calibration pad and the daily mean value at the observatory.

## 6. Homogeneity of the magnetic North direction

When establishing a new calibration pad, or if changes were made at or near an existing calibration pad, not only the magnetic North direction, but also its homogeneity has to be checked. The area of homogenous magnetic North direction is a circle with a radius corresponding to the largest distance between the aircraft sensor and the calibration pad center. Since positioning and leveling the theodolite exactly above a given point and making a measurement of magnetic North with a theodolite is time consuming, the number of magnetic North measurements can be reduced by choosing appropriate points where to measure magnetic North and by a complimentary use of total field measurements with a proton precession magnetometer (PPM).Two different types of magnetic anomalies have to be considered when assessing the inhomogeneity of the magnetic North direction:

The first type of magnetic anomaly is a magnetic object built some distance from the calibration pad, such as a building with steel construction material. In this case, it is advisable to measure the magnetic field in the center of the calibration pad, and at four positions around the edge of the calibration pad. The four positions at the edge should be 90° apart. These four positions should be chosen so that one of the points is closest to the building (or at the point where the strongest influence on the magnetic North direction is expected). Since the source of the magnetic anomaly is at some distance, it is likely that the magnetic North direction measured with the theodolite, at a height of approx. 1.3 meters, is similar to the magnetic North direction measured by the aircraft sensor. The influence of such an anomaly on the magnetic North direction is difficult to discern from total field measurements. If the anomaly changes the magnetic field in the east component, then it would also have a significant influence on the magnetic North direction. However, vector sum of the Earth's magnetic field and the anomaly field might be insignificantly different from the total field of the Earth's magnetic field alone, provided that the anomaly field is small.

The second type of magnetic anomaly is a small magnetic object buried in the ground on the calibration pad, such as a piece of steel rebar. If its dimension is very small compared to the calibration pad and its magnetic moment is not too strong, it can easily remain undetected when performing the 5 magnetic North determinations as described above. However, this anomaly could have an adverse influence on the aircraft sensor compensation procedure. To detect such anomalies with a wavelength small compared to the dimension of the calibration pad, a total field survey should be carried out. Measurements with a total field magnetometer are quick, taking only a few seconds each. A grid with a spacing of 1 or 2 meters should be established to cover the area where homogeneity is to be checked. The grid should be oriented with the four magnetic North measurement points at the edge of the calibration pad. Ideally, the total field magnetometer would be a gradiometer and can measure the vertical gradient as well. If possible, the total field values should be reduced with data from the nearest observatory to account for temporal variations. Local magnetic anomalies can easily be detected by plotting isoline maps of the total field or its vertical gradient for the investigated area. Should an anomaly be detected, then magnetic North measurements should be performed at its location. Since the object causing the magnetic anomaly is buried in the ground, it is likely that the magnetic North directions measured at the height of the theodolite are more inhomogeneous than those expected at greater height, where an aircraft's magnetic sensor is located.

**Acknowledgements**

Magnetic Maps of Germany are cooperatively produced by the German Geomagnetic Observatories in Niemegk and Wingst (GeoForschungs-Zentrum Potsdam) and Fürstenfeldbruck (University Munich). The fieldwork is carried out by Martin Fredow and staff of the Niemegk

Observatory (GeoForschungsZentrum Potsdam). The map for 2004 was calculated by Monika Korte (GeoForschungsZentrum Potsdam). The author is grateful to Martin Beblo (FUR), who made him familiar with magnetic measurements on calibration pads, and to Martin Feller (FUR), who joined him many times when doing the measurements. Martin Beblo stimulated and supervised the diploma thesis of Erhardt (1991).

The author would like to thank the reviewers for improving the manuscript.

## References

Beblo M., Best, A. und Schulz, G., 1995. Magnetische Karten der Bundesrepublik Deutschland. Scientific Technical Report STR95/22, Potsdam

Bolz, H., Kautzleben, H., Mundt, W. and Wolter, H., 1969. Die magnetische Landesvermessung der Deutschen Demokratischen Republik zur Epoche 1957.5; Ergebnisse und Auswertungen. Abhandlung Geomagnet. Inst. Potsdam, Nr. 41

Campbell, W., 2003. Introduction to geomagnetic fields. Cambridge University Press, 2nd Edition

Erhardt, U., 1991. Erstellung von Isolinienplänen der erdmagnetischen Elemente für beliebige Epochen. Diploma Thesis, Institute for Pure and Applied Geophysics, University of Munich, 146 p.

Federal Republic of Germany and the secular variation field from 1965 to 1992. Dt. Hydrogr. Z. 49, No. 1, pp 5 - 20

Jankowski, J. and Sucksdorf, C., 1996. Guide for magnetic measurements and observatory practice. IAGA, 235 pages

Korte, M. and Fredow, M., 2001. Magnetic Repeat Station Survey of Germany 1999/2000. Scientific Technical Report STR01/04, GeoForschungsZentrum Potsdam, 23 p.

Kring Lauridsen, E., 1985. Experience with the DI-fluxgate magnetometer inclusive theory of the instrument and comparison with other methods. Danish Meteorological Institute, Geophysical Papers R-57

Sabaka, T.J., Olsen, N. and Purucker, E., 2004. Extending comprehensive models of the Earth's magnetic field with Oersted and CHAMP data. Geophys. J. Int., 159, pp 521 - 547

Weingärtner, E., 1991. Rückblick auf 150 Jahre geomagnetische Vermessungen von Lamont bis heute. Münchner Geophys. Mitt., 5

# AIRPORT GEOMAGNETIC SURVEYS IN THE UNITED STATES

ALAN BERARDUCCI[19]
*US Geological Survey, Geomagnetism Program*

**Abstract.** The Federal Aviation Administration (FAA) and the United States military have requirements for design, location, and construction of compass calibration pads (compass roses), these having been developed through collaboration with US Geological Survey (USGS) personnel. These requirements are detailed in the FAA Advisory Circular AC 150/5300-13, Appendix 4, and in various military documents, such as Handbook 1021/1, but the major requirement is that the range of declination measured within 75 meters of the center of a compass rose be less than or equal to 30 minutes of arc. The USGS Geomagnetism Group has developed specific methods for conducting a magnetic survey so that existing compass roses can be judged in terms of the needed standards and also that new sites can be evaluated for their suitability as potentially new compass roses. First, a preliminary survey is performed with a total-field magnetometer, with differences over the site area of less than 75nT being sufficient to warrant additional, more detailed surveying. Next, a number of survey points are established over the compass rose area and nearby, where declination is to be measured with an instrument capable of measuring declination to within 1 minute of arc, such as a Gurley transit magnetometer, DI Flux theodolite magnetometer, or Wild T-0. The data are corrected for diurnal and irregular effects of the magnetic field and declination is determined for each survey point, as well as declination range and average of the entire compass rose site. Altogether, a typical survey takes about four days to complete.

**Keywords:** airport geomagnetic survey; compass rose survey; compass calibration pad

[19] To whom correspondence should be addressed at: US Geological Survey, Geomagnetism Program Denver Federal Center MS 966, Box 25046, Denver, CO 80225, USA. Email: berarducci@usgs.gov http://geomag.usgs.gov



*J.L. Rasson and T. Delipetrov (eds.), Geomagnetics for Aeronautical Safety,* 247–258.

## 1. Introduction

There are two primary objectives in performing airport geomagnetic surveys. The first is to determine the suitability of the site for a compass calibration pad (compass rose), and the second is to determine magnetic declination at a suitable site. Suitability is assessed using the standards outlined in FAA (Federal Aviation Administration) and DOD (Department of Defense) documents. Magnetic declination is determined using procedures developed by the US Geological Survey (USGS). Surveyors must have an understanding of geomagnetism and be trained to use several sophisticated magnetometers to acquire magnetic measurements. Personnel from geomagnetic observatories are especially suited to perform compass rose surveys. In the United States, the USGS and private companies provide airport geomagnetic surveys.

## 2. Requirements

The FAA and the DOD have specific requirements for design, location, and construction of a compass rose.

### 2.1. FAA REQUIREMENTS

FAA requirements for design, location and construction of a compass calibration pad are detailed in the FAA AC (advisory circular) 150/5300-13 Appendix 4. The advisory circular may be obtained at the following internet site: http://www.faa.gov/arp/pdf/5300-13p2.pdf. Highlights of sections 4 to 6 of the advisory circular are discussed below.

#### 2.1.1. *Section 4 - Design of Compass Calibration Pad*

Design details from section 4 are to be used as guidelines. Variations in design are acceptable if the general requirements are met. Design requirements of a compass rose are:

- Radials must be painted every 30 degrees beginning at magnetic north. (Figure 127)
- The compass rose must be built with non-magnetic materials.
- The size of the compass rose must be compatible with the size of aircraft using it (15 meters for small airplanes to 33 meters for large jets).

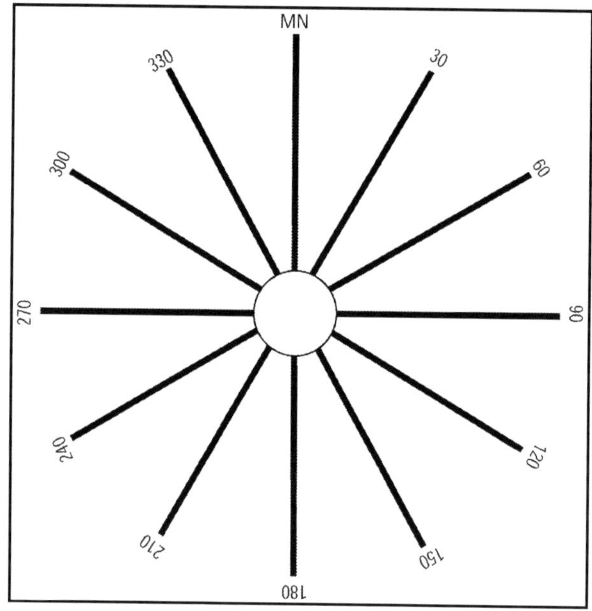


Figure 127. Compass rose radials.

### 2.1.2. *Section 5 - Location of compass calibration pad*

Section 5 of the FAA Advisory Circular provides guidelines for locating a suitable site for a compass rose. Section 5 also states the range of declination which is allowed over the compass rose area. Ideally, all criteria in the circular should be met. At many airports it is not possible to locate a compass rose following all the requirements. The general requirements for location of a compass rose are:

- Locate a compass rose 90 meters from power and communication cables and other aircraft.
- Locate a compass rose at least 180 meters from large magnetic objects, such as buildings, railroad tracks, high voltage transmission lines, or cables with direct current.
- Locate a compass rose off the side of a taxiway or runway to satisfy local clearances.

### 2.1.3. *Site location suitability*

To judge the suitability of a site a thorough magnetic declination survey must be made so that the criteria in 5d of the FAA AC are met. The criteria are:

- The difference between magnetic and true north must be uniform across the site.

- The range of declination must be less than one half degree (from 0.3 to 3 meters above the base and within 75 meters of the center).

To assess site suitability some subjectivity in judgment must be employed. For example, if a small magnetic object is located within 75 meters of the center of the compass rose, (but is not on the compass rose), and if the object will not affect compass calibration, the location is suitable, though a disclaimer should be included in the report. On the other hand, if a building is within 120 meters of the proposed site and has an effect on the compass rose (yet the pad still meets the one half degree criteria) the site should be deemed unsuitable because the natural magnetic field of the area is disturbed.

#### 2.1.4. *Section 6 – Construction of compass calibration pad*

Compass roses must be constructed according to the exact requirements detailed in Section 6 of the FAA AC. The requirements are:

- Use only non-magnetic materials for construction of a compass rose. Reinforcing steel, ferrous aggregates, and steel or reinforced concrete drainage pipe cannot be used. Many non-magnetic materials are available. Suspect materials must be checked before use.
- The radials of the compass rose must be oriented to within one minute of its magnetic bearing.
- The date, declination value, and annual change must be marked near the center of the compass rose. A permanent monument marking geographic north should be established at a remote location.
- A second magnetic declination survey must be made after construction of the compass rose is complete to confirm it meets FAA standards and to determine the average declination in order to paint the radials.
- Another magnetic declination survey must be made if major construction occurs within 180 meters of the center of the compass rose.
- Declination surveys must be made at least every five years to recertify the compass rose.

### 2.2. MILITARY REQUIREMENTS

There are only a few minor differences between the FAA and DOD requirements. The US Air Force may use the FAA or DOD requirements. The DOD documents may be found at the following internet sites: http://www.hnd.usace.army.mil/techinfo/UFC/UFC%203-260-01.pdf (see Attachment 11) and http://www.fas.org/nuke/intro/aircraft/afman32-1123.pdf (see Attachment 11).

## 3. Compass rose site selection

The USGS has developed specific methods for conducting magnetic surveys so that existing or new compass rose sites can be judged in terms of the FAA and DOD standards. The steps used to select a compass rose site are:

- A preliminary site assessment must be made. Airport drawings should be reviewed and airport personnel queried about potential sites.
- Total field magnetic surveys of potential sites must be made in order to evaluate the sites.
- A declination survey of the most favorable site must be made to determine that the site meets FAA guidelines, to determine the average declination over the site, and to determine where to locate the compass rose radials.
- When a new compass rose has been constructed, a second declination survey must be made to be certain that the compass rose still meets the FAA criteria.

### 3.1. PRELIMINARY SITE CHECKS

The airport management personnel usually have one or more sites in mind on which to build a compass rose. Managers should be encouraged to consider as many locations as possible and to take into consideration the ease of access of a site by aircraft that will use the compass rose, and traffic flow around the compass rose. The most common locations for compass roses are a taxiway, ramp, or a separate pad built specifically for a compass rose.

When considering potential sites, airport drawings should be reviewed for pipelines, conduits, and drainage pipes. Potential sites should be checked visually to scan for drainage grates, manhole covers, and any evidence of buried ferrous metals. Also, the airport maintenance supervisor should be interviewed to see if he knows of any magnetic items which may compromise sites.

### 3.2. TOTAL FIELD SURVEY

A magnetic total field survey of an existing or new compass rose is essential to determine the suitability of the site because preliminary checks may not show buried ferrous metals which can impact the magnetic field. A total field survey of a site can be made in 1-2 hours.

### 3.2.1. *Equipment for a total field survey*

Two total field magnetometers in general use are a Geometrics G-856 portable total field magnetometer or a GEM Systems GSM-19 portable total field magnetometer. Both magnetometers consist of an electronics console, cable, sensor, 2 meter staff, and carrying harness. Both consoles are able to store data points for later download.

### 3.2.2. *Procedure for a total field survey*

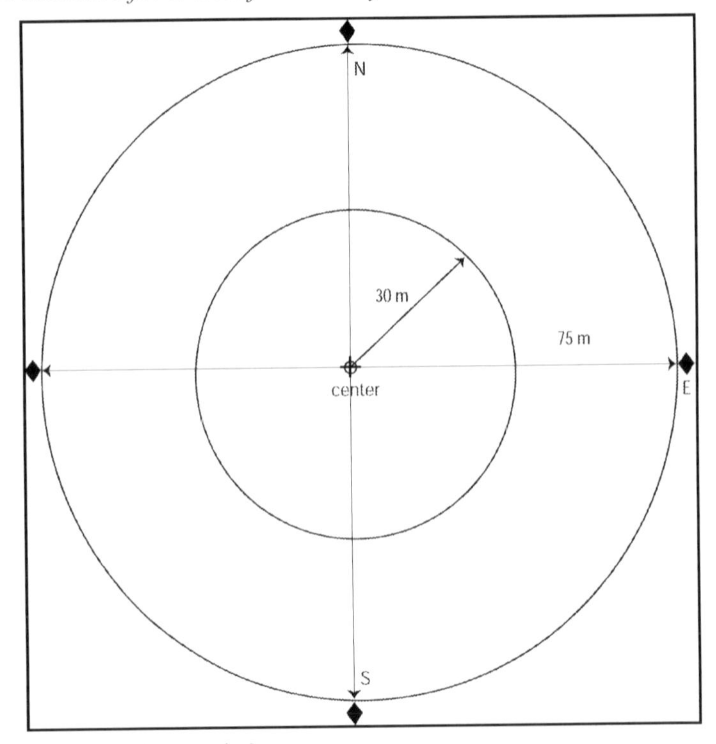


Figure 128. Total Field Magnetic Survey.

- Mark the center point and points 30 meters north, south, east, and west with temporary wooden stakes.
- Make total field readings at the center point and approximately every 3 meters along N-S and E-W lines. Exact locations are unnecessary. The distances between points may be approximated and established by simply pacing them.
- Where initial results look promising, make total field readings every 3 meters to cover a 30 meter radius area.
- From 30 meters out to 75 meters radius make total field readings every 10 meters. (Figure 128)

If the total field has a range of 75nT over a 75 meter area, it will meet the FAA and DOD requirements.

Experience and judgment must be employed to interpret the results of a total field survey. USGS interprets the FAA and DOD requirements to mean that if a slightly anomalous area (off the compass rose pad itself, but within 75meters of the center of the compass rose), does not affect declination on the compass rose pad, the pad may be certified (with a description of the anomaly included in the report).

## 3.3. SITE DECISION

Total field survey results are used to rank potential compass rose sites. Following the total field survey, a single site must be chosen for construction of the compass rose after which a detailed magnetic declination survey must be made to confirm the site meets the FAA requirements stated in 5d. The declination survey will take from one to three days to complete.

## 3.4. MAGNETIC DECLINATION SURVEY

The USGS has developed the procedure for performing a magnetic declination survey.

The steps for making a declination survey are:

- Set up test points in a grid pattern over the compass rose area.
- Determine geographic north.
- Measure magnetic declination at the test points.
- Apply diurnal and instrument corrections to the data.
- Create a final report for the airport.

### 3.4.1. *Equipment for a compass rose survey*

There are several declination magnetometers suitable for a compass rose survey. Discussed below are the Gurley transit, the DI Flux, and the Wild T0 magnetometers.

- The USGS uses a Gurley transit magnetometer to perform compass rose surveys as it is faster to use than the DI Flux and more accurate than the Wild T0. Since it is a mechanical compass, it has an instrument correction which must be determined. The Gurley has a unique reading eyepiece which allows compass readings to 15 seconds of arc. The Gurley is no longer manufactured.

- The DI Flux instrument is also suitable for compass rose surveys. The advantage in using a DI Flux is that it is an absolute instrument and has no instrument correction. A DI Flux theodolite reads to either 0.1 minute or 1 second of arc, depending on the model. The shortfall of the DI Flux is that it must be precisely leveled and four separate measurements are required to determine the magnetic meridian.
- Another suitable and reliable instrument is the Wild T0 Compass Theodolite. The T0 reads to 1 minute of arc. It is no longer manufactured.

3.4.2. *Test point and grid setup (Figure 129)*

To perform a declination survey over a compass rose area, many test points must be established. There are several ways to establish these points, but the best method is to use a grid pattern where the test points are spaced evenly over the compass rose. For example, a compass rose pad 36 meters in diameter should have main test points established in a grid pattern 6 meters apart. Grid spacing of 4 to 9 meters is also acceptable. The established grid is 7 lines by 7 lines. If the test points are on pavement, they should be marked with a 3 millimeter dot of marking paint. The paint marks should be labeled and broadly circled so that they are easily identified. When test points are on dirt or grass, they should be marked with a wooden stake which should have a 3 millimeter dot on top to mark the exact location of the test point. A theodolite or transit is suitable to set up the grid. The grid setup procedure follows:

- Mark the center of the compass rose with a wooden stake or paint mark. It should be labeled test point "D4". Set an azimuth stake 120-180 meters south of D4 for convenience. The azimuth stake should have a heavy vertical line marked on it and be labeled "D".
- Establish three test points north and three to the south on this line at 6 meters spacing. These test points are named D1 through D7. Establish auxiliary test points 75 meters along the same line north and south of D4. These points are named N75 and S75.
- Turn the transit exactly 90° east from the D azimuth and set an azimuth stake in this line 120-180 meters east of D4. Turn the transit 180° to the west and set an azimuth stake in this line 120-180 meters west of D4. The accuracy of the declination readings is dependant upon these azimuth stakes. Read and reset the azimuth stakes so that the final angle between them is 90°, plus or minus one half minute of arc.

- Establish six grid test points at 6 meters spacing on this line. Label these points A4 through G4. Establish auxiliary test points along this line 75 meters east and west of D4. Label these points E75 and W75.
- Move the transit to the C4 test point. Level the instrument, sight on the azimuth mark to the west, turn exactly 90° to the south, and set the "C" azimuth stake at 120-180 meters in this line. Establish six test points north and south of C4 on this line at 6 meters spacing and label them C1 through C7.
- Continue in the same manner until all 7 lines are established. Points A1, A7, G1, and G7 may be left off the grid.

The time required to establish a grid of this size ranges from three to six hours. A field assistant is needed to set up the grid stakes in the locations pinpointed by the surveyor.

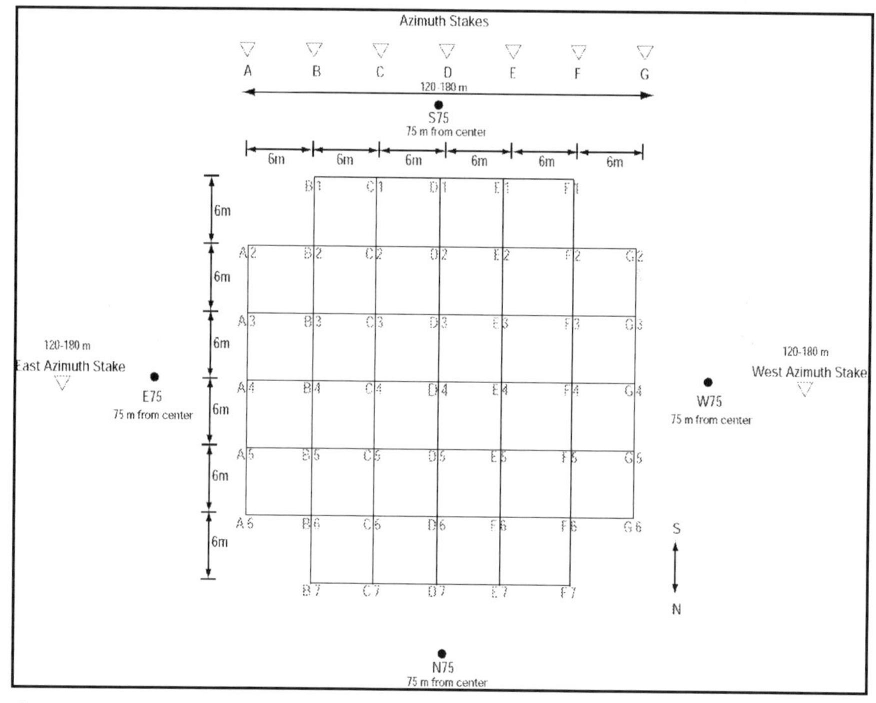


Figure 129. Typical grid layout.

## 3.4.3. *Determine geographic north*

Geographic north must be determined in order to convert the magnetic survey values to declination. The most common methods used to determine geographic north include:

- Astronomical (solar or star) observations. These are preferred as they are quick, easy, and accurate. A solar filter must be used for solar observations. A lighted mark and a light source are needed for star observations. Software is readily available for azimuth computations from solar or star observations. The drawback to using solar or star observations for locating geographic north is that the sun or stars must be visible.
- GPS. GPS is accurate, but the equipment is expensive.
- Gyrocompass. This equipment is heavy and lacks the accuracy of solar, stars, and GPS, but works in any weather.

A permanent geographic north azimuth marker should be established and azimuths to at least three equally spaced nearby prominent objects. A permanent azimuth marker and the prominent objects should be at least 75 meters from the center point of the compass rose. Prominent objects can be the airport beacon, the corners of nearby buildings, or any permanent, stationary object. The true azimuth of these objects is computed from the geographic north determination.

### 3.4.4. *Declination survey*

The declination survey must be done during quiet or unsettled magnetic conditions. Measurement procedure:

- Set the declinometer at station D4, the center point.
- Level the instrument.
- Set the horizontal circle to read 0° while sighted on the azimuth stake D. The azimuth mark should be set to 0° at each test point so the measurements are easily compared to each other. It will be evident if a reading at any test point is bad.
- Read the magnetic declination.
- Move declinometer to the next test point and repeat the measurement procedure.
- Continue in a methodical manner until the magnetic direction has been measured at all 45 test points and the 4 auxiliary test points.
- Make repeated measurements at the center point approximately every 30 minutes. These readings will allow the observer to apply diurnal corrections to each declination value.

The time required to make a survey of this size is 5-20 hours depending on size of the compass rose, observer's speed and instrument used.

### 3.4.5. *Apply diurnal and instrument corrections*

The diurnal (daily) variation of the magnetic field must be removed from the magnetic survey field data to negate the affect of the magnetic field changing during the measurement period. The diurnal variation may be determined either with data from a nearby observatory or by making repeated observations at the center point of the compass rose every 30 minutes. In either case, a chart is created with time on the horizontal scale and the declination on the vertical scale. The declination values measured at the center point every 30 minutes are plotted on the chart through the time of the survey. A straight line is drawn between each point. The times of the declination measurements at the other test points are plotted along the drawn line. The average declination during the measurement time is determined to the nearest minute of arc. The individual correction for each test point is determined by how many arc minutes above or below the average line is each test point measurement.

Instrument corrections must be applied to the data collected using the Gurley transit magnetometer and the Wild T0 Compass Theodolite. Instruments with a mechanical compass require an instrument calibration against a known standard to determine the instrument correction. DI Flux instruments are absolute and have no correction, but they must be compared to an instrument which has been compared to an international standard.

### 3.4.6. *Final report and recommendations*

Once the data have been collected, corrections applied, and azimuths determined, the compiled information is formatted into a report for the airport. The report includes:

- Airport name
- Date of the magnetic survey
- Geographic coordinates
- Description of the compass rose site
- Description of the grid layout
- True azimuths
- Diagram of the survey results
- Recommendations based on the magnetic measurements and professional judgment

## 4. Conclusions

The FAA and the DOD have requirements for design, location, construction, and recertification of compass roses, which have been developed through collaboration with USGS. These requirements are detailed in the FAA Advisory Circular AC 150/5300-13, Appendix 4, and in various military documents, such as Handbook 1021/1, but the major requirement is that the range of declination measured within 75 meters of the center of a compass rose be less than or equal to 30 minutes of arc. The USGS Geomagnetism Group has developed specific methods for conducting a magnetic survey so that existing compass roses can be judged in terms of the needed standards and also that new sites can be evaluated for their suitability as potentially new compass roses. First, a preliminary survey is performed with a total-field magnetometer. Differences of less than 75nT over the site area are sufficient to warrant additional, more detailed surveying. Next, a number of survey points are established over the compass rose and surrounding area, where declination is to be measured with an instrument capable of measuring declination to within 1 minute of arc, such as a Gurley transit, DI Flux theodolite, or Wild T-0 magnetometers. The data are corrected for diurnal and irregular effects of the magnetic field and declination is determined for each survey point, as well as declination range and average of the entire compass rose site. Once the data have been collected, corrections applied, and azimuths determined, the compiled information is formatted into a report for the airport.

**DISCUSSION**

Questions (Jurgen Matzka):

1. What is the reason that the compass instrument gives different readings than DI – flux (magnetic impurities or problem with needle)?
2. Why declination and not magnetic north at compass rose?

Answers (Alan Berarducci):

1. Most compass instruments have some ferrous parts in them, such as the pivot which the needle balances on. Also, many compass instruments are able to be adjusted so the correction is 0 or very small. Since the instrument can not be inverted and usually it is not read from either end of the needle such as with DI Flux measurements where the instrument is read in 4 positions, there are errors from the transit or theodolite.
2. Magnetic North is used on a compass rose.

# MAGNETIC REPEAT STATION NETWORK IN ITALY AND MAGNETIC MEASUREMENTS AT HELIPORTS AND AIRPORTS

ANGELO DE SANTIS[20]
GUIDO DOMINICI
*Istituto Nazionale di Geofisica e Vulcanologia*

**Abstract.** The Italian Magnetic Network is composed of 114 repeat stations and its last complete measurement took place in the period 1999-2001, i.e. centred at epoch 2000. Mathematical models and maps were produced for declination, horizontal and vertical components, together with the total intensity of the geomagnetic field. By the end of 2005 the Magnetic Network measurement will be repeated again completely. To the Italian sites have been added also 11 other stations in Albania. Those were measured in 2004 with the collaboration of the Albanian Academy of Science and the University of Tirana. Together with this activity also another one dedicated to airports or heliports measurements has been undertaken. This kind of measurements has the objective to provide absolute magnetic knowledge of the magnetic declination in airports and heliport swinging roses where it is possible then to calibrate aircraft compasses.

**Keywords:** repeat stations; magnetic declination; swinging rose; navigation

## 1. Introduction

The geomagnetic field is an important property of our planet. It has allowed the life to progress in the evolution up to our times, screening most of the electric charges coming from the sun and the cosmic radiation that otherwise would have hit more dangerously the surface of the Earth, causing health damage and allowing the atmosphere to be blown away from the planet. The presence of a geomagnetic field has also allowed human

[20]Address for correspondence: Istituto Nazionale di Geofisica e Vulcanologia, V. Vigna Murata 605, 00143 Rome, Italy. Email: desantisag@ingv.it



*J.L. Rasson and T. Delipetrov (eds.), Geomagnetics for Aeronautical Safety,* 259–270.

beings to use it as orientation information. In this regard, observing the geomagnetic field, in particular its angular difference in orientation with respect to the geographic meridian, the so called Declination, provides together with the use of a compass one of the cheapest and simplest ways to know the orientation for navigation purposes at the Earth surface and above. The Istituto Nazionale di Geofisica e Vulcanologia (I.N.G.V.) is the Italian Institution that has the duty to make continuous observations of the geomagnetic field over the Italian territory. It runs two Observatories in Italy, L'Aquila (centre of Italy, working from 1960) and Castello Tesino (north of Italy, working form 1965) and two in Antartica, Zucchelli Base (along the coast) and Concordia (internal to the continent), the latter in conjunction with the French institute of IPGP. I.N.G.V. maintains also two continuous magnetic stations in Belluno (North East Italy) and Gibilmanna (South Italy - Sicily) where a series of absolute measurements is made at least once a year. Also a National Magnetic Network composed of around one hundred of repeat stations is measured every 5 years approximately. In this paper we will illustrate some information about the Magnetic Network and then will describe some notes on the measurements made at the swinging rose platforms of airports and heliports for allowing compass calibration of aircraft.

## 2. The Italian Magnetic Network

In Italy a long history of magnetic measurements and practice can be traced back in time to the XVII century (see Cafarella et al., 1992a, b) but the continuity in regularly repeated magnetic measurements started with the national unification at the end of the XIX century. Since the 1930's a modern Italian Magnetic Network composed of repeat stations regularly distributed over the Italian territory and integrated by Magnetic Observatories, has allowed the determination of the spatial structure and the time variation of the Earth's magnetic field over Italy (Cafarella et al., 1992a, b).

Table 27 shows a brief history of the magnetic surveys made in the Italian territory from 1640 to present, indicating the magnetic elements observed together with the number of repeat stations (# sites) and the corresponding observers. The magnetic survey of 1891-92 represents the first 3-component magnetic survey in the then re-united nation of Italy: we present here the chart of the horizontal component published at that time (Figure 130)

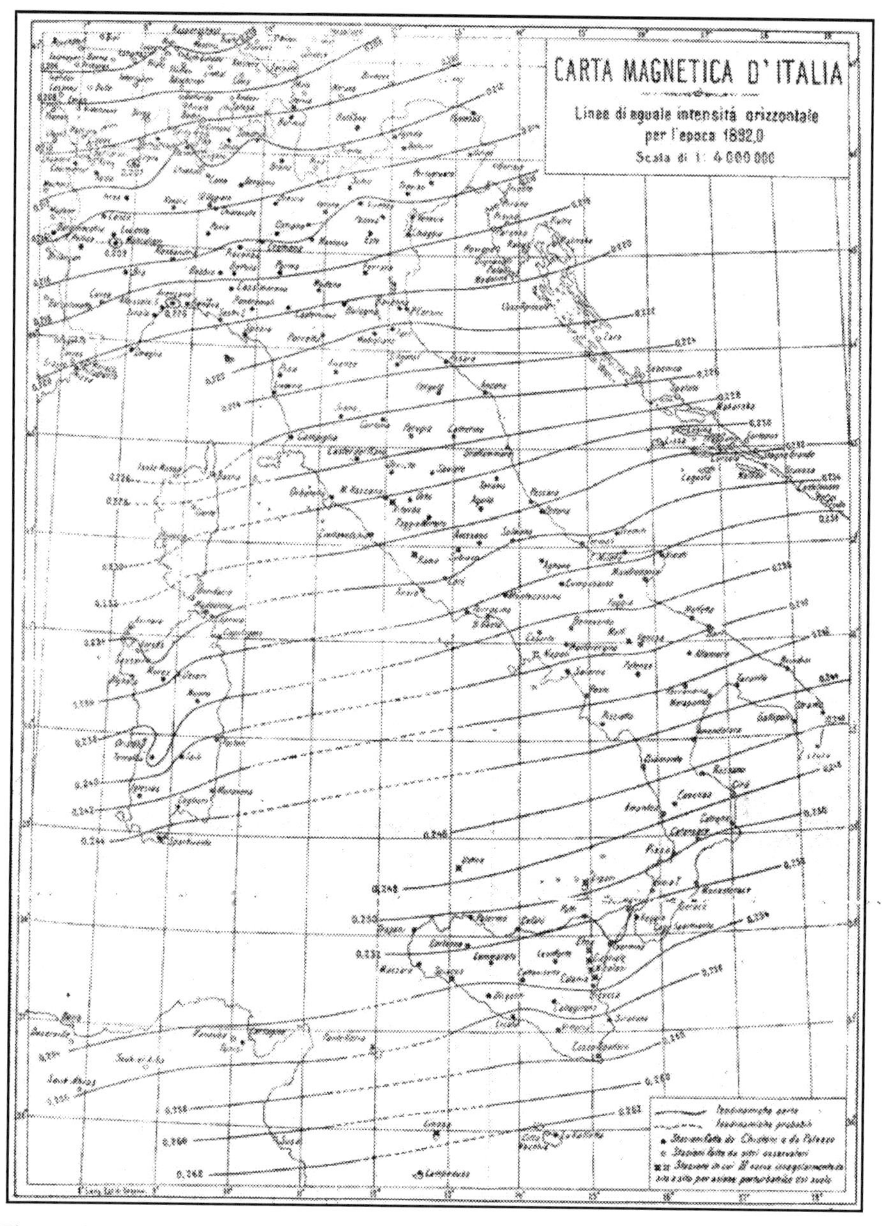


Figure 130. Chart of the horizontal magnetic component H for epoch 1892 as published by Chistoni and Palazzo after the 3-component magnetic survey of 1891-92 in Italy.

Table 27. History of Magnetic Surveys in Italy from 1640 to present. Acronyms are explained within the text. Additional acronym of UCMG stands for the Italian Central Office of Meteorology.

| Epoch | Mag. Elements | #Sites | Observer |
|---|---|---|---|
| 1640 | D | 21 | Fathers Borri-Martini |
| 1805-07 | I,H | 15 | Humboldt |
| 1845-56 | D,I,H | 24 | Kreil – Fritsch |
| 1875-78 | D,I,H | 77 | Father Denza |
| 1881-92 | D,I,H | 284 | UCMG(Chistoni-Palazzo)¶ |
| 1932-38 | D,I,H | 46(1496) | IGMI |
| 1948 | D,I,H | 46 | IGMI |
| 1959 | D,I,H | 46 | IGMI |
| 1965 | D,I,H | 28 | ING+Universities |
| 1973 | D,I,H | 50 | IGMI |
| 1979 | D,I,F | 106(2252) | ING+ Universities |
| 1985 | D,I,F | 106 | ING+IGMI |
| 1990 | D,I,F | 116 | ING |
| 1995 | D,I,F | 116 | ING |
| 2000 | D,I,F | 114 | INGV (Charts with IGMI) |
| 2005 | D,I,F | 114 | INGV (completed by end 2005) |

Numbers in parenthesis: second order measurements, i.e. only F,H,Z. ¶ For this Survey we present a chart for H (see Figure 130).

Repeat Stations of the National Magnetic Network must satisfy the following requirements:

i) absence of significant artificial and natural disturbance;
ii) representative of a quite large area;
iii) availability of targets for azimuths with geographic meridian.

The "magnetic selection" of the repeat station locations then follows the many years of experience that has led to the identification of areas with low magnetic crustal anomaly level, primarily in relation to the knowledge of the geological and tectonic environment.

A magnetic repeat station is materially constituted of a mark on a concrete pillar or on a $1m^2$ platform in the area of the location selected for the magnetic measurement procedures. The operators put the instrument tripod on top of this mark and, using several sightings on an azimuthally distributed landmark panorama, they can rely on a predefined geographical reference system for their measurements. Landmarks targets are generally materialized by churches bell towers, or crosses and recently also by antenna towers. Unfortunately the life time of a magnetic mark is not very long although a specific law protects these points of observation (Italian

Law n.1024, 3 June 1935): in fact the mark can be accidentally removed or a heavy magnetic interference can start with the edification of a new road or a building or other structures nearby. For this reason every repetition survey starts with an in depth inspection of the old marks and, if necessary, a new mark must be installed in the vicinity of the old unusable station.

Magnetic repeat stations are now well established and some of them show many years of "magnetic history" with measurements repeated on the same location for 80-90 years, thus providing information of great value about the secular variation of the geomagnetic field.

For each repeat station a monograph is prepared with the purpose not only to keep note of the magnetic measurements taken but also to allow a quick and correct finding of the place. For these requirements, the monograph consists of two parts: in the first we find indications about the relative area, e.g. field owner, and a brief description as to how reach the place with the help of a detailed map, coordinates and altitude of the site, together with the date of the first series of measurements. Also the Italian Institution that materialised the repeat station for the first time is indicated with the following acronyms:

- I.G.M.I. Istituto Geografico Militare Italiano, Florence;
- I.N.G.V. (I.N.G. before 2000) Istituto Nazionale di Geofisica e Vulcanologia, Rome;
- PADOVA Istituto di Fisica Terrestre, University of Padua;
- FERRARA Istituto di Mineralogia, University of Ferrara;
- GENOVA Ist. Geofisico e Geodetico, University of Genua;
- NAPOLI Osservatorio Vesuviano, Naples;
- BARI Ist. di Geodesia e Geofisica, University of Bari.

In the second part of the monograph, all targets with their azimuths are indicated. Finally all reduced magnetic field values are reported for all epochs of measurements.

In the repeat station survey of 2000, the number of repeat stations in Italy was in total 114, including two Observatories, L'Aquila (42°23'N, 13°19'E) and Castello Tesino (46°03'N, 11°39'E), corresponding to a density of about 1 station /3000 km$^2$ or about 58 km mean stations spacing.

For 2005, the survey included also the repetition of Albanian network: this sub-network consists of 10 repeat stations materialised and measured for the first time in 1994, in the framework of a joint project between the Center of Geochemistry and Geophysics of Tirana, the Physics Department of Tirana University and I.N.G.V. (I.N.G. at that time), and then repeated for the total intensity $F$ alone in August 2003 (Duka et al., 2004) and for all magnetic components in September 2004 with the addition of 1 station (Berat). Figure 131 shows the progress in the measurements for the unfinished 2005 survey. During the period of survey also two magnetometers were deployed in the seafloor of the Tyrrhenian Sea in the framework of the European Project GEOSTAR. The completion of the whole magnetic survey is foreseen by the end of 2005.

Figure 131. Points of Repeat Magnetic Stations of Italy and Albania and their progress in time so far. Completion of the whole survey (to reach a total of 114 Italian sites + 11 Albanian sites) is foreseen by the end of 2005. During the period of survey also two sites in the seafloor were deployed for magnetic measurements in the framework of the European project GEOSTAR.

Magnetic repeat station network measurements now generally include measurements of inclination *I*, declination *D* and total intensity *F*. From these measurements also the horizontal and the vertical components, *H* and *Z*, can be determined. The measurements should be repeated in time over the same stations respecting a 5-year average periodicity as suggested by IAGA (International Association of Geomagnetism and Aeronomy). Instruments used at repeat stations are now Proton Precession Magnetometers and DI Flux magnetometers. A gyroscopic theodolite for the determination of geographic north is used when necessary. The field magnetic measurements need an independent local time variation monitoring: this is generally undertaken at a magnetic observatory. When necessary, time variations of the geomagnetic field were recorded also in the field by means of a portable tri-axial variometer.

Measurements at each repeat station are generally made as 10 sets of measurements: five in the morning at about 7.30-9.30 am LT, and five in the afternoon at about 3.30-5.30 pm LT.

As is known, the value of declination changes slowly with diurnal and secular variation but abruptly during a magnetic storm. For this reason it is important to check if the magnetic field is quiet enough at the day of the measurement. Nowadays it is easy to do it by means of an internet connection: in Italy, the magnetic activity at the two magnetic observatories can be seen in real time by connecting to the INGV web site (www.ingv.it).

## 3. Data analysis, normal reference fields and cartography

For what concerns the data reduction, the following procedure was used: the magnetic elements $D$, $I$ and $F$ observed at the repeat stations are reduced firstly to 02 UT for the diurnal variation correction, with reference to digital data from L'Aquila or Castello Tesino Observatory, or from the portable variometric station when the station was installed. Secondly, data will be reduced with the secular variation of L'Aquila Observatory for the fixed epoch, i.e. 2005.0. Prior to the computation of the normal field coefficients, all values will be reduced to sea level considering only the dipolar contribution (see Meloni et al., 1994). In formulae, the value of element 'E' (i.e., D, I or F) at station's', (Es) reduced i.e., to epoch 2005.0, will be calculated following the two steps:

$$\mathrm{Es(02\ UT)} = \mathrm{Evar(02\ UT)} + [\mathrm{Es(t)} - \mathrm{Evar(t)}]$$

$$\mathrm{Es(2005.0)} = \mathrm{Eobs(2005.0)} + [\mathrm{Es(02\ UT)} - \mathrm{Eobs(02\ UT)}]$$

where:

- Es(02 UT)= Value of E at station s reduced at time 02 UT of day of measure
- Evar(02 UT)= Value of E at variometer of station s at 02 UT of day of measure
- Es(t) = Value of element E observed at station s at time t
- Evar(t)= Value of element E at variometer of station s at time t
- Eobs(t) = Value of element E measured at the Observatory at the same time t
- Es(2005.0)= Value of element at station s reduced to epoch 2005.0
- Eobs(2005.0)= Mean value of element at the Observatory for epoch 2005.0

The data so reduced (at sea level and at the centred epoch, i.e. 2005.0) are analysed by means of a least squares regression, in order to provide a normal reference field for each magnetic element *D*, *I* and *F*. The other magnetic elements are deduced from these by means the known formulae relating all magnetic elements. Each normal field is usually expressed as a second order polynomial in latitude and longitude, referred to a central point with coordinates of 12° E longitude and 42 ° N latitude. Usually a rejection criterion is applied in order to reject the least significant coefficients. Normal reference fields so produced are compared with IGRF and another national reference field determined from IGRF and L'Aquila Observatory annual means (Molina and De Santis, 1987). Final significant coefficients are then published together with the corresponding cartography for all magnetic elements.

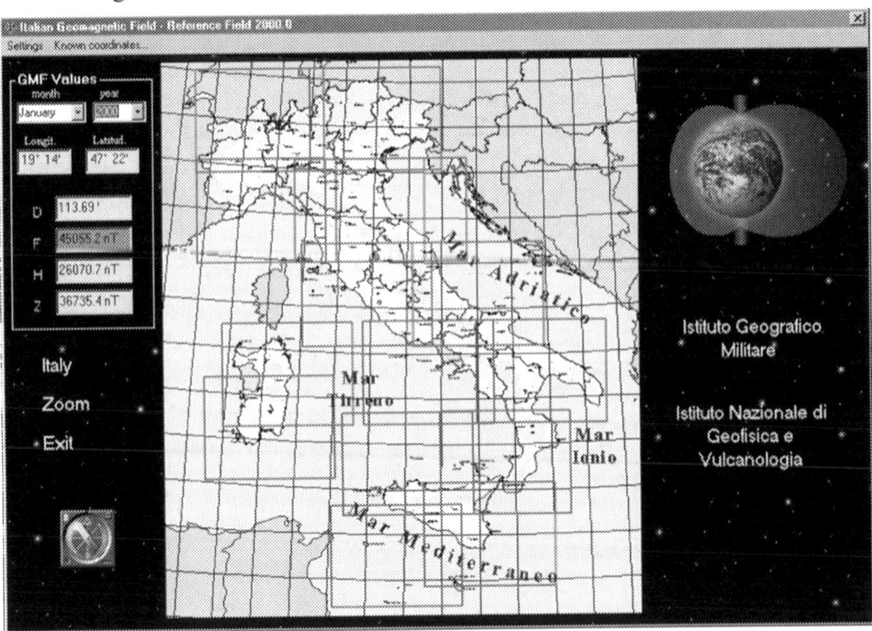


Figure 132. Snapshot of the interactive program that provides interactively the geomagnetic field values at any point of Italy in the epoch 2000.

The last complete survey made in Italy for magnetic maps production was undertaken in 1999-2001 and centred at 2000 (Coticchia et al. 2001). For that epoch in fact an edition of four magnetic maps to the scale 1/1,500,000, one for each element (D, H, Z and F), also with digital support (CD), was published in a joint collaboration between the Istituto Geografico Militare (I.G.M.) and the I.N.G.V.

The maps for F, H and Z were drawn with automatic graphic contouring programs after updating 2552 stations measured in the framework of the second order magnetic network undertaken in CNR Project of Research PFG at 1979.0, with the adopted secular variation model. For what concerns the map of declination the second order network was based on the old IGM 1529 stations data, updated with the usual method starting from the 1985 compilation.

A similar procedure will be followed for the new cartography of the magnetic survey of 2005.

An alternative analysis is also applied for providing a real Laplacian representation of the geomagnetic field over Italy: this is usually made by means of the Spherical Cap Harmonic Analysis (SCHA) (Haines, 1985; De Santis et al., 1990). In this case, no reduction in altitude nor in time (apart from the diurnal variation correction) is made, because this technique takes the proper altitude and time into account for any magnetic field observation. The corresponding references given at the end of this paper can be used to clarify all details of SCHA.

## 4. Declination measurements in heliports and airports

The regular condition of a periodic control of aircraft compasses is that the operator performs a procedure known as 'compass swing'. This operation consists in a rotation of the aircraft around a point (the centre of a platform also called the *swinging rose*) and compares the values of compass orientation with the horizontal direction of geomagnetic vector obtained by magnetic measurements.

In this section we report our experience in some activity specifically required by helicopter industries and aircraft companies for compass calibration (see Figure 133).

For a new pad dedicated to swinging rose, a visual inspection is important to check if the zone is:

- a level circular area,
- of sufficient strength to support the weight of aircraft,
- at least 180 m far from: buildings, railroad tracks, DC power lines, hangars causing possible big disturbances,
- at least 90 m far from other aircraft.

A total field survey with a proton magnetometer verifies that the area is magnetically quiet if the values measured do not scatter from the mean by more than 90 nT.

Particular attention is required for the construction of the platform so as not to use magnetic materials (e.g. reinforcing steel or ferrous aggregate), and to use non-magnetic material (e.g. PVC) if a drainage conduct is required within 90m.

Once the swinging rose is built, the azimuths of all possible geodetic target bearings must be measured with a gyro-theodolite having a precision not worse than 20 arc-seconds. Such a measurement is to be made only once. To avoid any possible confusion, it is suggested to paint the direction of the Geographic North with a different colour as the Magnetic North.

Magnetic angular measurements are made with a DI flux theodolite with precision of around 1-6'' (depending on the instrument). Each measurement is reduced with the closest Observatory and then the mean value of declination is computed. This is the value officially given to the Authorities.

Figure 133. Swinging rose at Brindisi (South-East Italy) heliport of Agusta (Manufacturer of helicopters).

After the determination of the direction of Magnetic North, radial lines (usually 24, one every 15 degrees) are drawn with non-magnetic paint, with precision in direction of 1 minute, and some circles are drawn, usually with radii of around 5 m, 15m and 25m.

The Italian law regulations require the determination of magnetic Declination at least every 5 years, for civil use, and every 1 year for military use. Nevertheless, it is strongly suggested to update it annually for any use.

## 5. Conclusions

Maps and models deduced from the repeat station surveys and magnetic observatories do not only represent the geomagnetic field in all its aspects but can also improve our knowledge of the field dynamics, which is the field evolution in time. Improving the studies of the geomagnetic field, providing better present and predictive models will allow in the next future to improve also the quality of our measurements in the heliports and airports.

As the technology makes important progresses to improve our quality of life, such as in transportation and communications, it usually becomes more complex and sensitive to possible disruptions. It is in those emergencies that we should be able to rely on more natural information such as the orientation through the magnetic compass. Measuring the magnetic declination all over the national territory, with particular attention in some important places, such as heliports and airports, becomes fundamental for improving navigation, in particular for the safety in all operations related to landing and take-off in airports, and flying.

## Acknowledgements

We would like to thank the organizers of the NATO ADVANCED RESEARCH WORKSHOP, held in Ohrid, Macedonia 18-20 May, 2005, in particular Dr. Jean Rasson, for inviting us to present this paper in the workshop. We also thank our colleagues of INGV, in particular Angelo Di Ponzio and Massimo Miconi, for their collaboration in many operations during the magnetic repeat station surveys of 2000 and 2005.

## References

Coticchia A., Dominici G., De Santis A., Di Ponzio A., Meloni A., Pierozzi M., Sperti, M., 2001. Italian Magnetic Network and Geomagnetic field maps of Italy at year 2000.0, *Bollet. Geodes. Scie Aff.*, **IV**, 261-291. (with 4 maps 1/2.000.000 and CD-ROM).

Cafarella L., De Santis A., Meloni A., 1992. Secular variation from historical geomagnetic field measurements, *Phys. Earth Planet. Inter.*, **73**, 206-221.

Cafarella L., De Santis A., Meloni A., 1992. *Il catalogo geomagnetico storico italiano, Publ. ING*, 160 pp..

De Santis A., O.Battelli, D.J. Kerridge, 1990. Spherical cap harmonic analysis applied to regional field modelling for Italy, *J. Geomag. & Geoelectr*, **42**, 1019-1036.

Duka B., Gaya-Piqué L.R., De Santis A., Bushati S., Chiappini M., Dominici G., 2004. A geomagnetic Reference Model for Albania, Southern Italy and Ionian Sea from 1990 to 2005, *Annals of Geophysics*, 47, 1609-1615

Haines, G.V., 1985. Spherical Cap Harmonic Analysis. *J. Geophys. Res.* **90** (B3): 2583-2591.

Meloni A., Battelli O., De Santis A., Dominici G., 1994. The 1990.0 magnetic repeat station survey and normal reference fields for Italy, *Annali di Geofisica*, Vol.XXXVII, **5**, 949-967.

Molina F., De Santis A., 1987.Considerations and proposal for a best utilization of IGRF over areas including a geomagnetic Observatory, *Physics Earth Plan. Inter.*, **48**, 379-387.

# AIRPORT CONDITIONS IN MACEDONIA: SEISMIC RISKS

LAZO PEKEVSKI*
*Seismological Observatory, Faculty of Natural Sciences and Mathematics. University "Sts Cyril and Methodius"*

**Abstract.** The territory of the Republic of Macedonia is located in the Balkan region – one of the most earthquake prone areas in Europe. Investigations of the seismicity of the Balkan Peninsula point out that this territory is exposed not only to autochthonous earthquakes but also to earthquakes from adjacent seismically active areas. The Macedonian Seismological network (SORM) and the network of the Institute of Earthquake Engineering and Engineering Seismology (IZIIS) detect, monitor and give detailed information on seismicity of the region. Their information directly impacts the accuracy of seismic hazard assessment. Application of statistical models of the "proposed future" seismic activity of the territory under investigation is of crucial importance in seismic hazard assessment. If the concept of "seismicity" indicates a measure with which natural seismic activity is determined in certain areas/sites (airports, flying fields etc.), it is of great importance in determining: 1) values of maximum expected intensity and ground accelerations of earthquakes and 2) given return periods (in years). The results of the research are presented on Seismic hazard maps for the areas with airports, airfields, and flying field sites. The parameter used as a measure of the seismic hazard for the areas under consideration, is the PGA, peak ground acceleration, (values of acceleration of *g* with a 64% probability to exceed).

**Keywords:** seismicity, seismic hazard, seismic risk

*To whom correspondence should be addressed at: Seismological Observatory, Faculty of Natural Sciences and Mathematics. University "Sts Cyril and Methodius", Skopje, Republic of Macedonia. Email: lpekevski@seismobsko.pmf.ukim.edu.mk



*J.L. Rasson and T. Delipetrov (eds.), Geomagnetics for Aeronautical Safety,* 271–279.

## 1. Introduction

Defining the parameters of ground motion during earthquakes (macroseismic intensity, displacement, velocity, acceleration etc.) and their application in determining seismicity parameters is of great interest.

Considering the random nature of an earthquake occurrence, modeling it is difficult without simplifications and assumptions related to the earthquake occurrence mechanism and structural behavior under seismic effects.

One way of presenting the earthquake loading for a given geographical region is with a seismic hazard map. Therefore, hazard mapping techniques must relate seismological and engineering parameters.

We present two hazard models, in order to show the effect of the mathematical/statistical approach to seismic hazard estimation. The maximum expected intensity and the peak ground acceleration (PGA) are used as a measure of the seismic hazard for the territory of Republic Macedonia.

## 2. Data

The territory of Republic of Macedonia was included in the seismic hazard investigations of the Euro-Mediterranean area. It is clear that the territory of the Republic of Macedonia and neighboring countries are within a prominent active seismic area of the Balkan Peninsula. This area is also exposed to earthquakes from adjacent seismically active areas (Figure 134).

The data for the study comes from records of seismic events from 1900 to 2000, isoseismal maps, and additional graphical and numerical results. These data are contributions of various research projects of the observed seismic activity on the territory of Republic of Macedonia. Besides the seismological data, available tectonic and engineering/seismological data for this area were also used. In this way, the existence of several well defined seismogenic zones (Vardar, Drim, Struma, and others), were defined. Extremely high activity has been identified in these zones where, during the neotectonic stage, intensive transformations of the old structures have taken place. The map of maximum calculated earthquake magnitudes is the result of such detailed investigations of the territory of the Republic of Macedonia. It is crucial to the seismic hazard investigations of the territory under consideration.

Comparing the seismological data and the tectonic conditions, a close relationship between the seismogenically active areas and the tectonic knots of faults (faults in different directions) is evident (Skopje, Valandovo, and other seismic areas). The magnitude of earthquakes in these zones depends upon whether the knots are created by an intersection of regional or local faults (Figure 135, Table 28, Jordanovski Lj, Pekevski, L. et al, 1998). This relationship is clearly presented on the map of maximum observed macroseismic intensities on the territory of the Republic of Macedonia for the period 1900-2000 (Figure 136).

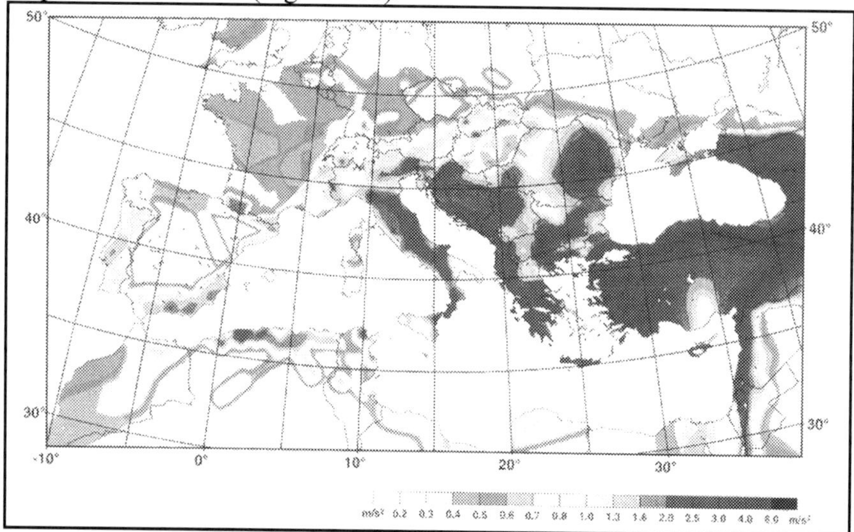


Figure 134. Seismic hazard assessment of the European-Mediterranean area in terms of peak ground acceleration for 90 % of non exceedance within a 50 year period. Red and brown colors indicate areas with the highest seismic hazard values. A significant part of the Republic of Macedonia belongs to the area with high seismic hazard values (Grunt Hal et al. 1999).

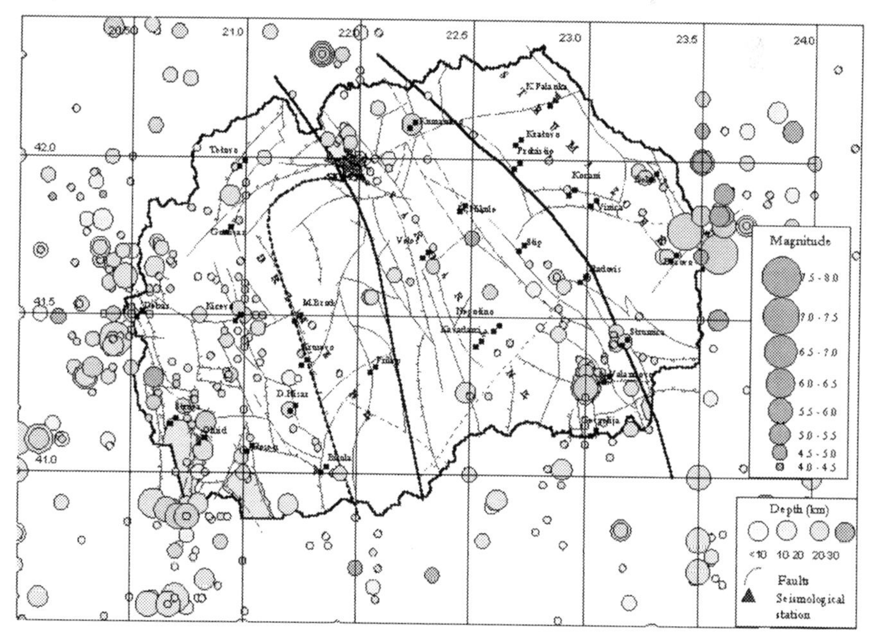


Figure 135. Epicenter map of earthquakes (1901-2000).

Table 28. The most significant historic and recent earthquakes occurred in the territory of the Republic of Macedonia and neighboring areas.

| *Year* | *M.* | *D.* | h min s (UTC) | $\varphi$ ($^0$) | $\lambda$ ($^0$) | *h* (km) | *M* | $I_0$ (MSK-64) | *Epicentral area* |
|---|---|---|---|---|---|---|---|---|---|
| (400) | | | | 41.50 | 22.00 | | 6.1 | 9 | *Gradsko (Stobi)* |
| 518 | | | | 42.10 | 21.40 | | 6.1 | 9 | *Skopje (Skupi)* |
| 527 | | | | 41.10 | 20.80 | | 6.1 | 9 | *Ohrid Lake* |
| 896 | 9 | 4 | | 41.70 | 23.00 | | 6.1 | 9 | *Pehcevo - Kresna* |
| 1555 | | | | 42.00 | 21.50 | | 6.1 | 9 | *Skopje* |
| 1755 | 2 | 26 | | 42.50 | 21.90 | | 6.1 | 9 | *Urosevac (Kacanik)* |
| 1904 | 4 | 4 | 10 2 38.1 | 41.78 | 22.93 | 25 | 7.3 | 9 | *Pehcevo - Kresna* |
| 1904 | 4 | 4 | 10 25 50.0 | 41.71 | 23.08 | 30 | 7.8 | 10 | *Pehcevo - Kresna* |
| 1905 | 10 | 8 | 7 28 51.4 | 41.80 | 23.10 | 30 | 6.5 | 8 | *Pehcevo - Kresna* |
| 1911 | 2 | 18 | 21 35 18.0 | 40.86 | 20.71 | 25 | 6.7 | 9 | *Ohrid Lake* |
| 1912 | 2 | 13 | 8 4 | 40.86 | 20.75 | 25 | 6.0 | 8 | *Ohrid Lake.* |
| 1921 | 8 | 10 | 14 10 40.0 | 42.30 | 21.40 | 20 | 6.1 | 9 | *Urosevac (Vitina)* |
| 1931 | 3 | 7 | 0 16 44.8 | 41.28 | 22.50 | 25 | 6.0 | 8 | *Valandovo* |
| 1931 | 3 | 8 | 1 50 24.0 | 41.28 | 22.50 | 10 | 6.6 | 10 | *Valandovo* |
| **1942** | **8** | **27** | **6 14 15.6** | **41.62** | **20.47** | **15** | **6.0** | **9** | ***Peskopia*** |
| 1963 | 7 | 26 | 4 17 11.7 | 42.02 | 21.42 | 5 | 6.1 | 9 | *Skopje* |
| 1967 | 11 | 30 | 7 23 49.9 | 41.42 | 20.43 | 20 | 6.5 | 9 | *Debar region* |

## 3. SEISMIC HAZARD MODEL FORMULATION

The construction of the seismic hazard models of the Republic of Macedonia is based upon the statistical evaluation of the past seismic activity combined with the geological setting. Different types of input data are needed to develop a reasonable model. They are:

- Geological setting. Location and behavior of the faults.
- Seismicity and source location.
- Maximum credible events for each source.
- Isoseismal maps of the selected earthquakes, for estimation of maximum expected magnitude and maximum intensity (according to the theory of extremes).
- Recurrence relationship.
- Attenuation relationship.
- Probabilistic model of seismic occurrences.

All information on the seismic history, geological structure, frequency of earthquake occurrences, and attenuation of ground motion with distance from the epicenter is used in the model to obtain the probable future seismic ground motion.

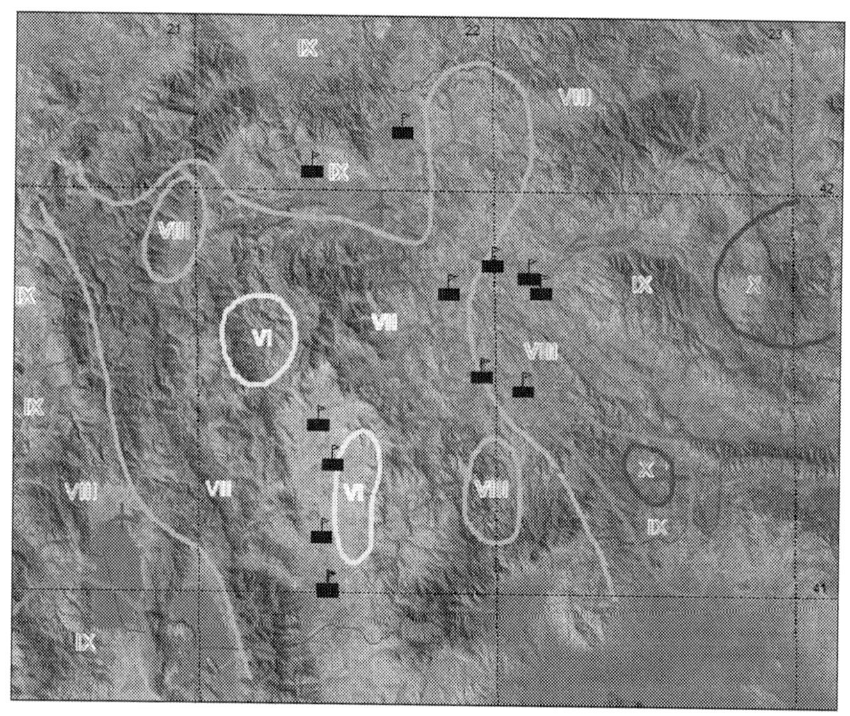


Figure 136. Map of maximum observed intensities (MCS) in the territory of the Republic of Macedonia (1901-2000).

Seismic hazard maps for the Republic of Macedonia for future time periods and different seismicity models are obtained using the above input and described procedures, (earthquake ground motion parameter forecasting). The hazard is measured in terms of maximum expected intensity and peak ground acceleration.

## 4. Results of hazard analysis

For enhanced flying and airport safety, seismic hazard maps (Figure 139) have been generated for areas where airports (AP, pink), airfields (AF, blue) and flying fields (FF, green) have been built. The parameters used as measures of the seismic hazard are the maximum expected intensity of the "future" earthquakes, and the PGA, values of acceleration of *g* with 64% probability to exceed in certain return periods (in years).

## 4.1. THEORY OF EXTREMES

Another seismic hazard parameter, as well as the characteristics of the regional seismicity, is the intensity of the maximum expected earthquakes macroseismic effect $I_{exp}$ in a certain return period T (in years). The extreme values method (Jenkinson's solution of the stability postulate) has been applied for the determination of $I_{exp}$ (Pekevski, 1992)

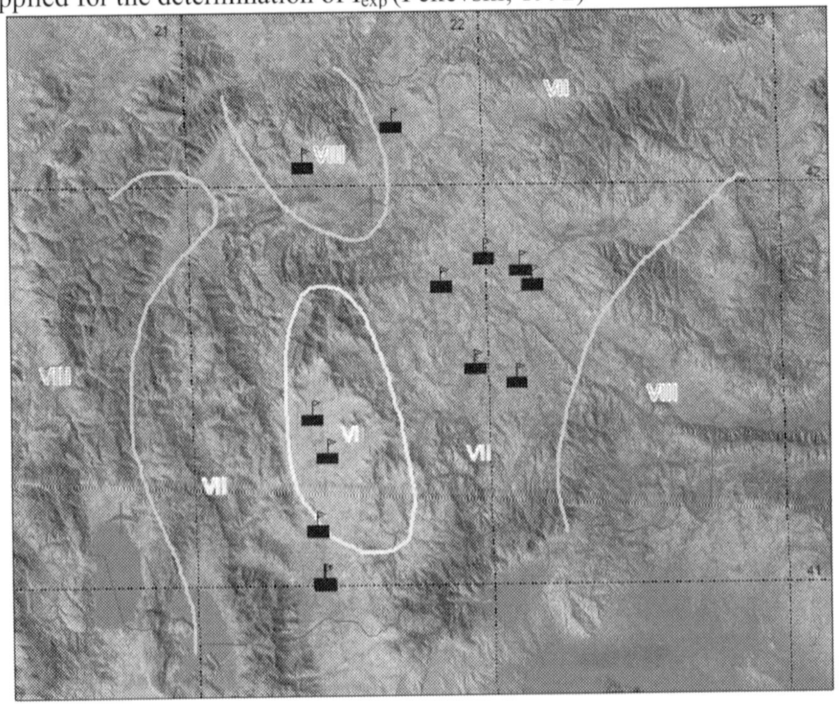


Figure 137. Theory of extremes (Jenkinson's solution) Expected intensity Imax in a 100 year period.

The results are presented on the Figure 137 and Figure 138. The seismogenic areas with high seismic activity are separated out by the isolines of certain MCS intensity. This procedure was used in research of the seismicity on the territory of former Yugoslavia (Pekevski, 1983, Ribaric et al., 1987).

## 4.2. PROBABILISTIC HAZARD

Figure 139 shows the seismic plane sources map of the Republic of Macedonia and AP, AF, and FF sites. Applying the model of plane seismic sources and the probabilistic approach, the calculated seismic hazard (for the grid points of the territory under investigation) in return period of 100 years, is presented on Figure 140.

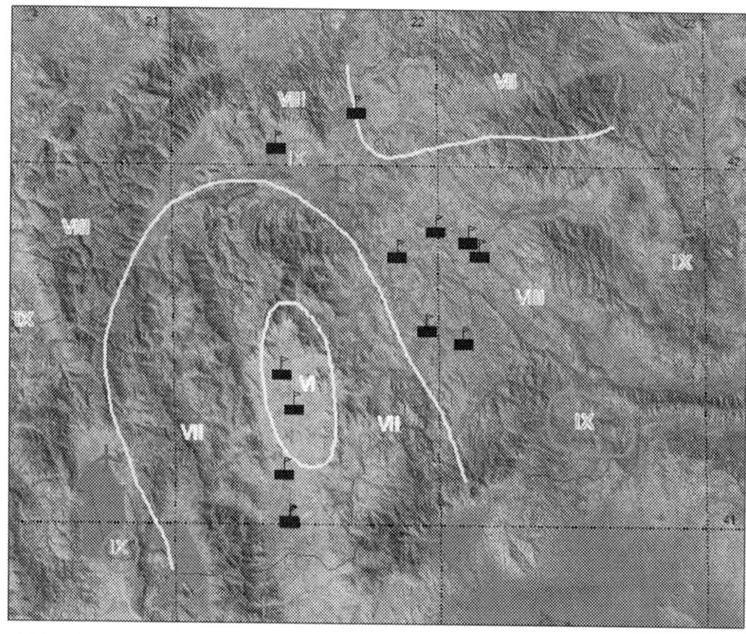


Figure 138. Theory of extremes (Jenkinson's solution): expected intensity Imax in a 200 year period.

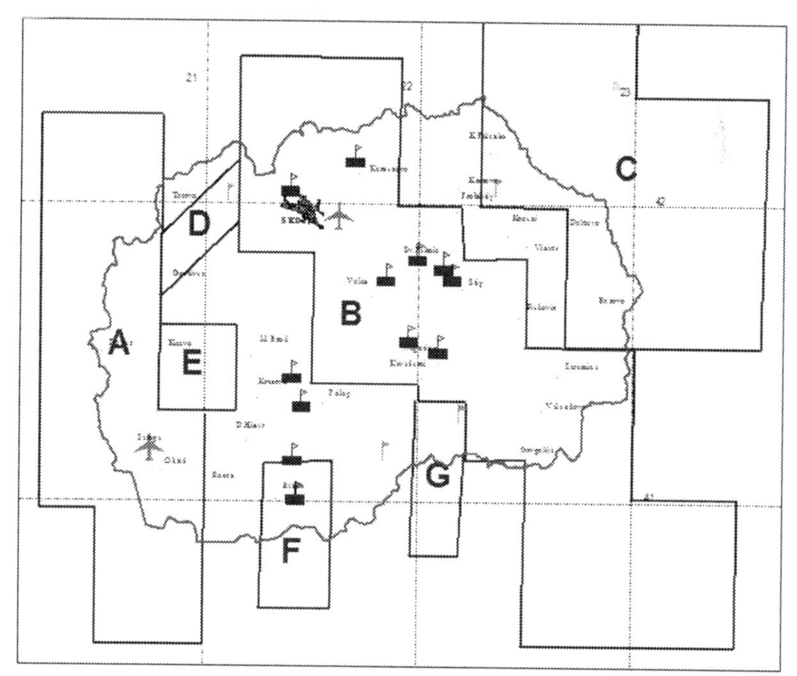


Figure 139. Map of plane sources (according to the seismological data) on the territory of the Republic of Macedonia.

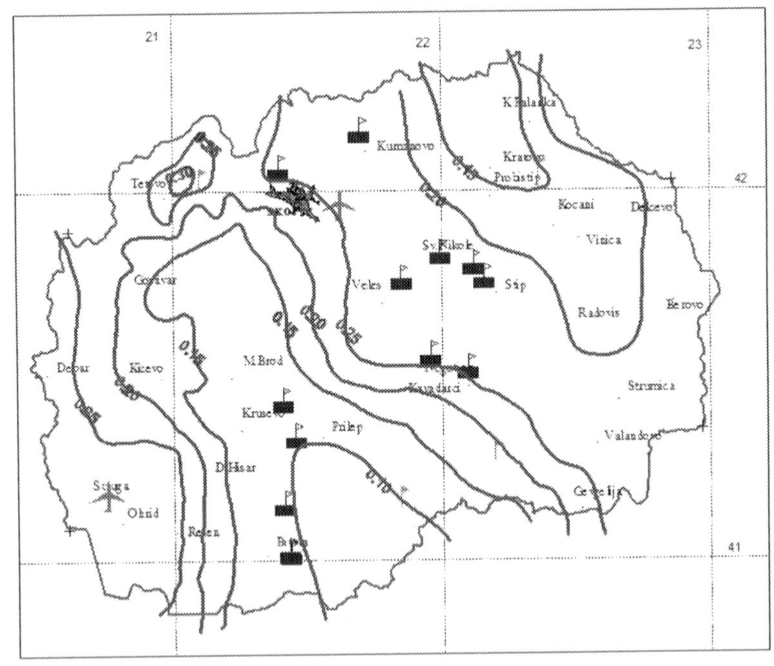


Figure 140. Probabilistic model: plane sources, return period 100 years.

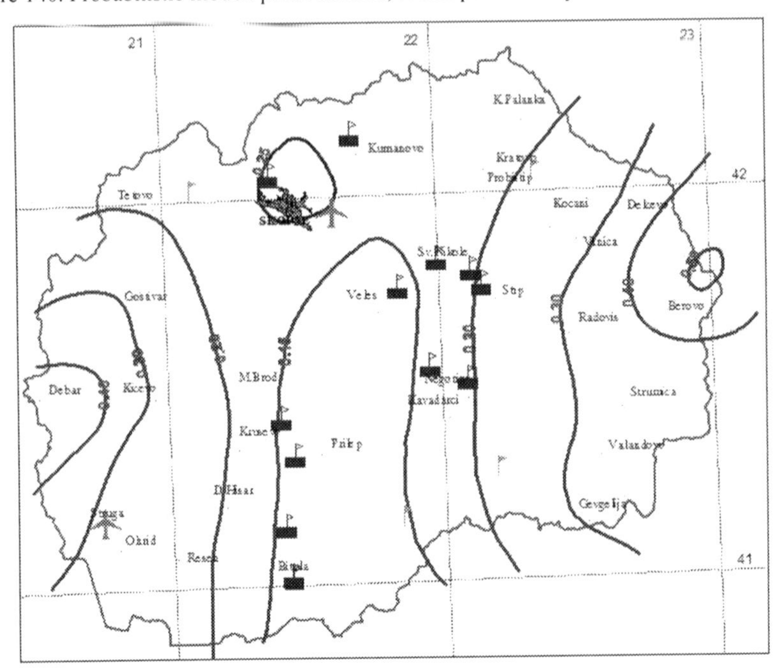


Figure 141. Probabilistic model: plane sources, return period 100 years.

Changing the model's seismic source type (line instead of plane) produces an evident difference in g-contour lines and in iso-acceleration contour lines as well, for the same return periods (100 years) and PGA values, with a 64% probability to exceed (Figure 141).

Comparison of these maps shows the effect of the *form/nature* of the seismic sources on the shape of the g-contour. The general shape of g contours depends on the groupings of earthquakes produced by different seismic sources (Mihailov D, Dojcinovski D., 1992).

## 5. Conclusion

The differences between the values of peak ground acceleration (PGA) for different types of seismic sources for the airports (AP), flying fields (FF), and airfields (AF) under investigation, show the need to study source modeling approaches, based not only on the historical seismological data but also on the specific tectonic conditions of the territory of the Republic of Macedonia, as well as on the analysis of recent seismic activity.

The results of this research will impact decision making on acceptable seismic risk levels for the safety of the AP, FF and AF sites.

## References

Jordanovski, Lj, Pekevski, L. et al., (1998): Basic characteristics of the seismicity on the territory of the Republic of Macedonia. Seismological Observatory, University "Sts Cyril and Methodius". Report 1998-01.

Mihailov V., Dojcinovski, D., (1992): Effects of uncertainty in seismic sources modeling on seismic hazard mapped parameters. Proceedings of the Workshop: Application of the artificial intelligence techniques in seismology and engineering seismology. Conseisl de L'Europe Cahiers du Centre Européen de Géodynamique et de Séismologie. Luxembourg, 1992. Vol. 6: 71-80.

Pekevski, L. (1992): An investigation of the seismicity applying the method of extremes values. Earthquake engineering, Tenth World Conference 1992. Publ. Balkema, Rotterdam: 283:286

# NONLINEAR TECHNIQUES FOR SHORT TERM PREDICTION OF THE GEOMAGNETIC FIELD AND ITS SECULAR VARIATION

ANGELO DE SANTIS[21]
ROBERTA TOZZI
*Istituto Nazionale di Geofisica e Vulcanologia*

**Abstract.** Recent studies appeared in literature on the chaotic behavior of the dynamical system producing the geomagnetic field, i.e. the geodynamo. They analyzed the secular variation as deduced from observatories annual means (Barraclough and De Santis, 1997; De Santis et al., 2002), as well as the information content of global models for the last century (De Santis et al., 2004), showing some interesting nonlinear properties. Suitable nonlinear techniques can be applied for short term prediction of the geomagnetic field, i.e. to extrapolate the field 1-2 years into the future. Using these methods it is possible to update geomagnetic field maps for navigational purposes and to improve the prediction in heliports and airports of the magnetic declination which is important for the safety and security of all operations related to landing and take-off.

**Keywords:** Nonlinear prediction; geomagnetic field; chaos; declination

## 1. Introduction

The geomagnetic field surrounding the Earth protects us from most of the outer space radiation. With its space and time variations it reveals many features of the dynamics of the outer terrestrial core, where the field is generated by means of the electric currents produced by the fluid convection of conductive iron alloys, a process called the geodynamo mechanism. Compasses provide the simplest way to know orientation in the

[21] To whom correspondence should be addressed at: Istituto Nazionale di Geofisica e Vulcanologia, V. Vigna Murata 605, 00143 Rome, Italy. Email: desantis@ingv.it



*J.L. Rasson and T. Delipetrov (eds.), Geomagnetics for Aeronautical Safety,* 281–289.

Earth reference frame. Exact knowledge of the compass pointing requires periodic monitoring of the magnetic declination, i.e. the angle between the direction of the magnetic field and true North. To this purpose national networks of magnetic stations are maintained and repeated every 3-5 years to collect magnetic measurements. Also, at airports and heliports the magnetic declination is measured periodically with particular attention for compass calibration.

Accurate measurements of the geomagnetic field are fundamental for the above mentioned reasons. But because of the unpredictable year by year change of the field, the secular variation becomes outdated as soon as a map is produced. To use the compiled maps some sort of short term prediction is needed, so prediction is not an option but a necessity. Usually prediction is simply linear extrapolation, that takes into account the mean change of the field over a certain period of time and then assumes the same change for the future. Commonly this linear prediction provides significant deviations from the real values, even after only one or two years. For this reason, a repetition of the magnetic measurement becomes necessary. Clearly, improving the prediction would allow prolonging the time interval between series of measurements.

The aim of this paper is to introduce and apply a new technique that should in principle improve the prediction results.

Some nonlinear techniques have been applied to magnetic data to find possible chaos or fractality of the geomagnetic field (Barraclough and De Santis, 1997; De Santis et al., 2002) with satisfying results. Thus, the idea here is to apply the same techniques to predict the geomagnetic field. After some generalities on the results obtained recently in terms of nonlinear features of the geomagnetic field, a nonlinear technique called the nonlinear forecasting approach (NFA) will be described and applied to make reasonable short term (1-2 years) predictions.

The NFA's most important points and possible future applications will be assessed as well.

## 2. Nonlinear features of the geomagnetic field

When a phenomenon shows that small or great changes of some initial conditions correspond to small or great changes of its evolution, respectively, its dynamics are said to be linear. Conversely, when small changes of some initial conditions involve unpredictable great changes in the future evolution, the dynamics are said to be nonlinear and this phenomenon is called sensitivity to initial conditions. In other words, in the latter case there is a nonlinear relation between the input (changes of initial conditions) and the output (future values of the signal under study). If this relation can be written in exponential form with positive exponent, the dynamics and the corresponding system are said to be chaotic. The chaotic nature of the geomagnetic field reflects the chaoticity of the system generating it, that is the chaoticity of the geodynamo.

In the recent years we have tried to find evidence of the chaoticity of the geomagnetic field. The vector power spectrum of the observatory annual means shows an almost power-law (linear) behaviour in the (log-log) plot for periods ranging from 6 years to around a century (De Santis et al., 2003). This can be explained as a consequence of the chaotic state of the magnetic field because when a phenomenon is chaotic it usually shows scaling spatial and temporal spectra with defined spectral exponents. Starting with simple assumptions about the spatial power spectra of the geomagnetic field and its secular variation, it was possible to predict the temporal power law spectrum with a specific scaling.

More recently (De Santis et al., 2004) some statistical concepts related to the Information Theory, such as the *information content* $I(t)$, have been applied to the last century (years 1900-2000) of IGRF (International Geomagnetic Reference Field) global models. $I(t)$ is a negative quantity which measures the knowledge of the state of the system when knowing only the probability distribution of all the possible states of the system. When a system is chaotic, $I(t)$ decreases linearly in time and the inverse of the (negative) slope defines a characteristic time of the dynamical system after which it is not possible to make any prediction at all. Linear plots with characteristic times of around 850 and 420 years were found when applying this concept to the geomagnetic field and its secular variation, respectively. From the application of L'Hôpital theorem to the definition of the probability used for the information content, the rough agreement of the two characteristic times was interpreted as a possible symptom of an impending geomagnetic reversal or excursion. The chaotic state of the geomagnetic field has then been considered a manifestation of this possible change of state of the field (De Santis et al., 2004).

The possible chaotic state of the geomagnetic field is also supported by the fractal magnetic potential at the core-mantle boundary as deduced from global models from around 1600 to present (De Santis and Barraclough, 1997).

The above considerations and results are indicative of nonlinear, possibly chaotic dynamics, of the geomagnetic field and suggest that a nonlinear technique is probably more reliable for making predictions than linear techniques.

In the following section, one of these nonlinear techniques will be introduced and some preliminary results will be shown.

## 3. Nonlinear forecasting

The application of nonlinear forecasting here described originates from some recent results supporting the idea that the geomagnetic field secular variation seems to be the result of a dynamical system (the fluid outer core of the Earth) possibly characterized by a chaotic behavior. Barraclough and De Santis (1997 and De Santis et al. (2002) apply a nonlinear forecasting technique (Sugihara and May, 1990) to discriminate deterministic chaos from randomness and periodicity in geomagnetic time series. These time series consist of the secular variation of the Cartesian components (X, Y and Z) of the Earth's magnetic field estimated as the first-differences from observatory annual means.

With random signals, the ability to predict future values is small and independent of the prediction interval, i.e. how far into the future the prediction is made. Periodic signals are characterized by high predictability and are independent of the prediction interval. With chaotic signals, predictability deteriorates as the prediction interval increases: it is good for short time predictions, but rapidly approaches zero after a certain characteristic time (related to the specific dynamics of the system generating the signal under study; Sugihara and May, 1990). The first step in nonlinear forecasting is to reconstruct the phase space starting with the time series and applying the Takens theorem (Takens, 1981). According to this theorem, the dynamics on each $n$-th axis of the space can be represented by the time series itself if shifted by ($n$-1) times a proper delay, $\tau$. The second step is to evaluate the so-called Largest Lyapunov exponent of the chaotic system which is related to the way the prediction ability deteriorates by increasing the prediction interval (Wales, 1991). In fact, for a chaotic system, two initially close orbits in the phase space diverge along a certain axis as $e^{\lambda t}$, where $t$ is time and $\lambda$ the so-called Lyapunov exponent associated with that axis. A three-dimensional dynamical system has three Lyapunov exponents, and if the largest exponent is positive, we say that the system is chaotic, because there is the tendency for the orbits to diverge at least in one direction of the phase space. For the Eastward component, Y, of the geomagnetic field the found largest Lyapunov exponent was around 0.2 year$^{-1}$ corresponding to a characteristic time of around 5 years, after which no reasonable prediction can be made. This value supports the practice of updating the International Geomagnetic Reference Field (IGRF) every 5 years.

This paper considers the nonlinear forecasting approach suggested by Fowler and Roach (1993). This approach allows us to determine the predicted value at a certain time $t$, termed the *predictee*, by comparing the ($t$-1) value with all past values. In fact, looking for the past numerical value, say the $n$-value, closest (thus termed the *similar*) to the ($t$-1) value, the *predictor* will be identified as the next value to the similar, i.e. the $n$+1 value. More generally in an $E$-dimensional phase space, the last known value (an $E$-dimensional point) is compared with all the other known values to find the $E$+1 nearest points (similar points) that are, therefore, characterized by the shortest distances from the last known point. Finally, the forecast is made by controlling the time evolution of these similar points after they have been inverse squared weighted.

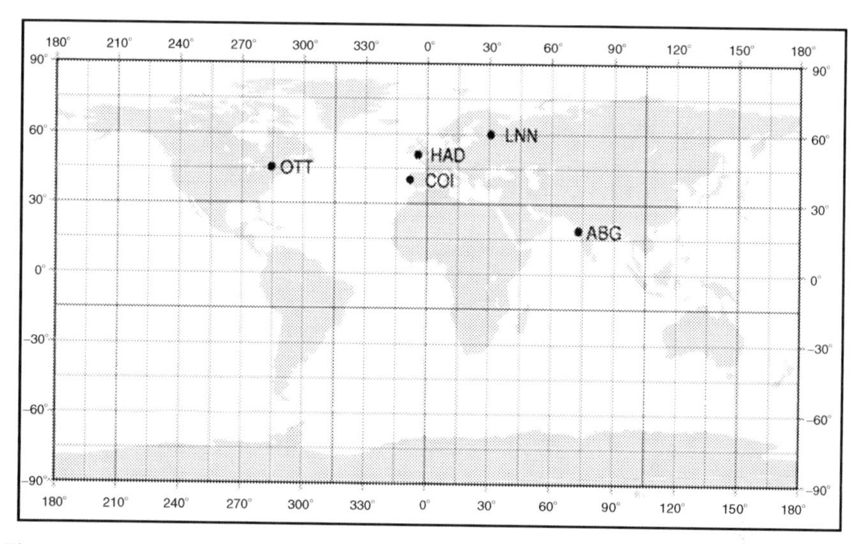


Figure 142. Geographical distribution of the five selected observatories indicated by their IAGA codes.

The number $n$ of axes necessary to reconstruct the dynamics represents the degrees of freedom of the system, and $n$+1=$E$ is also said the *embedding dimension* of the dynamics. For the secular variation of the geomagnetic field, a three-dimensional space ($n$=3 and $E$=4) is quite enough to get all the topological structure of the ideal phase space (Barraclough and De Santis, 1997). The appropriate delay $\tau$ can be estimated as the time when the autocorrelation function of the signal is close to zero. For observatory annual means, this value is about 1 year, that is the sampling itself of the time series.

Analyzed data come from five selected geomagnetic observatory time series of the Y component secular variation, whose geographical distribution is shown in Figure 142. The forecasting technique previously described was applied by averaging just the four points of the phase space closest to the most recent value in the time series. This technique was well able to predict the secular variation of the geomagnetic field 1-2 years into the future. In principle, a longer prediction interval would not be reliable because the ability of forecasting approaches zero after 5-6 years. Figure 143 shows the prediction (grey bold line) of the secular variation for the selected observatories together with the real values (black thin line). Since the technique is particularly suitable for prediction of the magnetic component Y, and therefore presumably, for the magnetic declination D, it can be used to update declination values at specific places, in particular at airports and heliports, where accurate measurements are critical to the safety of aircraft operations.

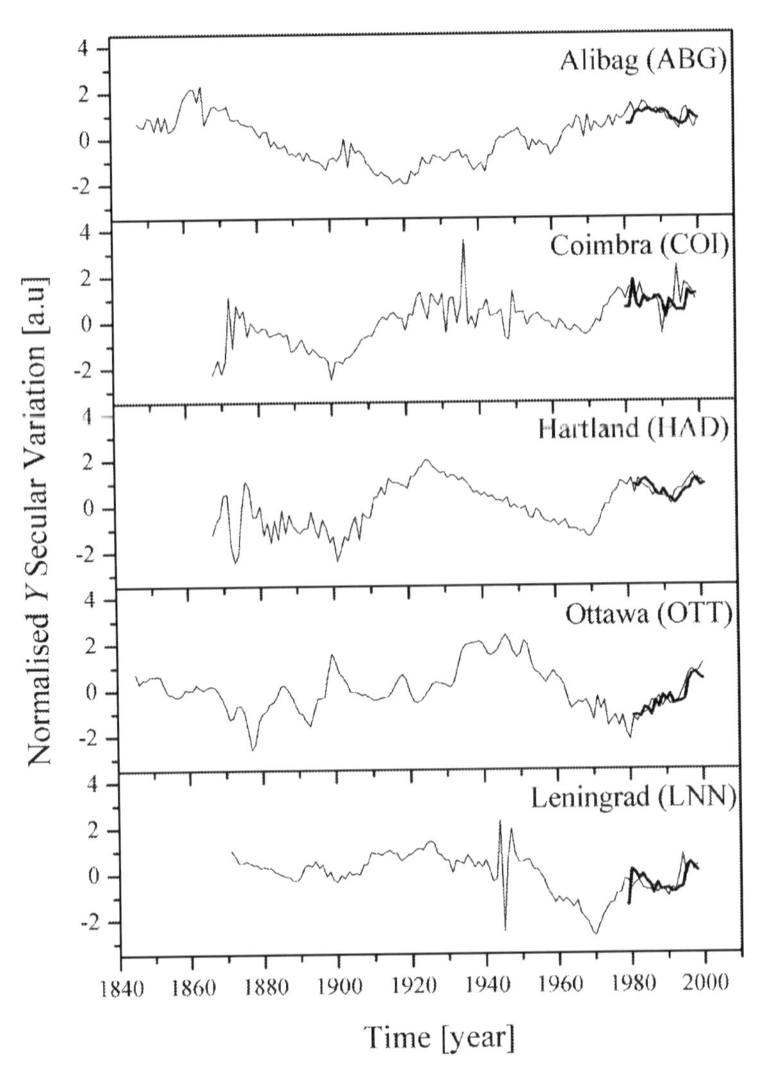


Figure 143. Normalized secular variation of the Eastward component of the geomagnetic field in arbitrary units: first-differences estimated from observatory annual means data (black thin line); secular variation predicted by means of nonlinear forecasting approach (grey bold line).

## 4. Conclusions

In this paper some nonlinear features of the geomagnetic field have been described. These, in turn, justify the application of a nonlinear technique to make reliable short-term predictions (1-2 years) of secular variation. Both the small number of time series to which the technique was applied, and the simple prediction scheme applied to the phase space, gave encouraging results and warranted further investigation. The application to a greater number of time series would be useful to search for a better scheme of phase space interpolation that would improve the final prediction. This kind of short term technique is potentially applicable to forecast future (1-2 years) values of the magnetic field elements at observatories. This would allow a better extrapolation of the geomagnetic secular variation of global models or at repeat stations. Therefore it would improve the regional maps of the geomagnetic field, in particular those of magnetic declination, which are so useful for navigation. Another application would be to make short term prediction of the declination at heliports and airports, where it is so important for the safety and security in all related operations.

## Acknowledgements

We would like to thank the organizers of the NATO ADVANCED RESEARCH WORKSHOP, held in Ohrid, Macedonia 18-20 May, 2005, in particular, Dr. Jean Rasson, for inviting us to present this paper at the workshop. Also, the discussions occurring during the meeting have been very stimulating for the preparation of the paper and for focusing on some aspects more than others. Part of this work has been supported by the REM project of the National Program of Research in Antarctica (PNRA).

## References

Barraclough, D.R., De Santis, A., 1997. Some possible evidence for a chaotic geomagnetic field from observational data. *Phys. Earth Planet. Inter.*, **99**, 207-220.

De Santis, A., Barraclough, D.R., 1997. A fractal interpretation of the topography of the scalar geomagnetic scalar potential at the core-mantle boundary, *Pure and Applied Geophys.*, **149**, No.4, 747-760.

De Santis, A., Barraclough, D.R., Tozzi, R., 2002. Nonlinear variability in the geomagnetic secular variation of the last 150 years, *Fractals*, **10**, 3, 297-303.

De Santis, A., Barraclough, D.R., Tozzi, R., 2003. Spectral and temporal spectra of the geomagnetic field and their scaling properties, *Phys. Earth Planet. Int.*, **135**, 125-134.

De Santis, A., Tozzi, R., Gaya-Piqué, L., 2004. Information Content and K-entropy of the Present Earth Magnetic Field, *Earth and Planetary Science Letters*, **218**, 269-275.

Fowler, A.D., Roach, D.E., 1993. Dimensionality analysis of time-series data: nonlinear methods. *Comput. Geosc.*, **19**, 41-52.

Sugihara, G., May, R.M., 1990. Nonlinear forecasting as a way of distinguishing chaos from measurement error in time series. *Nature*, **344**, 734-741.

Takens, F., 1981. Detecting strange attractors in turbulence. In: *Lecture notes in mathematics*, **898**. Ed.s: D.A. Rand, L.S. Young. Springer, Berlin, pp. 366.

Wales, D.J., 1991. Calculating the rate of loss information from chaotic time series by forecasting. *Nature*, **350**, 485-488.

## DISCUSSION

Question (Jean Rasson): Can you give the exact nature of the axes in the 3D phase space you use to establish your prediction?
Answer (Angelo De Santis): In general, a phase space is composed of a certain number of axes, each axis is a generalized coordinate characterizing the dynamics of the system we are studying and the number of axes corresponds to the degrees of freedom of the represented dynamic system. If the dynamics of the system is characterized by n independent differential equations of n unknowns, then n is the degrees of freedom and at each axis we can place the variation of each unknown. When the differential equations of the dynamics are not known, following to Takens' theorem (1981), from a signal f(t) it is possible to reconstruct the phase space placing at each axis f(t), f(t+τ), f(t+2τ) ..., with τ an appropriate delay time, usually corresponding to the first zero of the autocorrelation function of f(t). Each state of the dynamics of the system will visit one and only one site in the reconstructed phase space, and the topology of the 'shape' reconstructed by all orbits is specific of that system only, so that its study allows in principle to extract much information about the properties of the system and its dynamics.
What was said above is strictly valid when the signal characterizes a chaotic system. Necessary ingredients for a system to be chaotic are determinism, nonlinear differential equations of the dynamics, and initial condition sensitivity.
Question (Jürgen Matzka): How do data gaps affect NFA and bicoherence? Can the past be predicted (before the observatory was established)?
Answer (Angelo De Santis): Gaps have little effect on the NFA, since if it is the 'topology' that we are interested (to extract information such as degrees of freedom, divergence of orbits in the phase space, etc. or to infer some prediction) small gaps do not necessary change the gross properties of the phase-space in that sense. Also Bicoherence could be little affected, if we use some specific scheme of Fourier Transformation for irregularly distributed data, although it is probably more sensitive to gaps than the former technique. Of course, for obtaining positive results, in both cases gaps must be the exception in the time series and not the rule.

Regarding the second part of the question, in principle the answer is 'yes', however, and this is counterintuitive, our ability to predict the past of chaotic phenomena is worse than the ability to predict their future! This can be explained by the fact that total divergence of the orbits is larger when going back in time.

Question (Sanja Panovska): What is the semi-angle of SCHA for normal field for the territory of the Republic of Macedonia (within $\Delta\varphi=1^{\circ}31'$ and $\Delta\lambda=2^{\circ}35$ interval)

Answer (Angelo De Santis): In theory in your case a cap with half angle of around 1.5 degrees should be considered. However, in practice, such a small cap would imply basis functions having very high degrees $n_k$, entailing great difficulties in their computations. In my opinion, in the case of the Republic of Macedonia, it would be easier to apply some other technique for representing the geomagnetic field, for instance the rectangular harmonic analysis (Alldredge, 1981).

Question (Sanja Panovska): I know the theory for SCHA but I don't know how to put the temporal factor in equation (if I have data from 2003 which value for <t> to use)?

Answer (Angelo De Santis): For data distributed in a short time as one year, I think you could consider just a linear time behaviour of the field, therefore t=1.

# SPHERICAL CAP HARMONIC ANALYSIS OF THE GEOMAGNETIC FIELD WITH APPLICATION FOR AERONAUTICAL MAPPING

J. MIQUEL TORTA[22]
*Observatori de l'Ebre*

LUIS R. GAYA-PIQUÉ

ANGELO DE SANTIS
*Istituto Nazionale di Geofisica e Vulcanologia*

**Abstract.** The Spherical Cap Harmonic Analysis (SCHA) is a regional modeling technique based on appropriate functions which are solutions of Laplace's equation over a constrained, cap-like region of the Earth. The concept was introduced in 1985 in the context of geomagnetism as a local or regional extension of the classic global spherical harmonic analysis. Starting from the basic principles in which the analysis method is founded, this paper describes the latest applications for the modeling of the main magnetic field and its secular variation. Although examples of applications over small areas will be given, it will be shown that, in general, the bigger the region the more appropriate the technique. Therefore, this paper focuses on the results and perspectives over continental areas, like Antarctica or Europe. The possible application to the derivation of isogonic charts for navigational purposes with suitable time predictions will be emphasized. At the same time, the limitations of the method will be examined. Although recent revisions of the technique seem to solve some of the problems, our present research focuses on the quest for solutions to the still unanswered questions.

**Keywords:** SCHA; geomagnetic field modeling; spherical harmonics; declination

[22] To whom correspondence should be addressed at: Observatori de l'Ebre, Horta Alta 38, 43520 Roquetes, Spain. Email: jmtorta@obsebre.es



*J.L. Rasson and T. Delipetrov (eds.), Geomagnetics for Aeronautical Safety,* 291–307.

## 1. Introduction

The analytical representation of the geomagnetic field has been a topic of research for a long time, either for the aspects related with the definition of the origin of the field itself or for its variations. There is also great scientific interest in the phenomena that produce geomagnetic variations, like ionospheric and magnetospheric current systems, and those induced by them in the Earth's interior; along with the Earth's self-sustained dynamo, which is the origin of the main field and its secular variation. Modeling the sources that originate the crustal contribution, caused by differential magnetization of the rocks in the Earth's crust, are found in the same way.

From the beginning, such representations have been oriented towards global modeling or they have tended toward the representation of the phenomenon over a particular portion of the Earth's surface, because of a special interest in its study, or because of a denser distribution of the measurements in a particular region. A regional analysis tends to represent the field with better resolution, which is often a great advantage; but the mathematical algorithms that serve as a basis for such representations suffer frequently from restrictive constraints or impossible convergences. So, the algorithms have traditionally been better solved in the global case, given the quasi-spherical geometry of the Earth.

The Spherical Harmonic Analysis technique, introduced by Gauss in 1839, has resulted, by far, in the most popular method for modeling the main field and its secular variation at the global scale. Starting from Maxwell's equations, applied over the Earth's surface, it can be accepted with a good approximation that we are free from electric currents, so that the curl and the divergence of the field are null. The field can then be represented as the negative gradient of a magnetic potential $V$:

$$\mathbf{B} = -\nabla V \tag{1}$$

and such potential must then satisfy Laplace's equation:

$$\nabla^2 V = 0 \tag{2}$$

A solution for this equation in spherical coordinates may be obtained by the method of separation of variables (radial distance $r$ from the Earth's center, colatitude $\theta$ and longitude $\phi$) given as $V(r, \theta\ , \phi) = U(r)P(\theta)Q(\phi)$. Therefore, the problem is reduced to finding the solutions for these 3 differential equations, which depend on each of the variables:

$$\frac{d^2U}{dr^2} - \frac{n(n+1)}{r}U = 0 \tag{3}$$

$$\frac{1}{\sin\theta}\frac{d}{d\theta}\left(\sin\theta\frac{dP}{d\theta}\right)+\left[n(n+1)-\frac{m^2}{\sin^2\theta}\right]P=0 \tag{4}$$

$$\frac{1}{Q}\frac{d^2Q}{d\phi^2}=-m^2 \tag{5}$$

By adequately choosing the boundary conditions for the earth's sphere, the general solution of Laplace's equation can be expressed as a superposition of potential functions of this type:

$$\begin{aligned} V\,(r,\theta,\phi) &= a\sum_{n=1}^{Ni}\sum_{m=0}^{n}\left(\frac{a}{r}\right)^{n+1}\left\{g_n^{m,i}\cos m\phi+h_n^{m,i}\sin m\phi\right\}P_n^m(\cos\theta)+ \\ &+a\sum_{n=1}^{Ne}\sum_{m=0}^{n}\left(\frac{r}{a}\right)^{n}\left\{g_n^{m,e}\cos m\phi+h_n^{m,e}\sin m\phi\right\}P_n^m(\cos\theta) \end{aligned} \tag{6}$$

In this way, the represented potential consists of two parts: one produced by sources located within a sphere of radius $a$, and another by sources located outside this volume. The $P$ functions are the associated Legendre functions of first kind of degree $n$ and order $m$, which are integer parameters for this solution.

The product of the Legendre functions with the trigonometric functions in longitude forms the series of two-dimensional spherical harmonics. The $g$ and $h$ are the spherical harmonic coefficients, or Gauss coefficients. The general solution results in an infinite series of terms. In practice, it is truncated at finite indices $N_i$ and $N_e$.

The potential $V$, however, is not observable. According to equation (1), the cartesian components of the geomagnetic field are obtained as the partial derivatives of $V$ with respect to $r$, $\theta$, and $\phi$:

$$X\equiv -B_\theta=\frac{1}{r}\frac{\partial V}{\partial\theta} \tag{7}$$

$$Y\equiv B_\phi=\left(-\frac{1}{r\sin\theta}\right)\frac{\partial V}{\partial\phi} \tag{8}$$

$$Z\equiv -B_r=\frac{\partial V}{\partial r} \tag{9}$$

The most popular example of a global model for the main field (only internal long wavelength coefficients) is that known as the International Geomagnetic Reference Field (IGRF). The last up-to-date version of IGRF, known as the IGRF 10$^{\text{th}}$ generation (IAGA, 2005), includes models of the main field from 1900 to 2005 and a secular variation model for 2005-2010. The value of $N_i$ for the fixed field is equal to 10 (that is, 120 coefficients) for all models prior to 1995, while for years 2000 and 2005 it includes coefficients up to degree $n$=13 (i.e., 195 coefficients). The secular variation model is expanded up to degree $n$=8.

The most ambitious recent effort to model fields, not only from the Earth's core but also from the lithosphere, the quiet-day ionospheric sources and the magnetosphere, along with the associated induced currents, and interhemispheric field-aligned currents, is known as the comprehensive model of the near-Earth magnetic field (Sabaka et al., 2004). It includes about 2,000,000 data entries from the POGO, MAGSAT, Ørsted, and CHAMP satellites (for which the ionospheric fields are internal), and magnetic observatory hourly and annual means from 1960 to 2000, resulting in more than 25,000 parameters.

## 2. Regional techniques

When geomagnetic observations are known only over a small portion of the Earth's surface or the analysis is only required over a particular area, the above functions for the spherical analysis are not the most appropriate anymore. The different techniques for obtaining regional models can be subdivided into graphical and analytical (Haines, 1990).

The oldest models of the geomagnetic field, for which the maps were drawn by hand, and those which have used algorithms to generate uniform grids from non-uniformly distributed data by numerical interpolation were derived graphically.

The simplest analytical method uses a polynomial expression in latitude and longitude (e.g. De Santis et al., 2003). However, this technique, as with graphical methods, does not account for altitude variations, permits the possibility of geometrical inconsistencies, and does not guarantee the conditions imposed by the electromagnetic theory which requires that in regions free from magnetic sources and electric currents, the magnetic potential satisfies Laplace's equation.

Another procedure sometimes used consists in the application of spherical analysis to data in a restricted region. However this can generate numerical instabilities in the determination of the coefficients because the functions are not orthogonal over the limited area in which the analysis is developed.

So, instead of using basis functions which are orthogonal over the whole sphere, it is more natural to use appropriate functions for such regions. Two techniques employed for smaller regions on the globe are the Rectangular

Harmonic Analysis (RHA) and the Spherical cap Harmonic Analysis (SCHA).

In the RHA, the general solution for Laplace's equation is given by an expansion in terms of the ordinary Cartesian or rectangular coordinates, with the origin usually taken at the centre of the region where the data are located:

$$\begin{aligned} V = Ax + By + Cz + \sum_{m=1}^{M_0} \{ a_0^m \cos(mx) + b_0^m \sin(mx)\} \exp\{-k_x mz\} + \\ + \sum_{n=1}^{N_0} \{ a_n^0 \cos(ny) + c_n^0 \sin(ny)\} \exp\{-k_y nz\} + \\ + \sum_{m=1}^{M} \sum_{n=1}^{N} \{ a_n^m \cos(mx)\cos(ny) + b_n^m \sin(mx)\cos(ny) + \\ + c_n^m \cos(mx)\sin(ny) + d_n^m \sin(mx)\sin(ny)\} \exp\left\{- \sqrt{(k_x m)^2 + (k_y n)^2} z\right\} \end{aligned} \tag{10}$$

where $k_x=2\pi/L_x$ and $k_y=2\pi/L_y$, where $L_x$ and $L_y$, are the dimensions of the rectangular region in the $x$ and $y$ directions, respectively. In this way the dimensions in the horizontal coordinates are normalized to $2\pi$.

Although the components of the geomagnetic field obtained in this way are really derived from a potential that satisfies Laplace's equation, so that they suppose an analytical solution to the problem, the expansion does not converge uniformly over its interval of validity, but it is convergent only in mean square. This is because the functions used as a basis are periodic within such an interval; meanwhile the potential expanded in terms of such functions is not. In this way the termwise derivatives with respect to $x$ or $y$ are divergent. The effect can be appreciated by the exhibition of some ringing at the boundaries.

On the other hand, the terms $Ax$, $By$, and $Cz$ violate the boundary conditions for a potential only due to internal sources, which impose that it must be zero when $z$ tends to infinity. Their presence is explained by the fact that they tend to compensate the mentioned ringing, as well as the problems that appear (especially when the area is large) by the rectangular approximation of the spherical geometry.

## 3. Spherical Cap Harmonic Analysis

The Spherical Cap harmonic Analysis, or SCHA, developed by Haines (Haines, 1985), does not have the above mentioned problems and its basis functions give a convergent expansion both for the potential and for any of its derivatives.

In this case, when solving Laplace's equation, the boundary conditions are the same as those in the spherical case, except those in $\theta$ at the cap boundary. For a spherical cap the potential $V$ at $\theta_0$ and its derivative with respect to $\theta$ must satisfy the following boundary conditions, where $f$ and $g$ are arbitrary functions:

$$V(r,\theta_0,\phi) = f(r,\phi) \tag{11}$$

$$\frac{\partial V(r,\theta_0,\phi)}{\partial \theta} = g(r,\phi) \tag{12}$$

It has been demonstrated that the first condition is satisfied by choosing those values of $n$ such that the derivative of the potential with respect to the colatitude is zero:

$$\frac{dP_{n_k}^m(\cos\theta_0)}{d\theta} = 0 \tag{13}$$

Meanwhile, the second is satisfied for other values of $n$ such that the potential itself is zero:

$$P_{n_k}^m(\cos\theta_0) = 0 \tag{14}$$

These boundary conditions are satisfied by the associated Legendre functions with again a real, but not necessarily integer, degree.

Since the different real values of $n$ depend on $m$, they are described by $n_k(m)$, where $k$ is an integer index chosen to order the different roots $n$ for a given $m$. Thus defined, the $n_k(m)$ for which $k$-$m$ *are* even are the roots of equation (13), and those for which $k$-$m$ are odd are the roots of equation (14), when these equations are considered as equations in $n$.

By superposition and assuming the finite expansion approximation, the general solution of Laplace's equation for the spherical cap is:

$$\begin{aligned} V = a \sum_{k=0}^{KINT} \sum_{m=0}^{k} \left(\frac{a}{r}\right)^{n_k(m)+1} P_{n_k(m)}^m(\cos\theta) \sum_{q=0}^{QINT} \left\{ g_{k,q}^{m,i} \cos(m\phi) + h_{k,q}^{m,i} \sin(m\phi) \right\} t^q + \\ + a \sum_{k=1}^{KEXT} \sum_{m=0}^{k} \left(\frac{r}{a}\right)^{n_k(m)} P_{n_k(m)}^m(\cos\theta) \sum_{q=0}^{QEXT} \left\{ g_{k,q}^{m,e} \cos(m\phi) + h_{k,q}^{m,e} \sin(m\phi) \right\} t^q \end{aligned} \tag{15}$$

As it can be seen, equation (15) includes the possibility of adding a polynomial temporal dependence for the potential. Here the 2-D functions given by the product of the Legendre functions in colatitude with the trigonometric functions in longitude are called spherical cap harmonics, in analogy with the spherical harmonics in the global context.

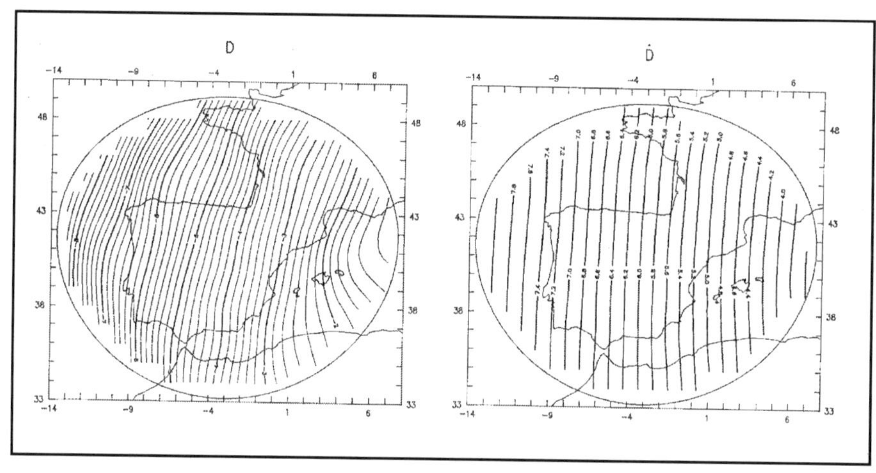

Figure 144. Contour maps of magnetic declination (left) and its annual change (right) for 1990 from a reference model for Spain (Torta et al., 1993).

The spatial wavelength over the Earth's surface represented by the model, for a harmonic of degree $n_k$, is simply given by the quotient between the Earth's perimeter and the harmonic degree. Table 29 shows the maximum (when $K$ is equal to 4) and minimum degrees associated with an SCHA performed over different cap sizes, and the corresponding wavelengths involved. So, for a 16° cap, which is roughly the size of Europe, the harmonics start at $n$ equals 6.1 and quickly reaches almost 25. When looking at the main field and its secular variation, which are characterized by degrees ranging from 1 to 10 or 12, the use of big caps is necessary; otherwise unrealistic detail is obtained in the maps. This did happen in a first attempt to apply the technique to the secular variation over Spain and adjacent areas (Torta et al., 1992), where, even though the data was kept over the original 16° cap, the model became more and more realistic as the size of the cap was increased. The boundary conditions (13) and (14) of the spherical cap harmonics were then defined in a realistic way at the border of the cap.

A similar procedure (Duka et al, 2004) was recently used for Albania and Southern Italy with data only restricted to a 3° cap but with the real area enlarged to a cap of 8°. Since this cap is still very small, the authors limited the expansion to a $K$ equal to 2.

With the model coefficients, it is possible to obtain maps for any of the magnetic elements and for any epoch within the interval of validity of the model; for instance, for the magnetic declination, the relevant element for aeronautical navigation. Figure 144 shows an example of the magnetic declination obtained for the Iberian Peninsula for 1990, in degrees East, and the annual change, in minutes per year, for the same epoch.

Table 29. Maximum and minimum degrees and wavelengths (in Km) associated with spherical cap harmonic analyses over different cap sizes.

| $\theta_0$ | $n_{max}$ (K=4) | $w_{min}$ | $n_{min}$ (K=4) | $w_{max}$ |
|---|---|---|---|---|
| 16° | 24.6277 | 1625 | 6.1481 | 6511 |
| 20° | 19.6044 | 2042 | 4.8432 | 8265 |
| 25° | 15.5864 | 2568 | 3.8056 | 10519 |
| 30° | 12.9083 | 3101 | 3.1196 | 12832 |
| 35° | 10.9958 | 3641 | 2.6347 | 15194 |
| 40° | 9.5619 | 4187 | 2.2754 | 17593 |
| 45° | 8.4471 | 4738 | 2.0000 | 20016 |

For all the above mentioned reasons the use of the SCHA technique has been growing among the geomagnetic field modeling community. Table 30 is a compendium of all the up-to-date English-written published papers and reports known to us about the technique and its applications.

Table 30. List of English language papers related to the SCHA as of May 2005.

| ABOUT THE TECHNIQUE |
|---|
| Haines, J. Geophys. Res. 90, 2583, 1985 |
| Haines, HHI-Rep. 21, 27, 1987 |
| Haines, Comput. Geosc. 14, 413, 1988 |
| Haines, J. Geomag. Geoelectr. 42, 1001, 1990 |
| Haines, Phys. Earth Planet. Inter. 65, 231, 1991 |
| Torta et al., Phys. Earth Planet. Inter. 74, 209, 1992 |
| Haines, Geophys. J. Int. 114, 490, 1993 |
| De Santis & Falcone, Proc. II Hot. Marus. Symp. 1994 |
| Torta & De Santis, Geophys. J. Int. 127, 441, 1996 |
| De Santis et al., Phys. Earth Planet. Inter. 97, 15, 1996 |
| De Santis et al., J. Geomag. Geoelectr. 49, 359, 1997 |
| De Santis & Torta, J. Geodesy 71, 526, 1997 |
| De Santis et al., Annali di Geofisica 40, 1161, 1997 |
| Lowes, Geophys. J. Int. 136, 781, 1999 |
| De Santis et al., Phys. Chem. Earth A 24, 935, 1999 |
| Düzgit & Malin, Geophys. J. Int. 141, 829, 2000 |
| Korte & Holme, Geophys. J. Int. 153, 253, 2003 |
| Thébault et al., Geophys. J. Int. 159, 83-103, 2004 |
| VARIATIONS OF THE TECHNIQUE |
| De Santis, Geophys. J. Int. 106, 253. 1991 |
| De Santis, Geophys. Res. Lett. 19, 1065, 1992 |
| REFERENCE FIELD MODELS |

| Haines & Newitt, J. Geomag. Geoelectr. 38, 895, 1986 |
|---|
| Nevanlinna et al., Deut. Hydro. Zeits. 41, 177, 1988 |
| Newitt & Haines, J. Geomagn. Geoelectr. 41, 249, 1989 |
| De Santis et al., J. Geomagn. Geoelectr. 42, 1019, 1990 |
| Newitt & Haines, Curr. Res. E, G.S.C., 275, 1991 |
| Nevanlinna & Rynö, HHI Rep. 22, 106, 1991 |
| Torta et al., J. Geomag. Geoelectr. 45, 573, 1993 |
| An et al., J. Geomag. Geoelectr. 46, 789, 1994 |
| An et al., Geomagn. Aeron. 34, 581, 1995 |
| Haines & Newitt, J. Geomag. Geoelectr. 49, 317, 1997 |
| Kotzé & Barraclough, J. Geomag. Geoelectr. 49, 452, 1997 |
| Chiappini et al., Phys. Chem. Earth 24, N.5, 433, 1999 |
| Kotzé, Earth Planets Space 53, 357, 2001 |
| De Santis et al., Geophys. Res. Lett. 29, N. 8, 33-1, 2002 |
| Gaya-Piqué, PhD Thesis, URL, 162 pp., 2004 |
| Duka et al., Ann. Geophys. 47, 1609-1615, 2004 |
| Gaya-Piqué et al., Earth Obs. with CHAMP. Berlin: Springer, 317-322, 2005 |
| SECULAR VARIATION MODELS |
| Haines, J. Geophys. Res. 90, 12563, 1985 |
| García et al., Phys. Earth Planet. Inter. 68, 65, 1991 |
| Miranda et al., J. Geomag. Geoelectr. 49, 373, 1997 |
| Torta et al., Tectonophysics 347, 179, 2002 |
| DETERMINATION OF THE NORTH MAGNETIC POLE |
| Newitt & Niblett, Can. J. Earth Sci. 23, 1062, 1986 |
| Newitt & Barton, J. Geomag. Geoelectr. 48, 221, 1996 |
| Newitt et al., EOS 83, 381, 2002 |
| ANOMALY FIELD MODELS |
| Haines, J. Geophys. Res. 90, 2593, 1985 |
| De Santis et al., NATO ASI Series C 261, 1, 1989 |
| Torta et al., Cahi. Cent. Eur. Geod. Seis. 4, 179, 1991 |
| An et al., J. Geomag. Geoelectr. 44, 243, 1992 |
| Duka, Annali di Geofisica 41, 49, 1998 |
| Zhen-chang et al., Chin. J. Geophys. 41, 42, 1998 |
| Rotanova & Odintsov, Phys.Chem. Earth A 24, 455, 1999 |
| Rotanova et al., Acta Geophys. Pol. 48, 223, 2000 |
| Korte & Haak, Phys. Earth Planet. Inter. 122, 205, 2000 |
| Kotzé, Geophys. Res. Lett. 29, N. 15, 5-1, 2002 |
| GEOMAGNETIC VARIATIONS EXTERNAL ORIGIN |
| Walker, J. Atmos. Terr. Phys. 51, 67, 1989 |

| Newitt & Walker, J. Geomag. Geoelectr. 42, 937, 1990 |
|---|
| Haines & Torta, Geophys. J. Int. 118, 499, 1994 |
| Torta et al., J. Geophys. Res. 102, 2483, 1997 |
| Walker et al., J. Atmos. Sol. Terr. Phys. 59, 1435, 1997 |
| Amm, Ann. Geophys. 16, 413, 1998 |
| IONOSPHERIC PARAMETERS |
| De Santis et al., Ann. Geophys. 9, 401, 1991 |
| De Santis et al., Adv. Space Res. 12, N. 6, 279, 1992 |
| De Santis et al., Comput. Geosc. 20, 849, 1994 |
| El Arini et al., Proc. Int. Beacon Sat. Symp. 358, 1994 |
| Dremukhina et al., J. Atmos. Sol. Terr. Phys. 60, 1517, 1998 |
| GEODESY |
| Jiancheng et al., Manuscr. Geod. 20, 265, 1995 |
| Hwang & Chen, Geophys. J. Int. 129, 450, 1997 |

Spherical Harmonic Analysis is well founded, but the SCHA is still in the dawn of its existence, so it must be used with some precaution. For instance, one must be aware of the problems concerning external-internal separation. Torta and De Santis (1996) performed an analysis of the daily variation over a cap of 18°, corresponding to the area represented by the European continent. They showed that while the fit to the total variation for any of the points of that area is excellent, the external and internal parts of such variation are not exactly the real ones, so that the errors in the external and internal fields are equal and opposite. The situation improves substantially with a cap of 30°, and further with larger caps, as soon as the intrinsic spectral content of the phenomenon to analyze coincides with that of the models basis functions. In any case, as the real and modeled separated fields are approximately in phase, the information about the ionospheric current systems generated by the magnetic variations is still valid.

These problems appear because the external-internal separation in reality implies the comparison of separate analyses for the horizontal and vertical components (Matsushita and Campbell, 1967). And, when the region becomes small enough, one of the first things that we appreciated (García et al., 1991) is that the potential for the horizontal and radial components cannot be simultaneously exactly represented. In fact, analyzing all components at the same time provides an approximation that attempts to fit both, but it is not as precise as fitting $Z$ separately from $X$ and $Y$. In any case, this problem is not unique to the SCHA; it appears with any method that attempts to analyze fields with wavelengths much larger than the area covered by the data (Lowes, 1995, 1999).

Even though the resulting potential represents a good approximation within the cap where the data are located and in their altitude range, it irremediably diverges for the opposite pole (because of the non-integer degree of the Legendre functions). So that, it is little by little intrinsically different form the real potential out of that area, and it prejudices the vertical extrapolations as well (Figure 145).

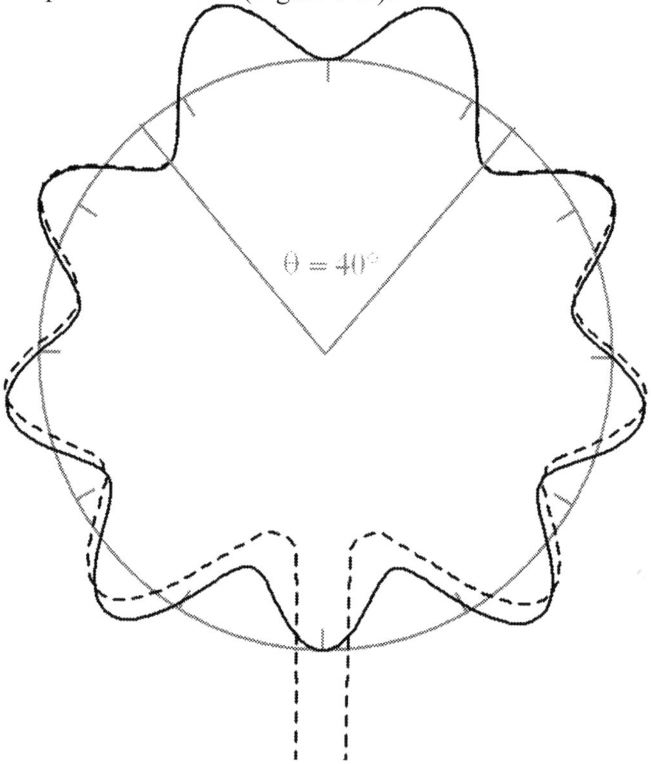

Figure 145. Solid line: associated Legendre function (n=9, $m$=2) at $\theta$=35° represented at all longitudes around a sphere. The same function has been computed on a regular grid inside a 40° spherical cap and fitted with an SCHA. Dashed line: the result of such SCHA over the same circular path around the sphere.

A revision of the technique has been recently presented (Thébault et al., 2004) in which the potential expansion is expressed as complex Legendre (conical) functions in colatitude and log-trigonometric series in longitude. The local potential $Vc$ expanded in the local basis (a spherical cap defined within $r$=a and $r$=b) is given as $Vc = V_1 + V_2$, where $V_1$ *is* the same potential as given by Haines (1985) for the even-set and $V_2$ is defined as:

$$V_2 = a \sum_{p \geq 1} \sum_{m > 0} R_p(r) \left\{ G_p^m \cos(m\phi) + H_p^m \sin(m\phi) \right\} K_p^m(\theta) + \\ + a \sum_{m > 0} R_0 \left\{ G_0^m \cos(m\phi) + H_0^m \sin(m\phi) \right\} P_0^m(\theta) \tag{16}$$

where $R_0$ is the square root of $1/(e^S-1)$ (with $S = \ln(b/a)$),
and

$$R_p(r) = \frac{(2S)^{-1/2}}{\sqrt{\left(\frac{2\pi p}{S}\right)^2 + 1}} \sqrt{\frac{a}{r}} \left\{ \frac{2\pi p}{S} \cos\left[\frac{\pi p}{S} \log\left(\frac{r}{a}\right)\right] + \sin\left[\frac{\pi p}{S} \log\left(\frac{r}{a}\right)\right] \right\} \tag{17}$$

where $p$ is an integer belonging to the imaginary part of a complex harmonic degree $n$ (see Thébault et al., 2004). $K^m{}_p$ are conical functions, i.e. Legendre functions with $n$ = complex.

The advantage of this proposal is that it provides a better fit for the radial variation, even for the $X$, $Y$, and $Z$ components, with respect to the classical SCHA. However, the internal-external separation is not made in $r$, but with respect to the cap region (i.e. internal or external to the cap).

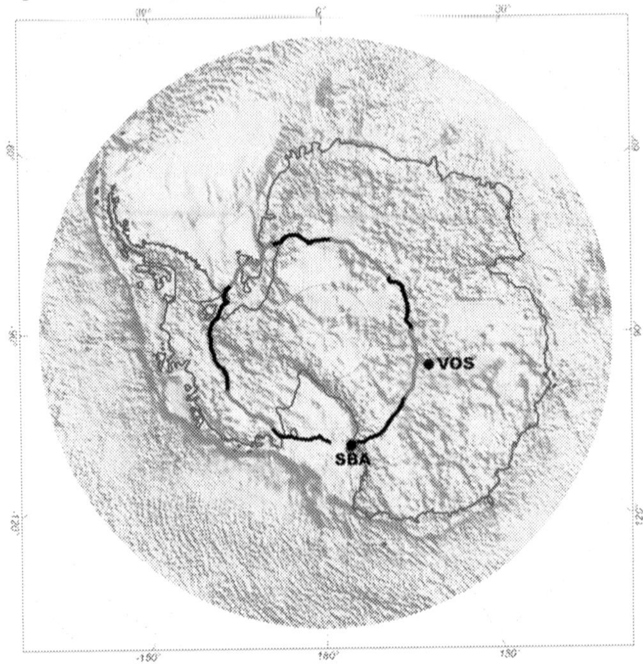


Figure 146. Path followed by the balloon (anticlockwise sense) carrying the magnetic instrumentation. Alternate colors mean different days.

## 4. Antarctic Reference Model

Assuming very carefully all the above mentioned limitations, the last application of the technique was revised (Gaya-Piqué, 2004; Gaya-Piqué *et al.*, 2006). It is called the Antarctic Reference Model (ARM) and to our knowledge it is the first full main field (i.e. main field plus its secular variation) geomagnetic model for the Antarctic. A model in this region is very important not only for scientific studies but especially for navigating with a compass there. It would be almost impossible without correct knowledge of large differences in declination typical of Antarctica.

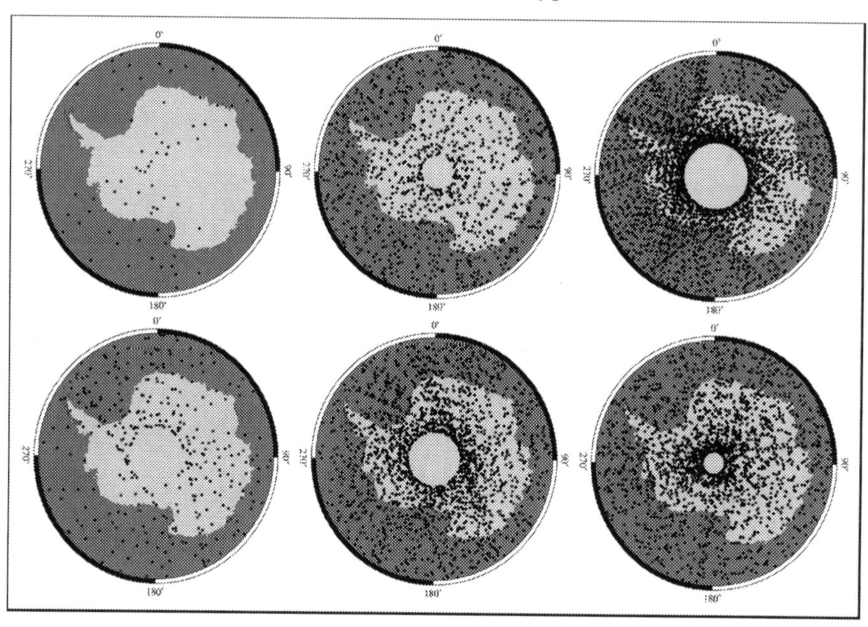

Figure 147. Spatial distribution of the satellite magnetic measurements used to develop the Antarctic Reference model. From left to right and from top to bottom: OGO-2, OGO-4, OGO-6, MAGSAT, ØRSTED and CHAMP.

The model has been developed using the most recent data sets available for the region. The annual means from 1960 on from all Antarctic magnetic observatories south of 60°S were used. However, the number and the extent of gaps in the data are important to note, because the time derivatives simply taken as the first differences tend to provide non-realistic values of the secular variation. To overcome this problem we took differences relative to a fiducial observation, in particular with respect to the mean over all data at each observatory. In this way, both the main field and the crustal anomaly are removed, obtaining for the $i^{th}$ measurement of the field at the $\nu^{th}$ observatory at epoch $t_{i\nu}$ (Haines, 1993):

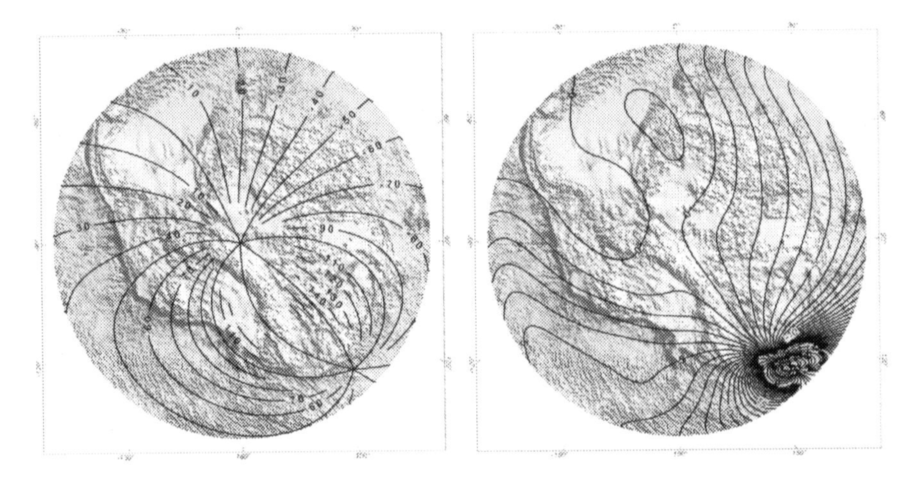

Figure 148. Contour maps of declination (left, units: degrees East) and its annual change (right, units: minutes/year) for 2005 from the Antarctic Reference model.

$$\vec{B}_{iv}(t) - \overline{\vec{B}}_{v} = \sum_{q=1}^{Q} \vec{b}_{qv}\left(t_{iv}^{q} - \overline{t_{v}^{q}}\right) + \vec{\varepsilon}_{iv} - \overline{\vec{\varepsilon}}_{v} \tag{18}$$

where $\vec{b}_{qv}$ are spatial functions (expressed as series expansions in spherical cap harmonics) evaluated at the position of the $v^{\text{th}}$ observation, and $\vec{\varepsilon}_{iv}$ is the measurement error. In equation (18) the temporal basis functions are expressed as power functions, but they can be chosen as Legendre, Fourier, or any other appropriate set of functions.

Secondly, a balloon mission was undertaken to obtain a data set of magnetic measurements from stratospheric altitude (Figure 146). Special attention was given to this mission since its magnetic measurements were used for the first time in this work.

Finally, magnetic data from six satellite missions were used, covering epochs over the 40 years of validity of the model. These data have been selected according to different criteria to model only values corresponding to magnetically quiet periods (Figure 147).The model parameters follow:

- A 30° half-angle Spherical Cap centered at the South Pole
- A maximum spatial degree expansion of $K$=8, which means $n \approx 25$, or a wavelength of approximately 1,600 km
- A variable maximum temporal degree expansion using cosine series

One hundred sixty three statistically significant coefficients were obtained by means of a stepwise regression procedure (see Haines and Torta, 1994 for details; an alternative to this procedure based on a regularized method has been recently presented by Korte and Holme, 2003). The regression procedure allows for the determination of the field for epochs between 1960 and 2005 (Figure 148). The fit to the secular variation of the observatory annual means is better than those of global models like IGRF or the Comprehensive model (Gaya-Piqué *et al*.., 2006), and in Figure 149 one can see its validity, with good fit of the magnetic elements at four different observatories being obtained.

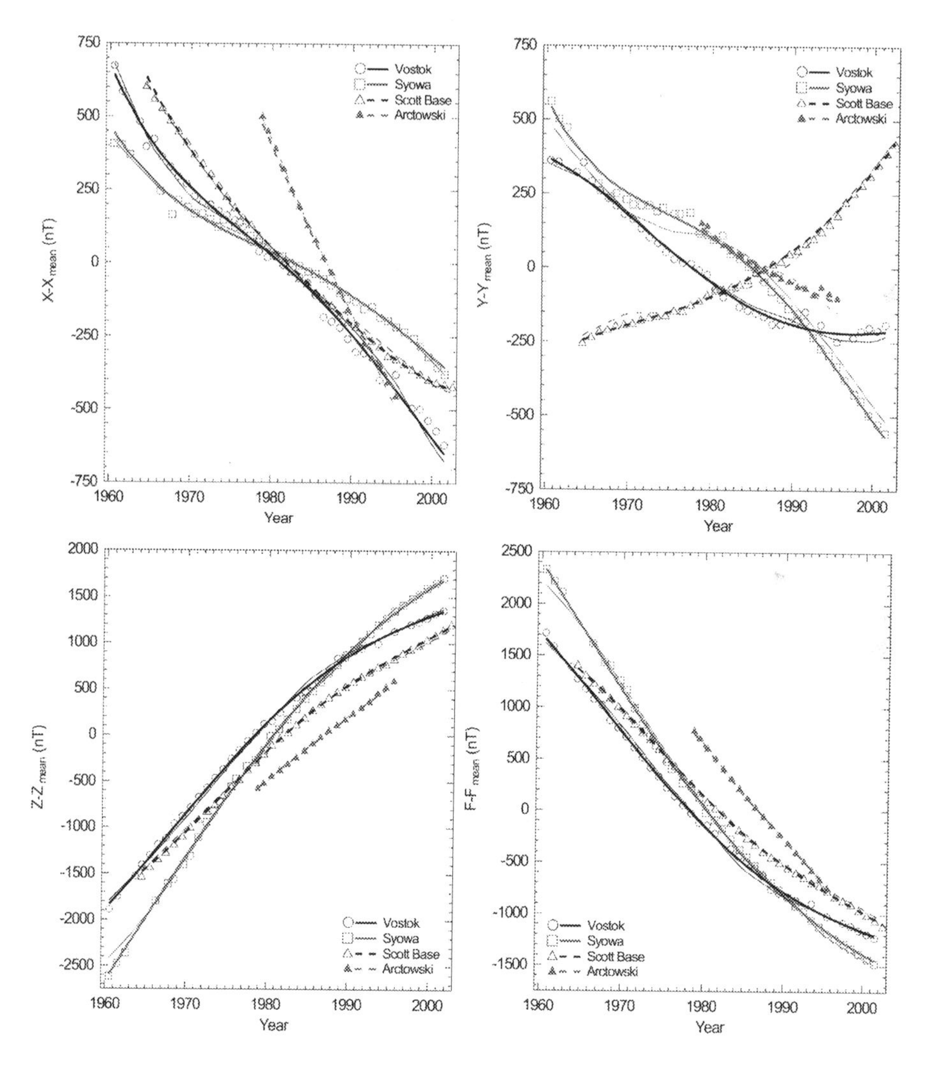


Figure 149. X (top left), Y (top right), Z (bottom left), and F (bottom right) annual means registered at ARC (solid triangles), SBA (open triangles), SYO (squares), and VOS (circles) observatories relative to their respective mean values over the time period. The thick lines show the fit given by ARM, and the thin lines that by IGRF 9th generation.

## 5. Conclusions

It has been shown how it is possible to develop models that reflect the spatial and temporal variations of the geomagnetic field in a restricted region with more detail and precision than is usually possible using the standard global models. Thus, regional models of secular variation allow for a better integration of disparate magnetic surveys and uniform digital anomaly charts can be obtained in this way. The spatial precision can be achieved using the Spherical Cap Harmonic Analysis technique. The temporal dependence of differences in the main field relative to the means at each observatory has been shown by analysis to be definitively more robust than the fit to variations obtained by numerical differentiation.

It was not the intent of this paper to present the SCHA method as the definitive technique for the analysis of the geomagnetic field in a restricted region of the Earth's surface. The intention has been to demonstrate how the SCHA can be of great value in some applications, once its drawbacks are analyzed, and whenever it is used conscientiously with its limitations recognized.

## References

De Santis, A., Gaya-Piqué, L. R., Dominici, A., Meloni, A. Torta, J. M., Tozzi, R., 2003, Italian geomagnetic reference field (ITGRF): update for 2000 and secular variation model up to 2005 by autoregressive forecasting, *Annals of Geophysics* **46** (3): 491-500.

Duka, B., Gaya-Piqué, L. R., De Santis, A., Bushati, S., Chiappini, M. and Dominici, G., 2004, A geomagnetic reference model for Albania, Southern Italy and Ionian Sea from 1990 to 2005, *Annals of Geophysics* **47** (5): 1609-1615.

García, A., Torta, J.M., Curto, J.J., Sanclement, E., 1991, Geomagnetic Secular Variation over Spain 1970-1988 by means of Spherical Cap Harmonic Analysis, *Phys. Earth Planet. Inter.* **68**: 65-75 [see Correction, *Phys. Earth Planet. Inter.*, 1992, **72**: 135].

Gaya-Piqué, L.R., 2004. *Analysis of the Geomagnetic field in Antarctica from near-surface and satellite data*, PhD Thesis, Observatori de l'Ebre, Universitat Ramon Llull, Spain.

Gaya-Piqué, L.R., Ravat, D., De Santis, A., Torta, J.M., 2005, Improving the main magnetic field representation in Antarctica, *Antarctic Science* (submitted).

Gaya-Piqué, L.R., Ravat, D., De Santis, A., Torta, J.M., 2006, New Model Alternatives for Improving the Representation of the Core Magnetic Field of Antarctica, Antarctic Science (in press)

Haines, G.V., 1985. Spherical Cap Harmonic Analysis. *J. Geophys. Res.* **90** (B3): 2583-2591.

Haines, G.V., 1990, Regional magnetic field modelling: a review, *J. Geomagn. Geoelectr.* **42** (9): 1001-1018.

Haines, G.V., 1993, Modelling geomagnetic secular variation by main-field differences. *Geophys. J. Int.* **114**: 490-500.

Haines, G.V., Torta J.M., 1994, Determination of Equivalent Currents Sources from Spherical Cap Harmonic Models of Geomagnetic Field Variations. *Geophys. J. Int.* **118**: 499-514.

International Association of Geomagnetism and Aeronomy (IAGA), Division V, Working Group VMOD: Geomagnetic Field Modeling, 2005, The 10th-Generation International Geomagnetic Reference Field, Geophys. J. Int **161**: 561-565.

Korte, M., Holme, R., 2003, Regularization of spherical cap harmonics. *Geophys. J. Int.* **153**: 253-262.

Lowes, F.J., 1995: The application of Spherical Cap Harmonic Analysis to the separation of the internal and external parts of Sq type fields (Abstract), *IUGG 95 Abstract Book, XXI General Assembly of IUGG*, p. B133.

Lowes, F.J., 1999: A problem in using Spherical Cap Harmonic Analysis to separate internal and external fields (Abstract), *IUGG 99 Abstract Book, XXII General Assembly of IUGG*, p. B324

Matsushita, S., Campbell, W.H., 1967. *Physics of geomagnetic phenomena*, Academic Press, New York, 2 vols., 1398 pp.

Sabaka, T.J., Olsen, N., Purucker, M.E, 2004, Extending comprehensive models of the Earth's magnetic field with Ørsted and CHAMP data, *Geophys. J. Int.* **159**: 521-547.

Thébault, E., Schott, J.J., Mandea, M., Hoffbeck, J.P., 2004, A new proposal for spherical cap harmonic modeling, *Geophys. J. Int.* **159**: 83-103.

Torta, J.M., De Santis, A., 1996, On the Derivation of the Earth's Conductivity Structure by means of Spherical Cap Harmonic Analysis. *Geophys. J. Int.* **127**: 441-451.

Torta, J.M., García, A., De Santis A., 1992, New representation of geomagnetic secular variation over restricted regions by means of SCHA: application to the case of Spain. *Phys. Earth Planet. Inter.* **74**: 209-217.

Torta, J.M., García, A., De Santis, A., 1993, A geomagnetic reference field for Spain at 1990. *J. Geomag. Geoelect.* **45**: 573-588.

## DISCUSSION

Question (Spomenko J. Mihajlovic): For which area do you think that SCHA is an optimal method (How wide should the area be)?

Answer (Miquel Torta): The area where SCHA is optimal depends on what kind of field one wants to represent. The spatial wavelength over the Earth's surface represented by the model, for a harmonic of degree $n_k$, is simply given by the quotient between the Earth's perimeter and the harmonic degree. And the different values that take the $n_k$ harmonics depend on the cap size, the bigger the cap the smaller the harmonics, and so the larger the wavelengths associated with each harmonic. Therefore, if one is interested in representing the main field and its secular variation, which we know are characterized by degrees going from 1 to 12 or 13, the use of big caps (say of continental size) are necessary; otherwise we will have unrealistic detail in our maps. If the model is aimed at representing smaller scale features (e.g. lithospheric anomalies), a small cap (e.g. of few degrees cap half-angle) can be suitable. As a rule of thumb, the area of the existence of the feature (or features) of the field that is going to be represented must be at least to some extent coincident with the cap-like region defining the analysis.

# ACTIVITIES COMPLETED TOWARD ESTABLISHING A GEOMAGNETIC OBSERVATORY IN THE REPUBLIC OF MACEDONIA

MARJAN DELIPETROV[23]
*Faculty of Mining and Geology*

**Abstract.** This paper presents the activities carried out in the territory of the Republic of Macedonia toward establishing a geomagnetic observatory. It gives the geographic location of the repeat station at Mount Plackovica, where construction of the geomagnetic observatory is planned. The paper also presents a proposal for the construction of the observatory.

**Keywords:** Geomagnetic Observatory, Republic of Macedonia, Geomagnetic field, Geomagnetic buildings

## 1. Activities completed so far toward establishing a geomagnetic observatory in the Republic of Macedonia

Since the declaration of independence in 1991, the Republic of Macedonia has lacked a geomagnetic observatory. Prior to 1991, during the existence of FR Yugoslavia, all geomagnetic measurements and permanent observations of the geomagnetic field were performed by Grocka observatory personnel in Serbia.

After independence, the Department of Geology and Geophysics of the Faculty of Mining and Geology in Štip purchased the first magnetometers for the investigation of some anomalies of the geomagnetic field. Of interest were ore deposits, archaeomagnetism, and the structural composition of some terrains.

The first steps in establishing the geomagnetic observatory started with the international project "Establishing a geomagnetic observatory in the Republic of Macedonia according to the standards of INTERMAGNET".

[23] Address for correspondence: Faculty of Mining and Geology, Department of Geology and Geophysics 2000 Štip, Macedonia. Email: marjan@rgf.ukim.edu.mk



*J.L. Rasson and T. Delipetrov (eds.), Geomagnetics for Aeronautical Safety,* 309–323.

The project was undertaken in collaboration with the Royal Meteorological Institute - Geomagnetic observatory in Dourbes, Belgium under the leadership of Dr. Todor Delipetrov, Dr. Jean Rasson, and the Faculty of Natural Sciences and Mathematics, Seismological observatory in Skopje.

The first measurements of the geomagnetic field were completed at the beginning of the project in 2002 at a number of locations in the country in order to select the most suitable terrain for construction of the geomagnetic observatory.

After the initial field survey in 2002, a grid of 15 repeat stations was established. Laboratory processing of measured results was carried out. Since these investigations aimed to define the location for construction of a geomagnetic observatory in the country, detailed analyses were performed. Bearing in mind the INTERMAGNET standards and the local conditions, the aim was to define an area that would meet the following criteria:

- The horizontal and vertical gradients of the geomagnetic field surrounding the location should not exceed 5 nT/m. At the locations for the absolute and variometer huts, the gradients should be within 1 nT/m;
- The wider area and the observatory should be located on a tectonically inactive block, meaning not in the zone of active tectonic dislocations,
- No seismic activity should be detected,
- The ground should be physically stable (no landslides, dip, bulging or similar);
- Observation pillars should be anchored on an undisturbed rock mass;
- Risk of floods or sudden changes in ground water levels should be minimal or nonexistent;
- Only moderate risk of strong winds and thunderstorms should exist;
- The surrounding area should not be populated or occupied by industrial facilities or be in an area where growth is expected. Infrastructure that would have negative impacts on observatory activities should be avoided.
- The possibility of acquiring surrounding property (5 – 10 hectares) for the needs for the observatory should be considered;
- The site should have the necessary infrastructure: road access, electricity, etc;
- The land should not be arable;
- From an economic aspect, the value of the area should be minimal, possible state land of low agricultural value;

- The site should be accessible year-round.

The locations in Ponikva, Galicica, and Plackovica (Figure 150)[2] completely satisfy the magnetic standards. However, bearing in mind other parameters such as property prices, area infrastructure, possible future growth, state ownership, and a favorable location with regard to neighboring observatories, it was decided that the best location for the construction of the geomagnetic observatory in the Republic of Macedonia is Mount Plackovica.

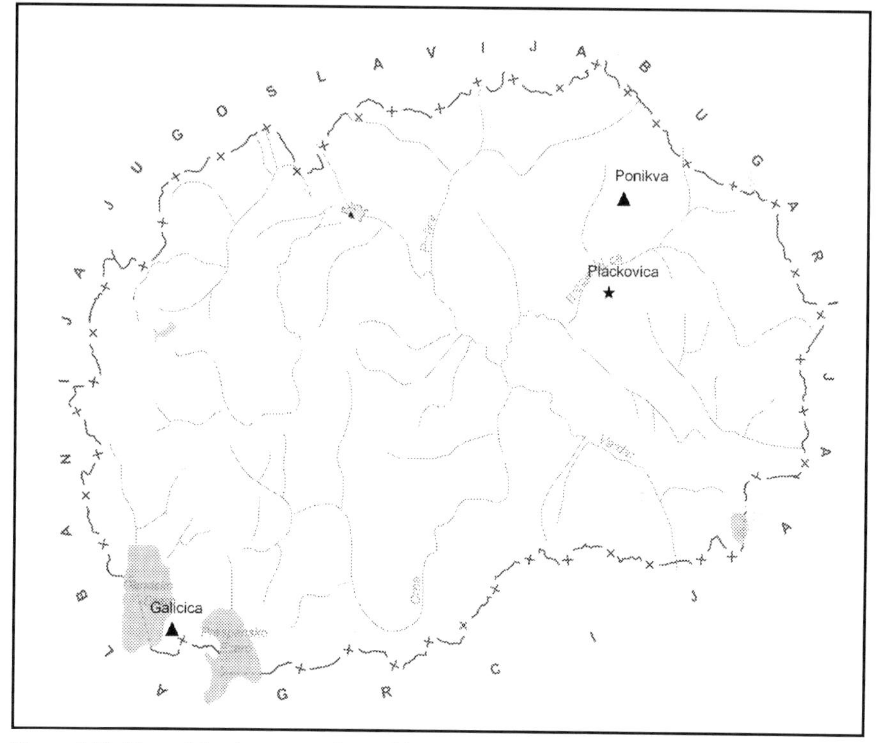


Figure 150. Map of the three most favorable locations for the observatory.

After deciding on the location of the geomagnetic observatory, the Department of Geology and Geophysics and the Seismological observatory in Skopje continued the work needed for starting construction. The National Government and the Ministries gave support for implementation of the idea because it was of scientific and state interest. Information about geomagnetic field changes should be announced in public media similar to weather forecasts in developed countries.

Changes in the geomagnetic field affect navigation and wireless communication. Airports in particular require accurate, up-to-date geomagnetic information. Those who use compasses for orientation also require this information.

Understanding the geomagnetic field was of interest to and was supported by the Ministry of Education and Science and also by the Ministries for Defense and Transport.

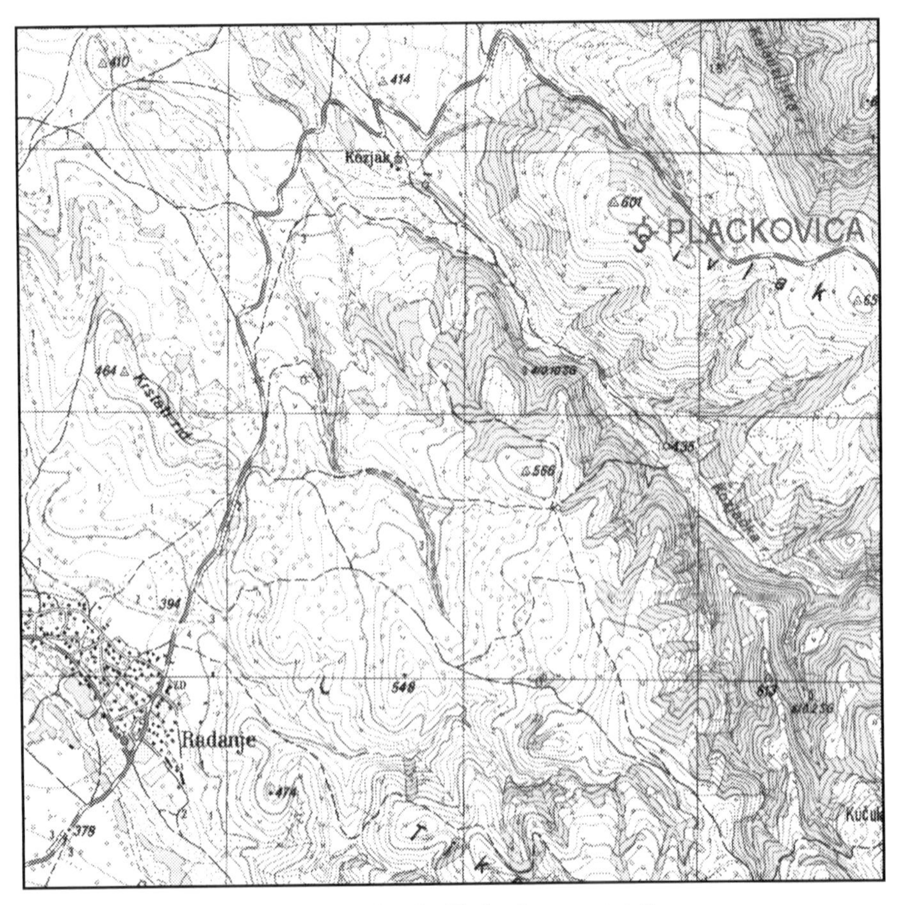


Figure 151. Topography of access road to the Plackovica repeat station.

The Department of Geology and Geophysics and the Seismological observatory completed a proposal for a geophysics activities law which is in the process of discussion and adoption.

A geomagnetic observatory needs adequate equipment, so contacts were established with several top institutions from countries in the European Union, within the Tempus Project "Geomagnetic Measurements and Quality standards". The European Training Foundation is financing the observatory project. The Tempus project provided significant funds for modern geomagnetic equipment. The equipment was tested at the observatory in Dourbes, Belgium and training was organized for a team of people for measuring the geomagnetic field.

## 2. Plackovica - location for construction of the geomagnetic observatory

The coordinates of the repeat station in Plackovica are: Longitude: 22°18'13"; Latitude: 41°47'41", Altitude: 677 m. This repeat station is situated on Mount Plackovica near the town of Štip and the village of Radanje (Figure 151). The local geology consists of micaschists and gneisses[1]. The plan is to install the magnetic observatory at the site of the repeat station.

During the past several months, architect Ljubica Velkovska[3] completed a proposal for the construction of the buildings for the geomagnetic observatory. The proposal was requested by the Department of Geology and Geophysics and the Seismological Observatory.

According to the requirements and consistent with the terrain conditions, a complex of structures for the observatory was designed.

This proposal gives schematics for the huts and structures for the actual buildings making up the geomagnetic observatory (see drawings in Annex 1).

## References

1. Delipetrov, T., Report: "Establishing geomagnetic observatory in the Republic of Macedonia according to INTERMAGNET standards", Stip, R. Macedonia, 1991
2. Rasson, JL., Delipetrov, M., "Republic of Macedonia: Magnetic Repeat Station Network Description, Dourbes, Belgium", 2004
3. Velkovska, Lj., Proposal project for geomagnetic observatory in the Republic of Macedonia, Skopje, R. Macedonia, 2005

**DISCUSSION**

Question (Valery Korepanov): What is the minimal distance from the main road to the observatory buildings?
Answer (Marjan Delipetrov): The minimal distance from the main road to the observatory buildings is more than 500m. But from the office building to observatory buildings the distance is about 150m.

## ANNEX 1: DRAWINGS FOR THE PROPOSED BUILDINGS OF THE MAGNETIC OBSERVATORY ACCORDING TO ARCHITECT LJUBICA VELKOVSKA

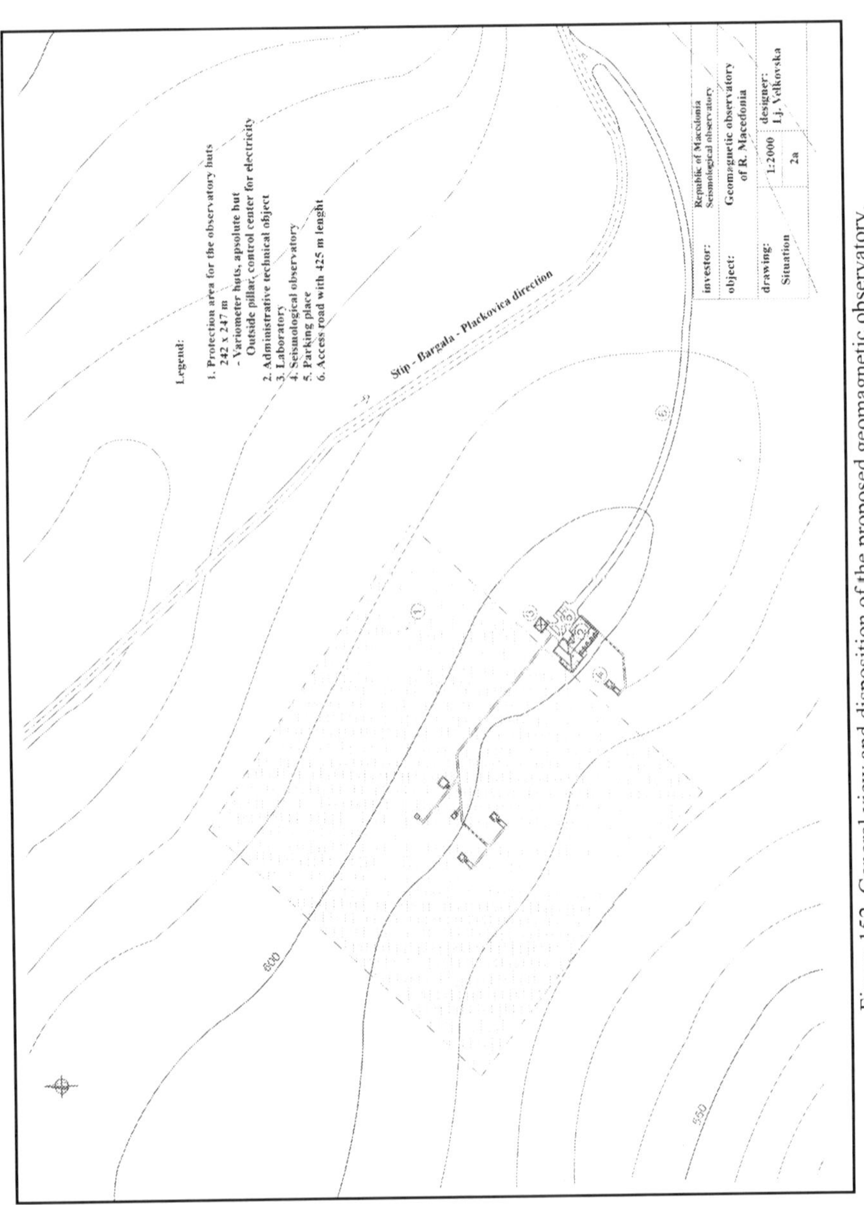


Figure 152. General view and disposition of the proposed geomagnetic observatory.

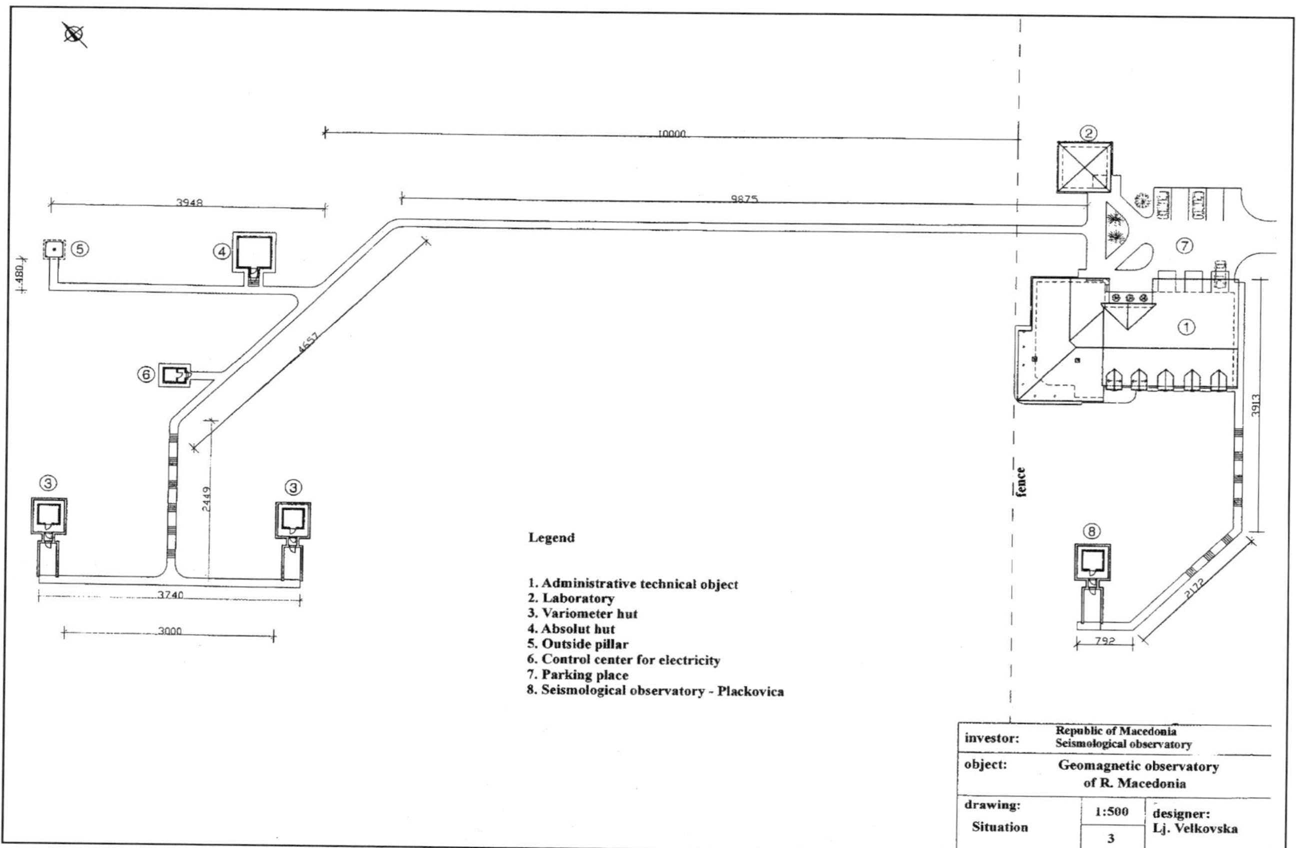


Figure 153. Close-up on the buildings for the proposed geomagnetic observatory.

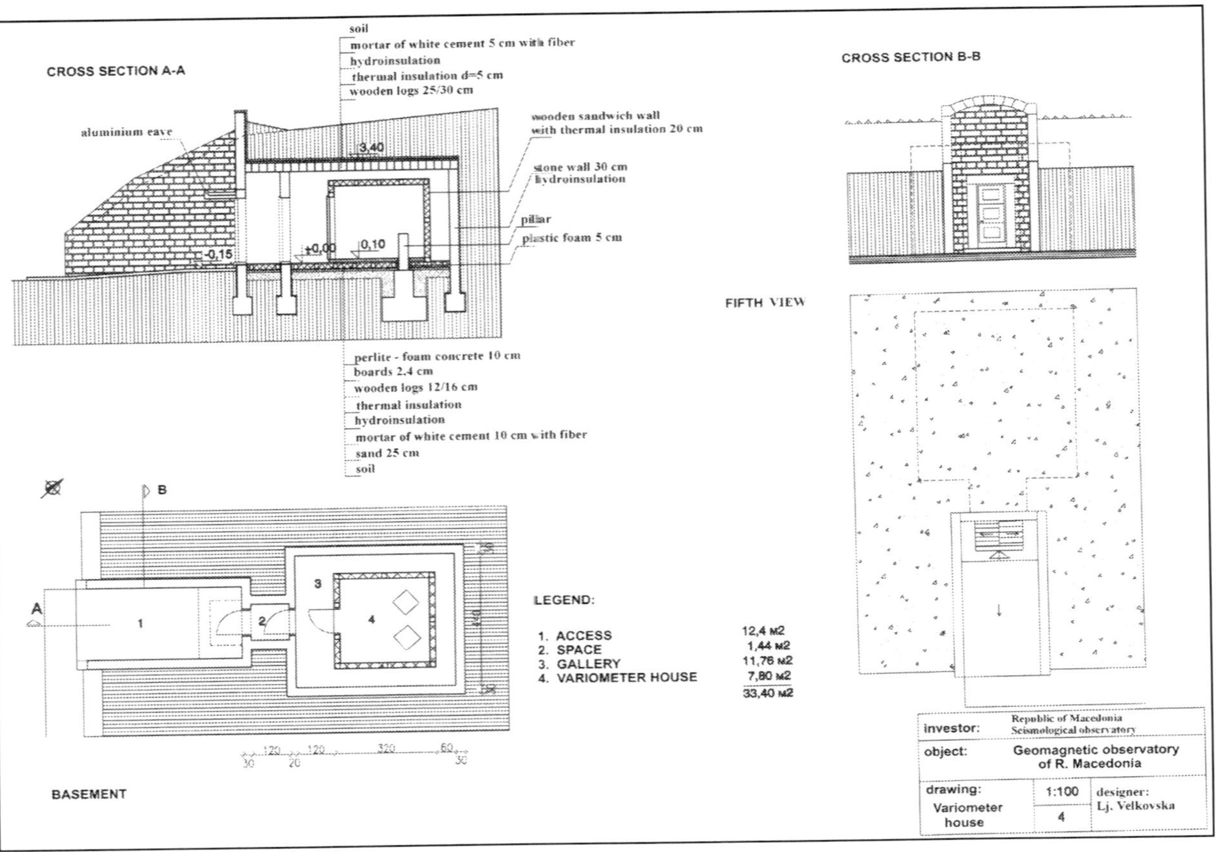


Figure 154. Variometer house.

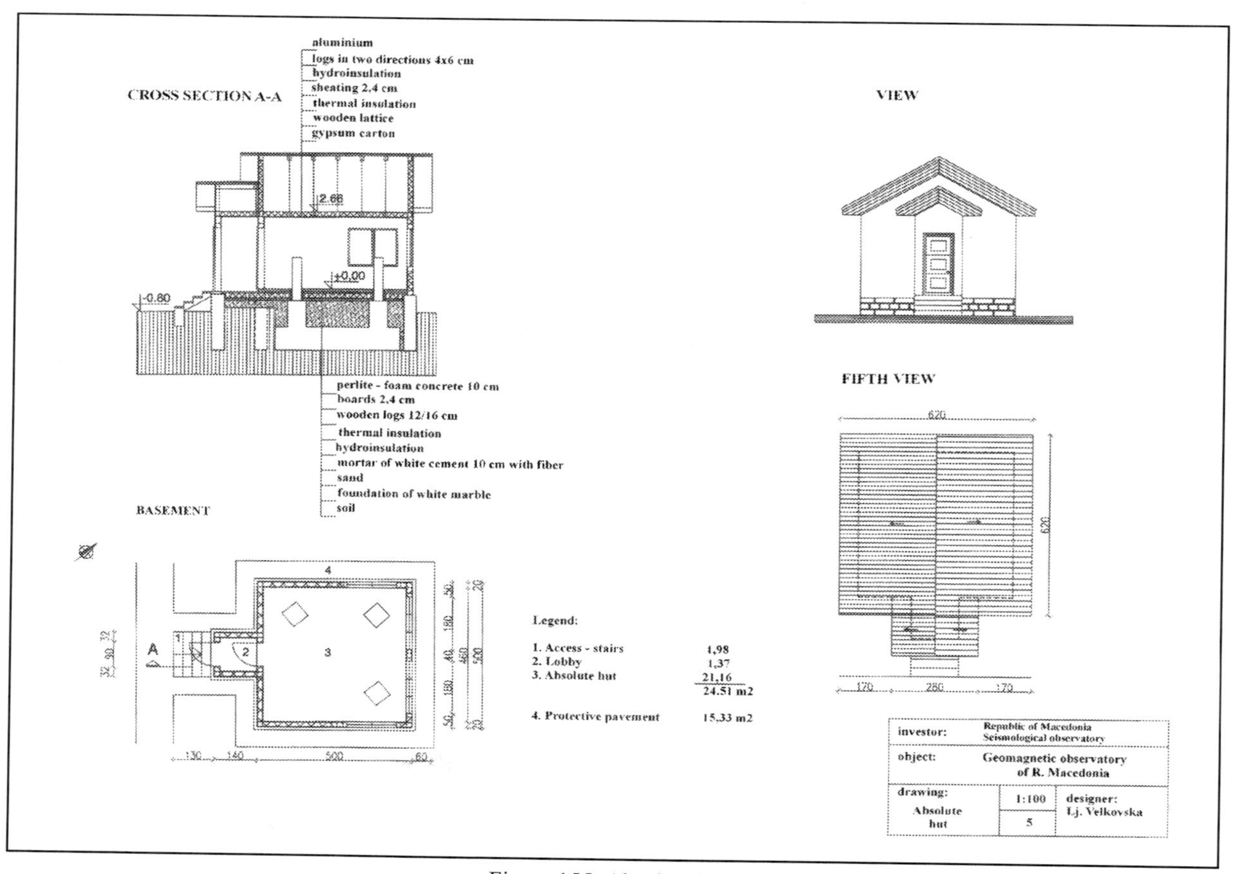


Figure 155. Absolute hut.

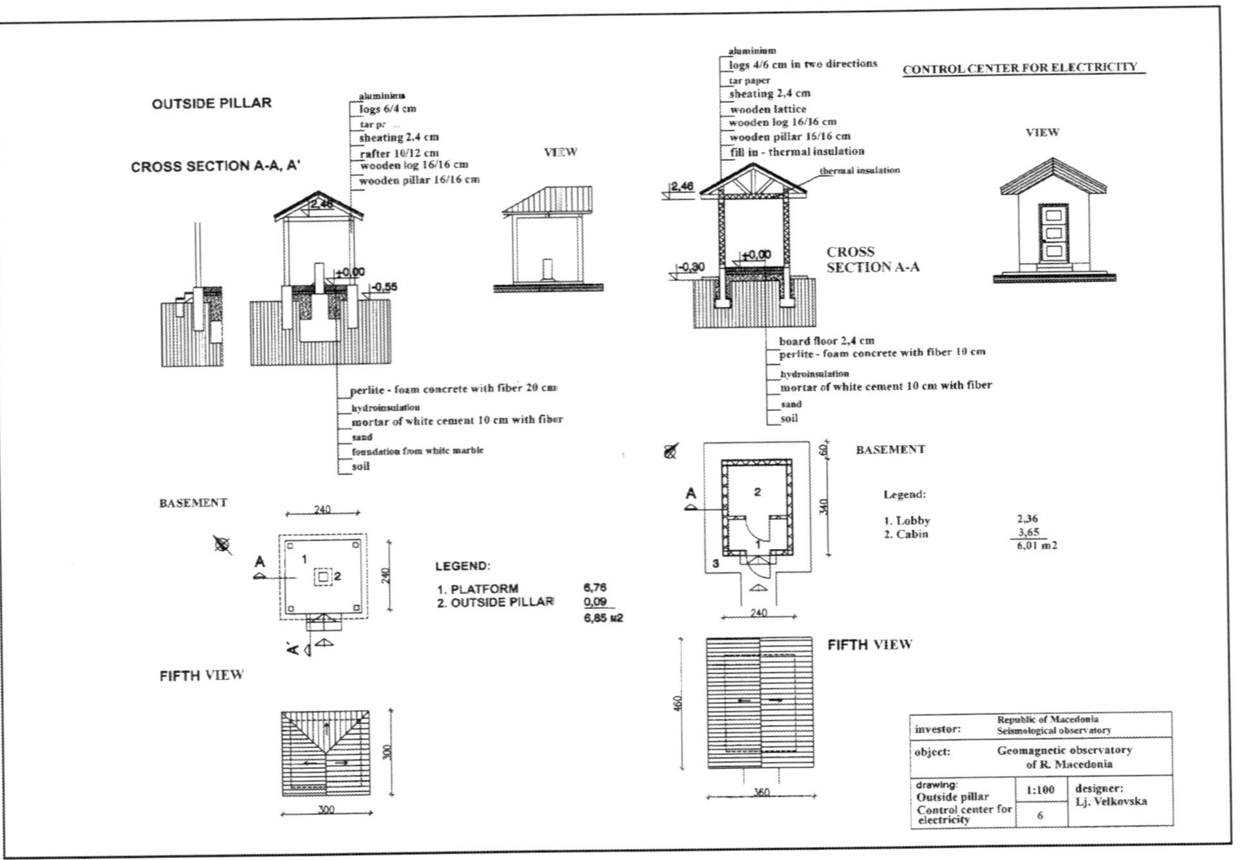


Figure 156. Outside pillar and control center for electricity.

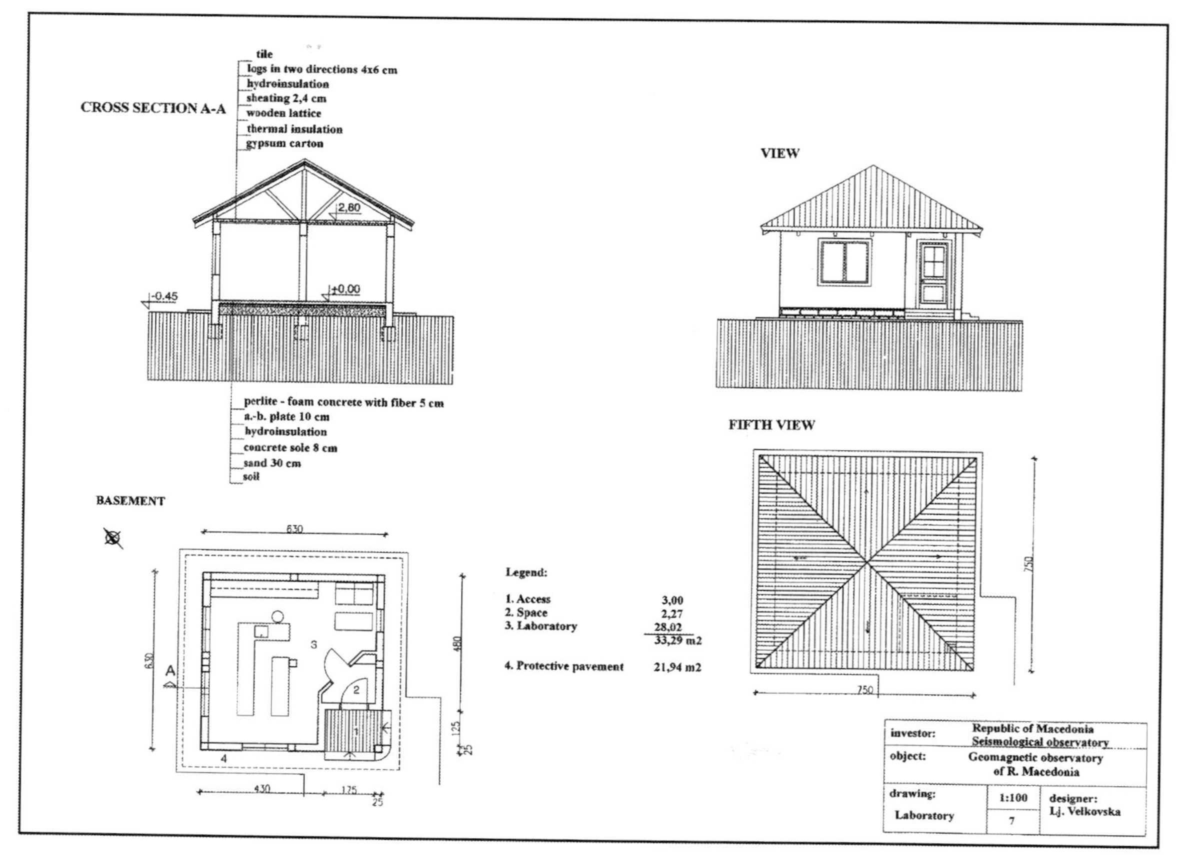


Figure 157. Laboratory.

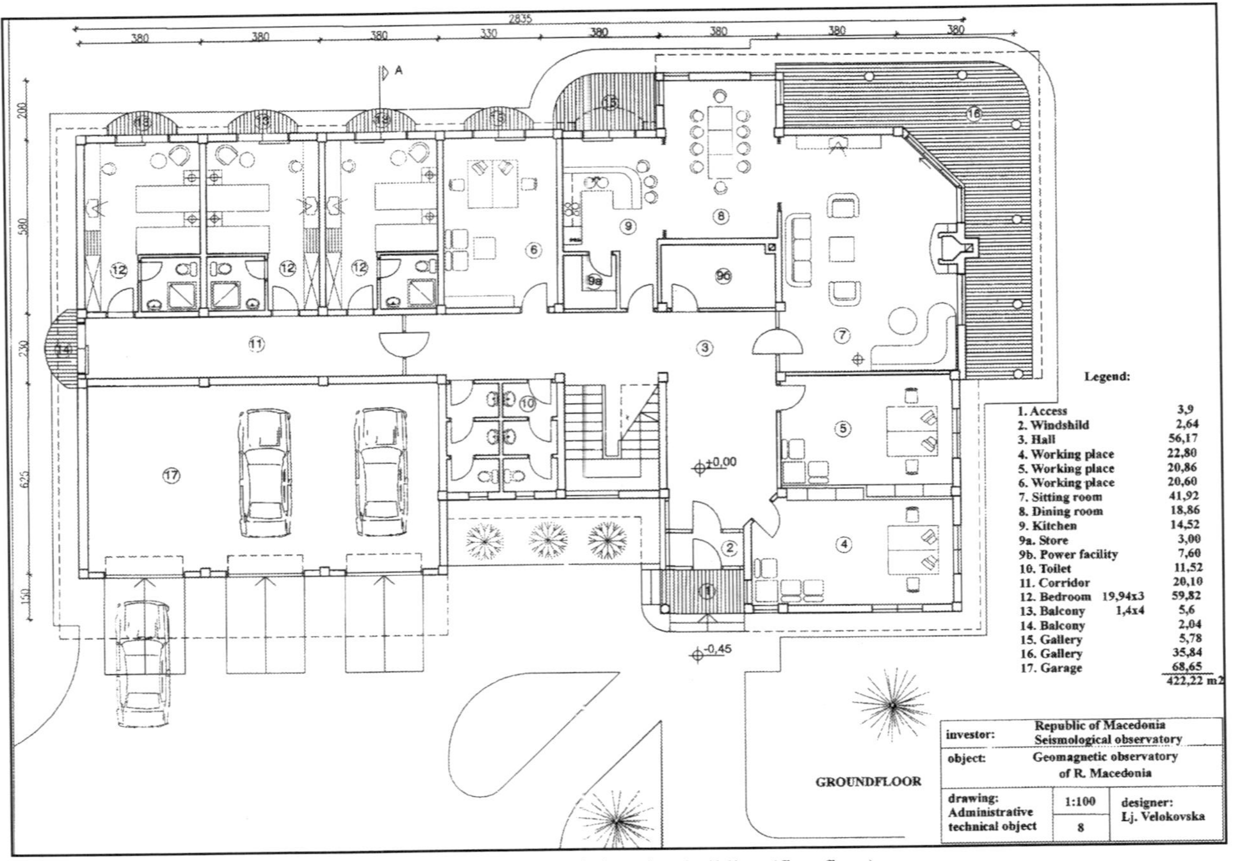


Figure 158. Administrative building (first floor).

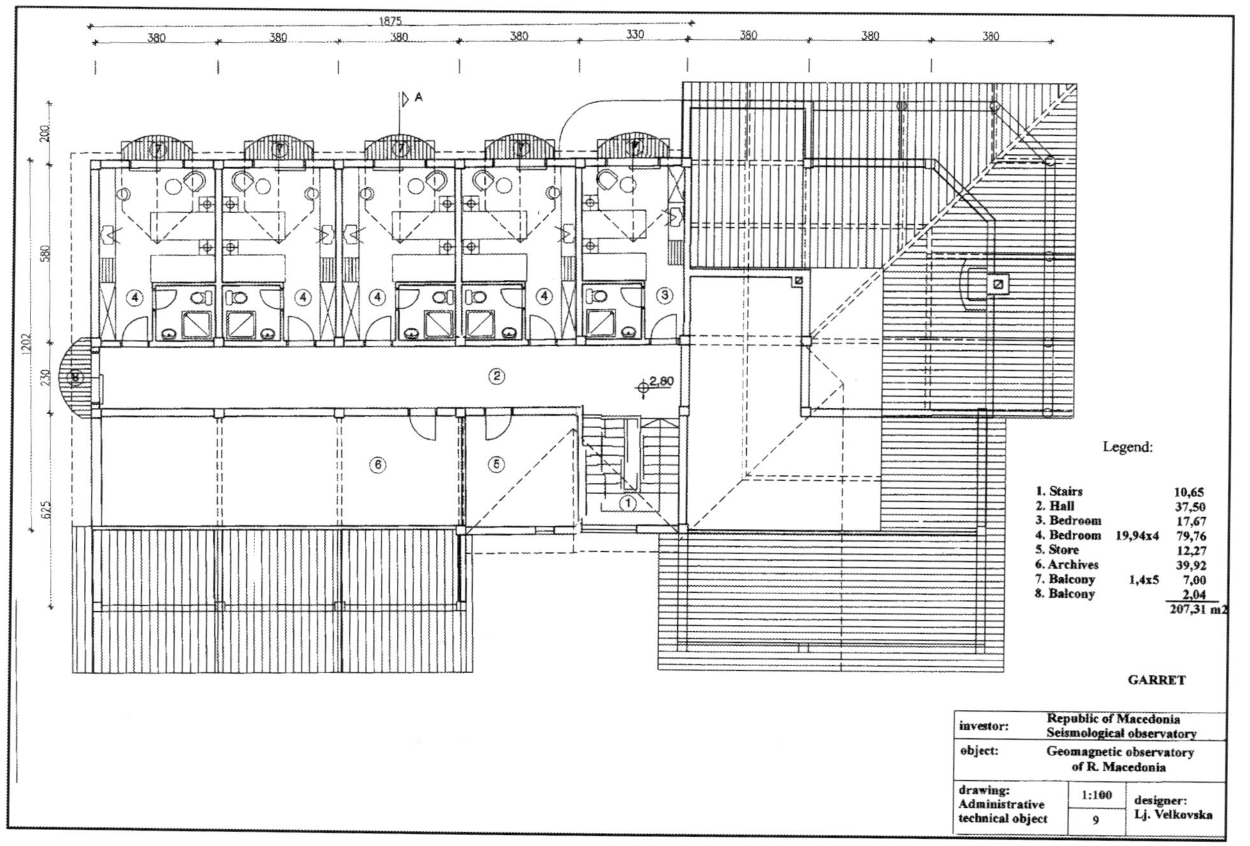


Figure 159. Administrative building (second floor).

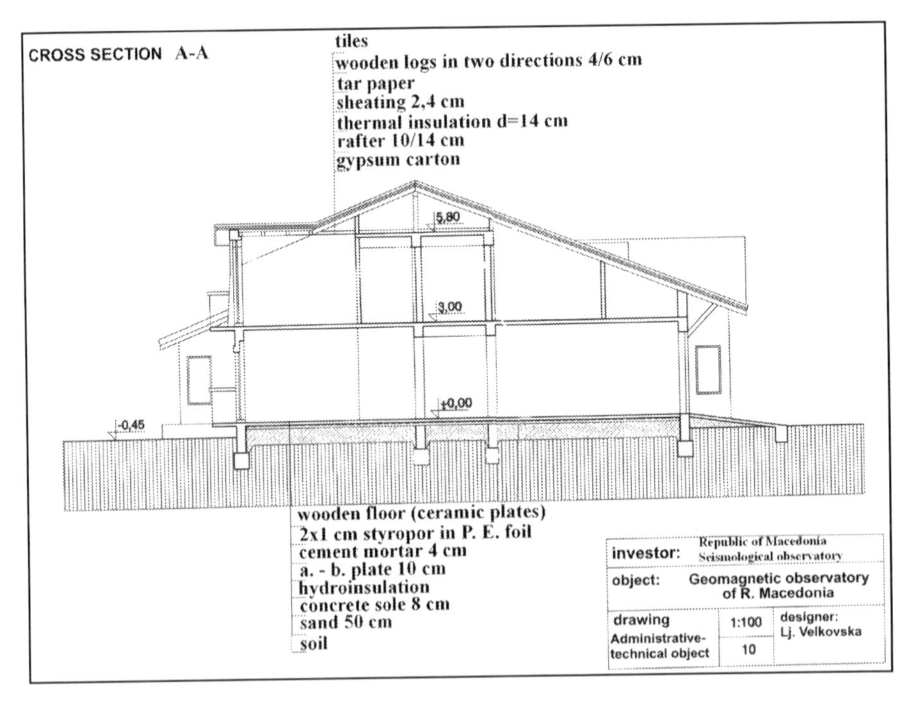


Figure 160. Cross section A – A of the administrative building.

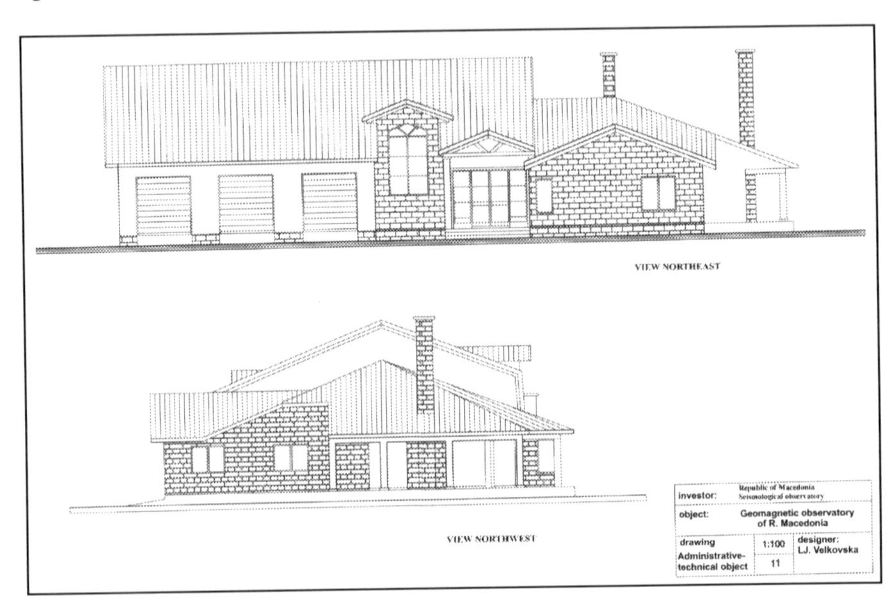


Figure 161. Northeast and northwest view of administrative building.

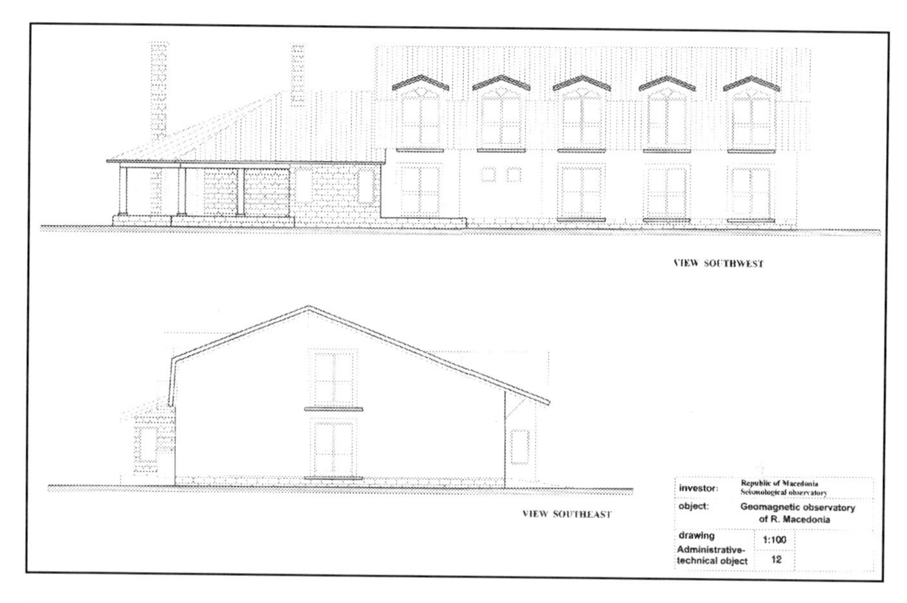


Figure 162. Southwest and southeast view of administrative building.

# ANALYSIS OF GEOMAGNETIC FIELD DATA FROM MEASUREMENTS DURING 2003 IN MACEDONIA

SANJA PANOVSKA[24]
TODOR DELIPETROV
*Faculty of Mining and Geology*

**Abstract.** Maps were compiled on the basis of the geomagnetic field measurements (I, D, F) carried out in 2003 in the territory of the Republic of Macedonia. Values of I, D, and F were calculated using the IGRF model. In 2003 geomagnetic measurements were performed at 15 repeat stations. Isolines of the geomagnetic elements were drawn with the SURFER program package. The normal dependences of the geomagnetic field on geographic latitude and longitude and the altitude of repeat stations were not taken into consideration during the processing. Approximations drawn in this manner make it possible to analyze the results relative IGRF model results. Such an approach is possible if one bears in mind that the territory of Macedonia is within $\Delta\varphi = 1^{\circ}31'$ and $\Delta\lambda = 2^{\circ}35'$ interval, its surface area is $P = 25713$ km$^2$, and the highest repeat station is 1684 meters above sea level.

## 1. Normal value of the magnetic field of a homogenously magnetized Earth

If we suppose that the Earth is a homogenously magnetized sphere, then the magnetic moment (M) of the sphere can be expressed as the product of the volume of the sphere (V) and the magnetic moment of volume unit (J), as shown by Delipetrov (2003). In other words, the magnetic moment is

$$M = V \cdot J = \frac{4}{3} R^{3} \cdot \pi \cdot J$$

where R is the radius of the sphere.

[24] To whom correspondence should be addressed at: Faculty of Mining and Geology, Department of Geology and Geophysics, 2000 Štip, Macedonia. Email: panovskasanja@yahoo.com.



*J.L. Rasson and T. Delipetrov (eds.), Geomagnetics for Aeronautical Safety,* 325–346.

The potential of the magnetic field of the sphere can be expressed as:

$$U = \frac{M}{r^2}[\sin\varphi \cdot \sin\varphi_m + \cos\varphi \cdot \cos\varphi_m \cdot \cos(\lambda - \lambda_m)]$$

When the coordinates of northern magnetic pole $N_m(\varphi_m, \lambda_m)$ are steady, it can be written that

$$g_1^0 = \frac{4}{3}\pi \cdot J \cdot \sin\varphi_m$$

$$g_1^1 = \frac{4}{3}\pi \cdot J \cdot \cos\varphi_m \cdot \cos\lambda_m$$

$$h_1^1 = \frac{4}{3}\pi \cdot J \cdot \cos\varphi_m \cdot \sin\lambda_m$$

and in this case, the expression for the potential is:

$$U = \frac{R^3}{r^2}[g_1^0 \cdot \sin\varphi + (g_1^1 \cdot \cos\lambda + h_1^1 \cdot \sin\lambda) \cdot \cos\varphi]$$

Derivatives of the potential of the magnetic field, with respect to certain directions, are field components or the field in those directions.

$$X = -\frac{1}{r} \cdot \frac{\partial U}{\partial \varphi}$$

$$Y = -\frac{1}{r \cdot \cos\varphi} \cdot \frac{\partial U}{\partial \lambda}$$

$$Z = -\frac{\partial U}{\partial r}.$$

The X component refers to the northern direction, Y refers to the eastern direction, and Z to vertical component of the magnetic field of a homogenously magnetized sphere.

If the expressions for the potential are differentiated to variable values $\varphi$,$\lambda$, and r and, in the case of analysis of the magnetic field components of the Earth's surface, we say that the distance r corresponds to the radius of the Earth $r = R$ then the following expressions are obtained:

$$X = g_1^0 \cos\varphi - (g_1^1 \cos\lambda + h_1^1 \sin\lambda) \cdot \sin\varphi$$

$$Y = -g_1^1 \sin\lambda + h_1^1 \cos\lambda$$

$$Z = 2[g_1^0 \sin\varphi + (g_1^1 \cos\lambda + h_1^1 \sin\lambda)\cdot\cos\varphi]$$

The present-day magnetic moment of the Earth is $M_{Earth} = 7.8 \times 10^{15}$ $Tm^3$. The average radius of the Earth is 6375 km. The coordinates $(\varphi_m, \lambda_m)$ of the north magnetic pole in 2003 were 82.0° N latitude and 112.4° W longitude. It is possible to calculate coefficients $g_1^0, g_1^1, h_1^1$ in units nT, given in Table 31.

Table 31. The coefficients $g_1^0, g_1^1, h_1^1$ in units nT for epoch 2003.

| Epoch | $g_1^0$ | $g_1^1$ | $h_1^1$ |
|---|---|---|---|
| 2003 | 29813.1 | -1596.67 | -38738.1 |

If we assume that the axis of the magnet coincides with the rotation axis of the Earth, and therefore that the magnetic and geographic poles overlap then the coefficient $g_1^1 = 0$. If, on the other hand, we take for the starting meridian the meridian that passes through the point $N_m$ or the point in witch the magnetic axis passes through the Earth's surface, we have $h_1^1 = 0$. From this, the following is obtained for the components X, Y, and Z:

$$X = g_1^0 \cos\varphi$$

$$Y = 0$$

$$Z = 2g_1^0 \sin\varphi$$

If the angle φ also represents the angle of magnetic latitude, then only the north (X) component is present, and it is also the horizontal component of the field.

When calculated for one point in the Republic of Macedonia at geographic latitude φ = 42° N, these components, equal:

$$X = H = \frac{M}{R^3}\cos\varphi = 22155.4 \text{ nT}$$

$$Z = \frac{2M}{R^3}\sin\varphi = 39897.7 \text{ nT}$$

For the intensity of the total vector (T), which is the vectorial sum of horizontal (H) and vertical (Z) components we obtain the value:

$$T = \frac{M}{R^3}(1 + 3\sin^2\varphi)^{1/2} = 45636.5 \text{ nT}$$

## 2. Values of the gradient of the magnetic field of the axial dipole

The values of the gradient of the field of the axial dipole, expressed in nT/km can be calculated. The horizontal gradients of the magnetic field of the axial dipole would be:

$$\frac{\partial Z}{r \cdot \partial \varphi} = \frac{2M}{r^4} \cos \varphi = Z \cdot \frac{1}{r} \text{ctg}\varphi$$

$$\frac{\partial H}{r \cdot \partial \varphi} = -\frac{M}{r^4} \sin \varphi = -H \cdot \frac{1}{r} \text{tg}\varphi$$

These gradients represent the increase in of the north direction component. The vertical gradient would be:

$$\frac{\partial Z}{\partial r} = -\frac{6M}{r^4} \sin \varphi = -3Z \cdot \frac{1}{r}$$

$$\frac{\partial H}{\partial r} = -\frac{3M}{r^4} \cos \varphi = -3H \cdot \frac{1}{r}$$

If, for example we consider a point of the territory of Macedonia, with the geographic latitude $\varphi = 42°N$, then the values of the gradient of the magnetic field in the approximation of the axial dipole would be:

$$\frac{\partial Z}{r \cdot \partial \varphi} = \frac{Z}{r} \text{ctg}\varphi = 6.11 \text{ nT / km}$$

$$\frac{\partial H}{r \cdot \partial \varphi} = -\frac{H}{r} \text{tg}\varphi = -3.43 \text{nT / km}$$

$$\frac{\partial Z}{\partial r} = -3\frac{Z}{r} = -16.50 \text{ nT / km}$$

$$\frac{\partial H}{\partial r} = -3\frac{H}{r} = -11.43 \text{ nT / km}$$

## 3. Normal values of the geomagnetic field for Macedonia in 2003

The normal magnetic field of a territory can be approximated by the expression (Stefanovic,1978):

$$E(\Delta\varphi, \Delta\lambda) = a_1 + a_2 \cdot \Delta\varphi + a_3 \cdot \Delta\lambda + a_4 \cdot \Delta\varphi^2 + a_5 \cdot \Delta\lambda^2 + a_6 \Delta\varphi \cdot \Delta\lambda$$

where

$E(\Delta\varphi, \Delta\lambda)$ is the value of the normal field at the point with geographic coordinates $\varphi_1$ and $\lambda_1$;

$\varphi_1$ and $\lambda_1$ – geographic latitude and longitude of the point;

$\varphi_0$ and $\lambda_0$ – geographic latitude and longitude of the reference point;

$\Delta\varphi = \varphi_1 - \varphi_0$ – difference of geographic latitude in minutes;

$\Delta\lambda = \lambda_1 - \lambda_0$ – difference of geographic longitude in minutes;

$a_i$ – coefficients of the differences in nT/minutes, minutes/minutes, or nT and minutes. Commonly, the differences in geographic latitude and longitude are calculated relative to the coordinates of a geomagnetic observatory in that territory.

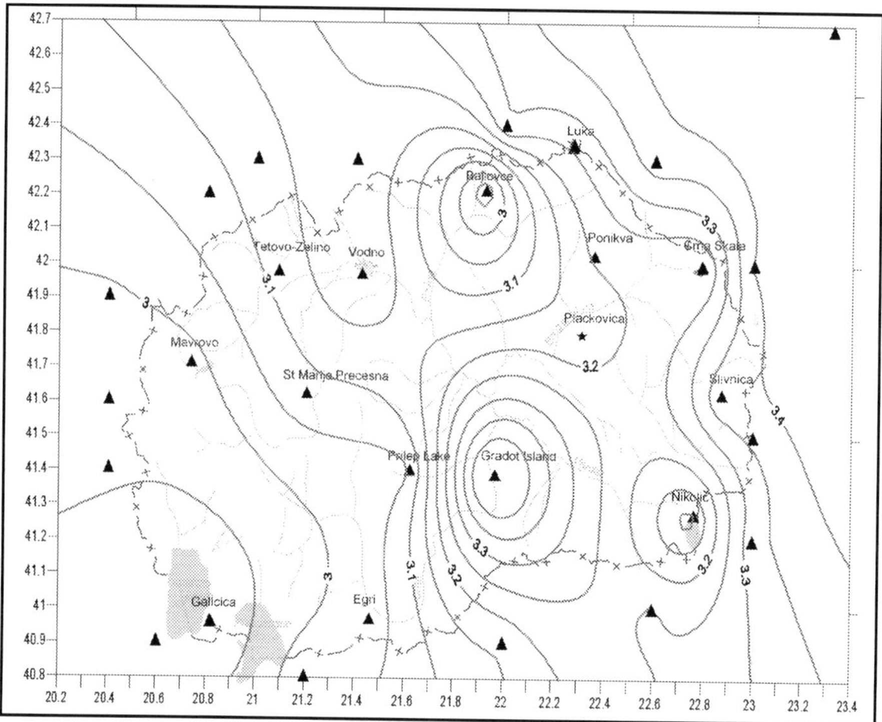


Figure 163. Map of measured values of declination of the geomagnetic field, epoch 2003.5, in the territory of the Republic of Macedonia.

The normal field in the territory of the Republic of Macedonia is calculated on the basis of measurements of absolute values of total intensity (T), declination (D), and inclination (I) in 2003 at 15 repeat stations. From these values the values of other components of the geomagnetic field were calculated. The equation for the normal field was calculated relative to the point with geographic coordinates latitude = 40° N and longitude = 20° E. The coefficients are expressed in nT/minute or minutes, the declination and inclination in minute differences except for coefficient $a_1$ which is expressed in nT or minutes. In the calculation, the values of the magnetic field at the Tetovo –Zelino repeat station were suppressed because the measurement data of the target azimuth was missing.

Since the density of the grid of points on which the measurements were performed is small, and the geologic structure of the territory on which coefficients are calculated is complex with the presence of magnetic rocks, there are deviations in the calculated coefficients. These coefficients for 2003 for every element of the magnetic field gave the values shown in Table 32.

Table 32. The coefficients for 2003 for every element of normal magnetic field in Macedonia.

| El | $a_1$ | $a_2$ | $a_3$ | $a_4$ | $a_5$ | $a_6$ |
|---|---|---|---|---|---|---|
| T | 46088.482 | 2.5629 | 0.6066 | 0.014872 | 0.0009347 | 0.000000281123 |
| D | 1.753810 | 0.021659 | 0.007858 | -0.00012336 | -0.000034789 | 0.000000000862 |
| I | 57.154877 | 0.005649 | -0.00036 | 0.000072412 | 0.000010130 | -0.00000000039 |
| H | 24980.448 | -1.93367 | 0.579256 | -0.04556586 | -0.006420438 | 0.000000402028 |
| X | 24900.809 | -0.80533 | 1.512406 | -0.059514413 | -0.01389987 | 0.000000848647 |
| Y | 770.00284 | 9.240089 | 3.456552 | -0.055937766 | -0.01561722 | 0.000000405225 |
| Z | 38715.402 | 4.819595 | 0.339774 | 0.0422145 | 0.00516986 | 0.000000076729 |

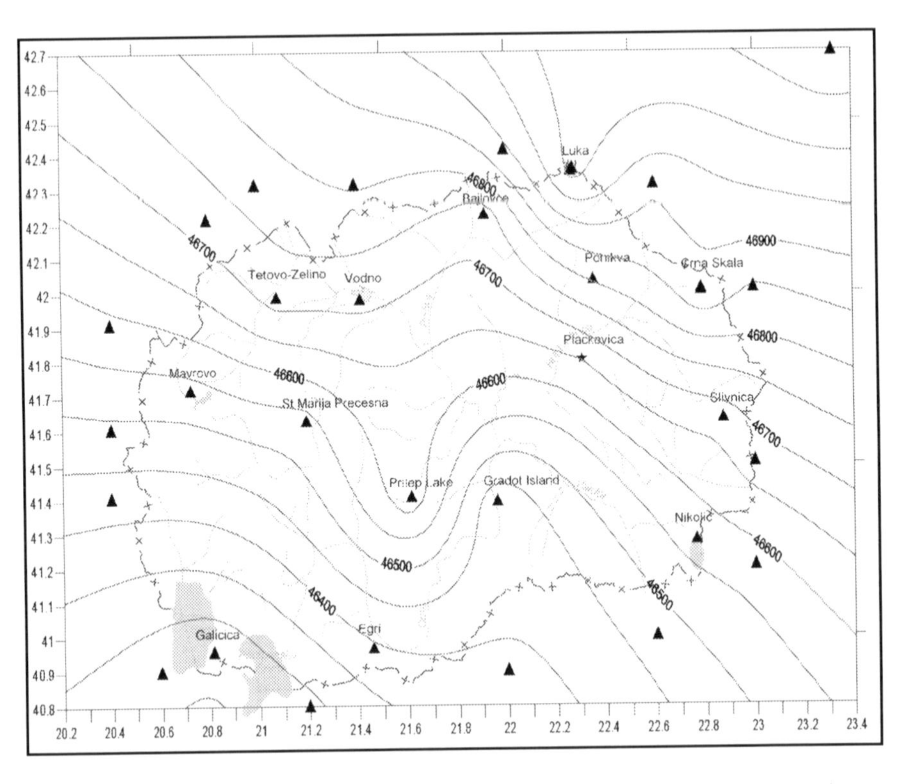


Figure 164. Map of measured values of total intensity of the geomagnetic field, epoch 2003.5, in the territory of the Republic of Macedonia.

Table 33 shows normal values of all components of the magnetic field calculated with coefficients for 14 repeat stations in the Republic of Macedonia.

Table 33. Normal values of all components of magnetic field for 14 repeat stations in Macedonia.

| Repeat station | T | D | I | H | X | Y | Z |
|---|---|---|---|---|---|---|---|
| Egri | 46354 | 3,0366 | 57,7601 | 24728 | 24702 | 1310,3 | 39207 |
| Mavrovo | 46544 | 2,9702 | 58,4998 | 24319 | 24244 | 1257,6 | 39685 |
| Plackovica | 46701 | 3,2683 | 58,6592 | 24290 | 24255 | 1385,5 | 39887 |
| Slivnica | 46686 | 3,2566 | 58,5099 | 24387 | 24349 | 1385,3 | 39811 |
| Vodno | 46689 | 3,0911 | 58,8473 | 24153 | 24082 | 1300,9 | 39956 |
| Bailovce | 46843 | 3,0966 | 59,1939 | 23990 | 23926 | 1294,5 | 40233 |
| Gradot Island | 46515 | 3,2271 | 58,1861 | 24522 | 24488 | 1380,7 | 39527 |
| Nikolic | 46538 | 3,1727 | 58,1542 | 24555 | 24502 | 1358,2 | 39532 |
| Ponikva | 46808 | 3,2343 | 58,9474 | 24144 | 24110 | 1362,9 | 40100 |
| St. Marija Precesna | 46542 | 3,1190 | 58,4034 | 24386 | 24333 | 1326,2 | 39643 |
| Crna Skala | 46849 | 3,2659 | 58,9338 | 24175 | 24156 | 1378,3 | 40130 |
| Luka | 46947 | 3,1021 | 59,3765 | 23915 | 23870 | 1293,2 | 40399 |
| Galicica | 46319 | 2,8982 | 57,7208 | 24736 | 24706 | 1250,8 | 39161 |
| Prilep Lake | 46495 | 3,1943 | 58,1771 | 24517 | 24484 | 1366,3 | 39506 |

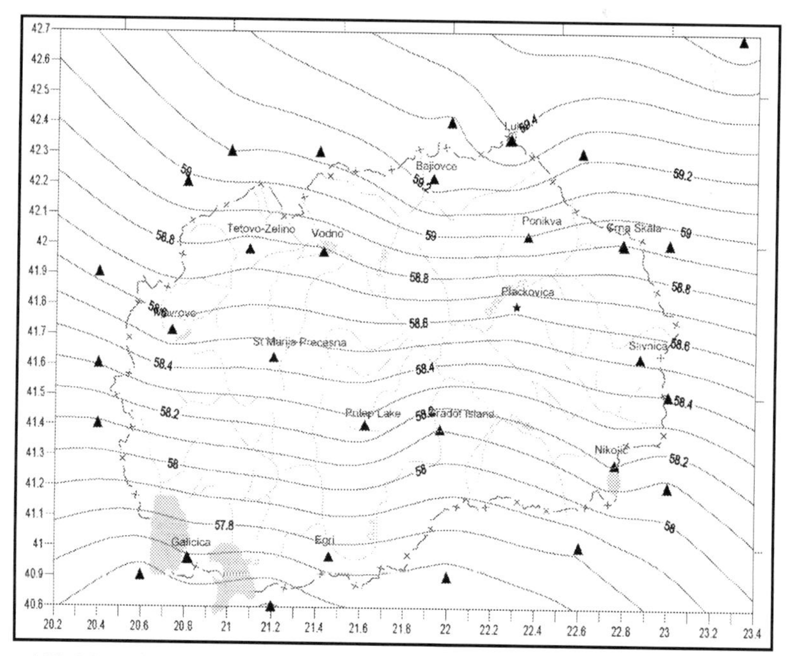


Figure 165. Map of measured values of inclination of the geomagnetic field, epoch 2003.5, in the territory of the Republic of Macedonia.

## 4. Models of components of the geomagnetic field in the Republic of Macedonia for 2003

In 2003, field measurements of the geomagnetic field were performed at 15 repeat stations, as shown by Rasson (2004). The measured results are shown in Table 34. Calculated values of the normal field are shown in Table 35.

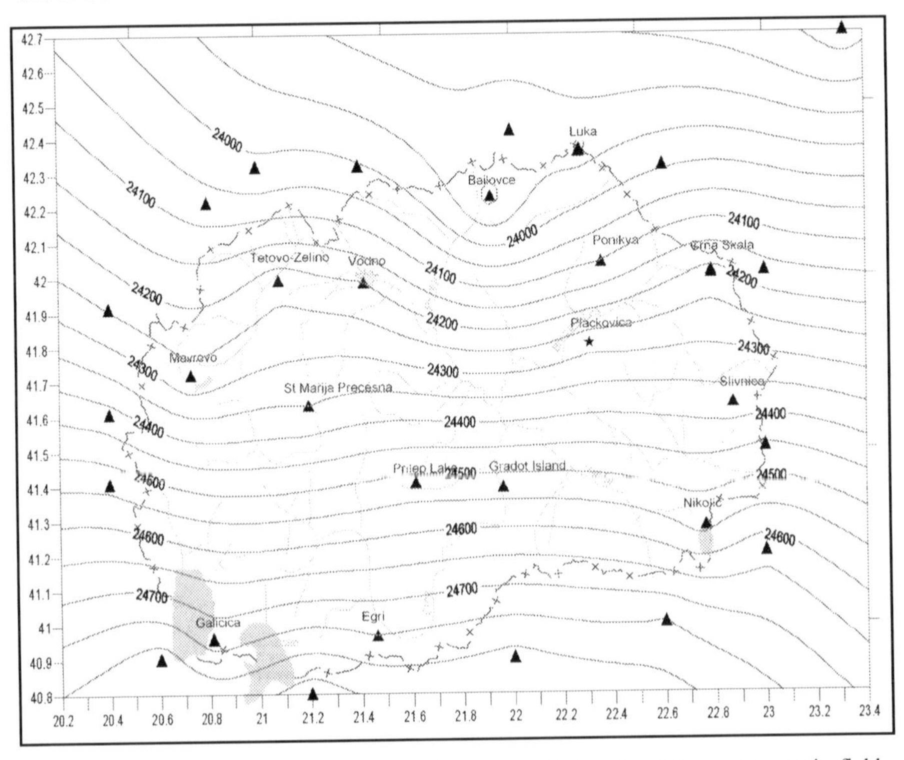


Figure 166. Map of the measured values of horizontal intensity of the geomagnetic field, epoch 2003.5, in the territory of the Republic of Macedonia.

Table 34. Measured values of 15 repeat stations in Macedonia.

| | | Measured | | |
|---|---|---|---|---|
| No | Repeat station | D(°) | I(°) | F(nT) |
| 1 | Tetovo | - | 58.754 | 46718.0 |
| 2 | Egri | 3.017 | 57.756 | 46391.7 |
| 3 | Mavrovo | 2.984 | 58.570 | 46531.4 |
| 4 | Plackovica | 3.163 | 58.620 | 46645.3 |
| 5 | Slivnica | 3.384 | 58.504 | 46665.2 |

| | | Measured | | |
|---|---|---|---|---|
| No | Repeat station | D(°) | I(°) | F(nT) |
| 6 | Vodno | 3.201 | 58.787 | 46712.7 |
| 7 | Bailovce | 2.915 | 59.257 | 46722.5 |
| 8 | Gradot island | 3.514 | 58.095 | 46414.8 |
| 9 | Nikolic | 3.077 | 58.198 | 46567.1 |
| 10 | Ponikva | 3.208 | 58.990 | 46800.2 |
| 11 | St. Marija | 3.083 | 58.445 | 46531.9 |
| 12 | Crna skala | 3.191 | 58.886 | 46885.1 |
| 13 | Luka | 3.258 | 59.393 | 47014.3 |
| 14 | Galicica | 2.898 | 57.689 | 46261.9 |
| 15 | Prilep Lake | 3.040 | 58.277 | 46632.6 |

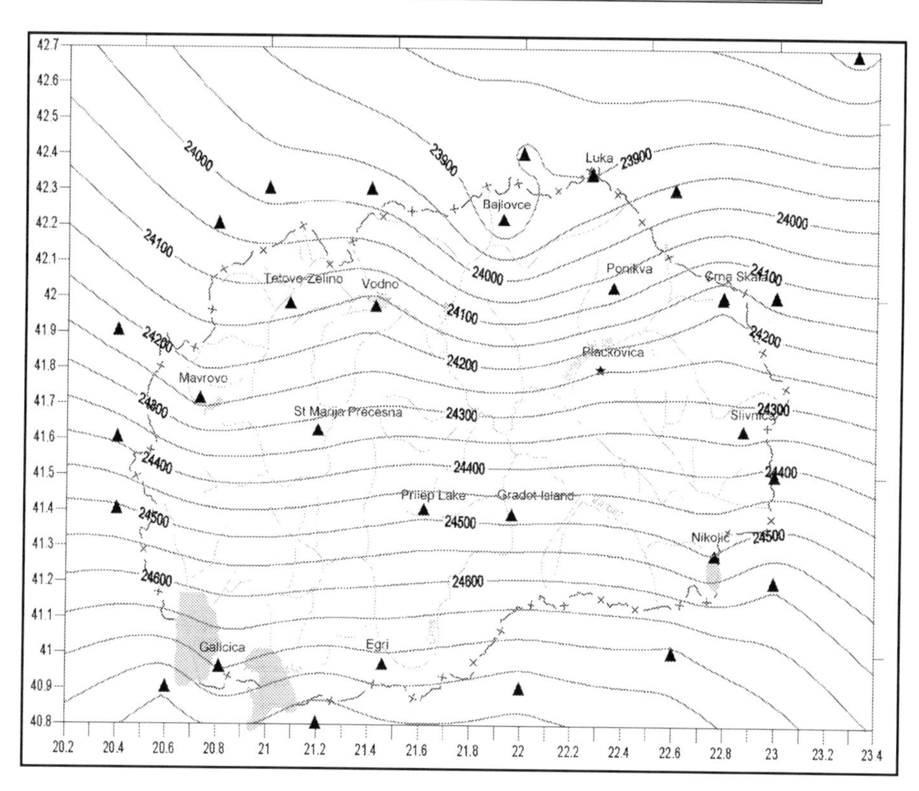


Figure 167. Map of measured values of the northern (X) component of the geomagnetic field, epoch 2003.5, in the territory of the Republic of Macedonia.

Table 35. Calculated values from 15 repeat stations in Macedonia.

| No | Repeat station | Calculated H(nT) | X(nT) | Y(nT) | Z(nT) |
|---|---|---|---|---|---|
| 1 | Tetovo | 24233.26 | - | - | 39941.46 |
| 2 | Egri | 24751.17 | 24716.87 | 1302.71 | 39237.34 |
| 3 | Mavrovo | 24264.10 | 24231.20 | 1263.12 | 39704.21 |
| 4 | Plackovica | 24288.75 | 24251.75 | 1340.17 | 39822.61 |
| 5 | Slivnica | 24379.72 | 24337.21 | 1439.08 | 39790.33 |
| 6 | Vodno | 24207.50 | 24169.74 | 1351.72 | 39950.88 |
| 7 | Bailovce | 23883.98 | 23853.08 | 1214.61 | 40156.53 |
| 8 | Gradot Island | 24530.80 | 24484.68 | 1503.55 | 39402.71 |
| 9 | Nikolic | 24540.18 | 24504.80 | 1317.27 | 39576.18 |
| 10 | Ponikva | 24110.89 | 24073.10 | 1349.27 | 40111.39 |
| 11 | St. Marija Precesna | 24350.92 | 24315.68 | 1309.65 | 39651.61 |
| 12 | Crna skala | 24227.53 | 24189.96 | 1348.62 | 40140.25 |
| 13 | Luka | 23937.17 | 23898.48 | 1360.40 | 40464.26 |
| 14 | Galicica | 24727.66 | 24696.04 | 1250.18 | 39098.67 |
| 15 | Prilep Lake | 24520.03 | 24485.53 | 1300.37 | 39665.69 |

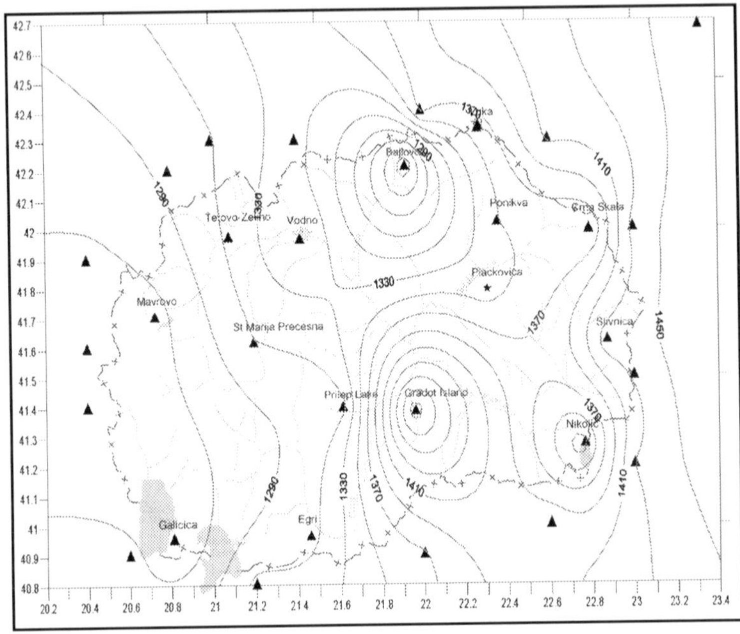


Figure 168. Map of measured values of the eastern (Y) component of the geomagnetic field, epoch 2003.5, in the territory of the Republic of Macedonia.

Digital models for each component of the geomagnetic field were calculated based on the data in Table 34 and Table 35. Using the SURFER software package, maps were compiled as shown in the following figures. The geomagnetic maps were based on the digital map of the Republic of Macedonia compiled at the Faculty of Mining and Geology, Stip and DATAMAP, Sofia in the MAPINFO program package. The values of the points that are outside the boundary of the Republic of Macedonia were calculated according to the IGRF 2000 model (Figure 163 – Figure 169).

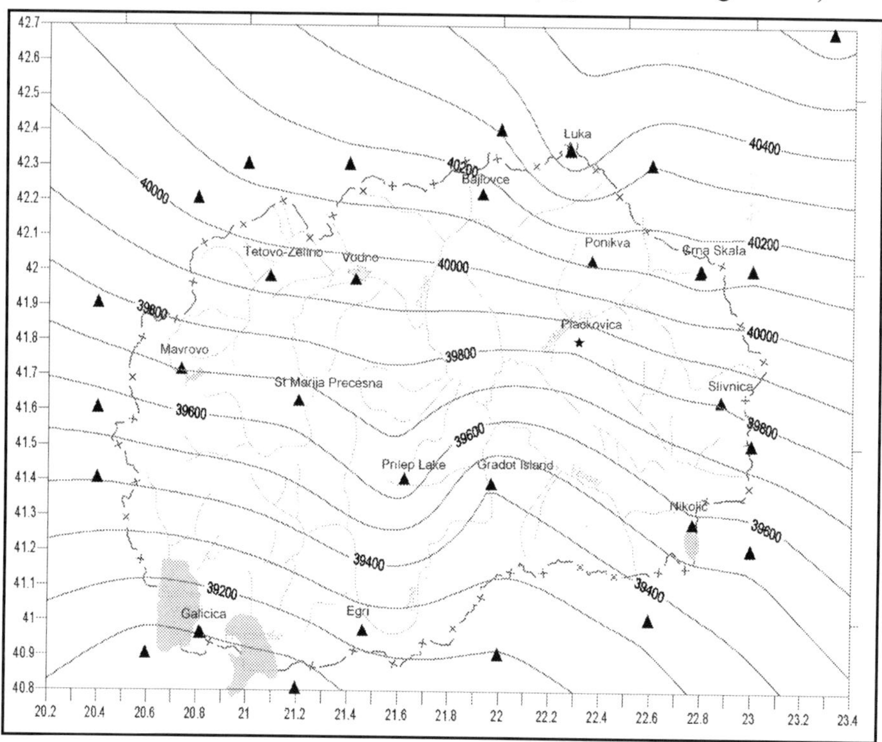


Figure 169. Map of measured values of the vertical (Z) component of the geomagnetic field, epoch 2003.5, in the territory of the Republic of Macedonia.

## 5. IGRF model for 2003.5 for the territory of the Republic of Macedonia

Models for all components of the geomagnetic field, for epoch 2003.5, were compiled for the territory of the Republic of Macedonia using the IGRF model. The USGS on-line calculator was used (available at the web site: http://geomag.usgs.gov/geomag/geomagAWT.html)

The values so obtained for all components of the geomagnetic field are shown in Table 36 and Table 37:

Table 36. The values of declination, inclination, and total intensity in Macedonia obtained from the IGRF 2000 model.

| | | IGRF 2003.5 | | |
|---|---|---|---|---|
| No | Repeat station | D (°) | I (°) | F (nT) |
| 1 | Tetovo | 3.1091 | 58.7707 | 46671.47 |
| 2 | Egri | 3.0811 | 57.6867 | 46360.66 |
| 3 | Mavrovo | 3.0292 | 58.4521 | 46532.72 |
| 4 | Plackovica | 3.2826 | 58.6581 | 46709.02 |
| 5 | Slivnica | 3.3486 | 58.5062 | 46684.85 |
| 6 | Vodno | 3.1616 | 58.7873 | 46696.06 |
| 7 | Bailovce | 3.2623 | 59.0841 | 46817.57 |
| 8 | Gradot Island | 3.1937 | 58.1910 | 46552.38 |
| 9 | Nikolic | 3.3003 | 58.1185 | 46580.13 |
| 10 | Ponikva | 3.3102 | 58.9081 | 46768.00 |
| 11 | St. Marija Precesna | 3.0967 | 58.3925 | 46555.70 |
| 12 | Crna skala | 3.3748 | 58.9086 | 46812.76 |
| 13 | Luka | 3.3275 | 59.2391 | 46873.96 |
| 14 | Galicica | 2.9818 | 57.6248 | 46279.39 |
| 15 | Prilep Lake | 3.1419 | 58.1804 | 46515.68 |

Table 37. The values of H, X, Y, and Z components in Macedonia obtained from the IGRF 2000 model.

| | | IGRF 2003.5 | | | |
|---|---|---|---|---|---|
| No | Repeat station | H(nT) | X(nT) | Y(nT) | Z(nT) |
| 1 | Tetovo - Zelino | 24197.47 | 24161.85 | 1312.40 | 39908.75 |
| 2 | Egri | 24782.00 | 24746.17 | 1332.03 | 39181.16 |
| 3 | Mavrovo | 24346.42 | 24312.40 | 1286.59 | 39655.34 |
| 4 | Plackovica | 24295.39 | 24255.53 | 1391.18 | 39893.19 |
| 5 | Slivnica | 24388.49 | 24346.85 | 1424.57 | 39808.00 |
| 6 | Vodno | 24198.64 | 24161.81 | 1334.61 | 39936.80 |
| 7 | Bailovce | 24053.92 | 24014.94 | 1368.83 | 40165.83 |
| 8 | Gradot Island. | 24537.26 | 24499.15 | 1367.00 | 39560.67 |
| 9 | Nikolic | 24601.94 | 24561.13 | 1416.31 | 39553.17 |
| 10 | Ponikva | 24151.58 | 24111.28 | 1394.55 | 40049.31 |
| 11 | St. Marija Precesna | 24399.72 | 24364.10 | 1318.08 | 39649.55 |
| 12 | Crna skala | 24174.36 | 24132.44 | 1423.08 | 40087.83 |
| 13 | Luka | 23974.01 | 23933.59 | 1391.54 | 40279.21 |

| No | Repeat station | IGRF 2003.5 | | | |
|---|---|---|---|---|---|
| | | H(nT) | X(nT) | Y(nT) | Z(nT) |
| 14 | Galicica | 24780.85 | 24747.30 | 1289.07 | 39085.69 |
| 15 | Prilep Lake | 24525.20 | 24488.34 | 1344.19 | 39524.96 |

The following graphs (Figure 170 – Figure 176) show components of the geomagnetic field of Republic of Macedonia.

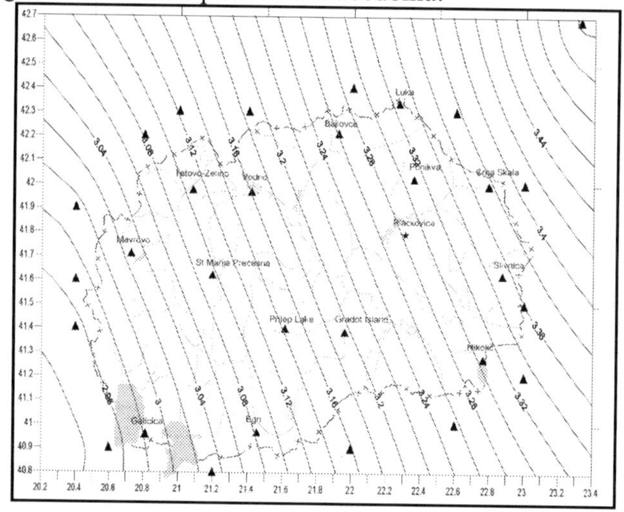


Figure 170. Map of declination (D) from the IGRF 2000 model for epoch 2003.5, in the territory of the Republic of Macedonia.

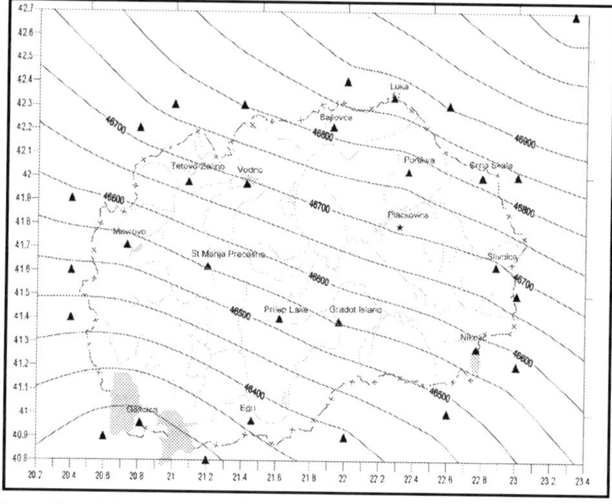


Figure 171. Map of total intensity (T) from the IGRF 2000 model for epoch 2003.5, in the territory of the Republic of Macedonia.

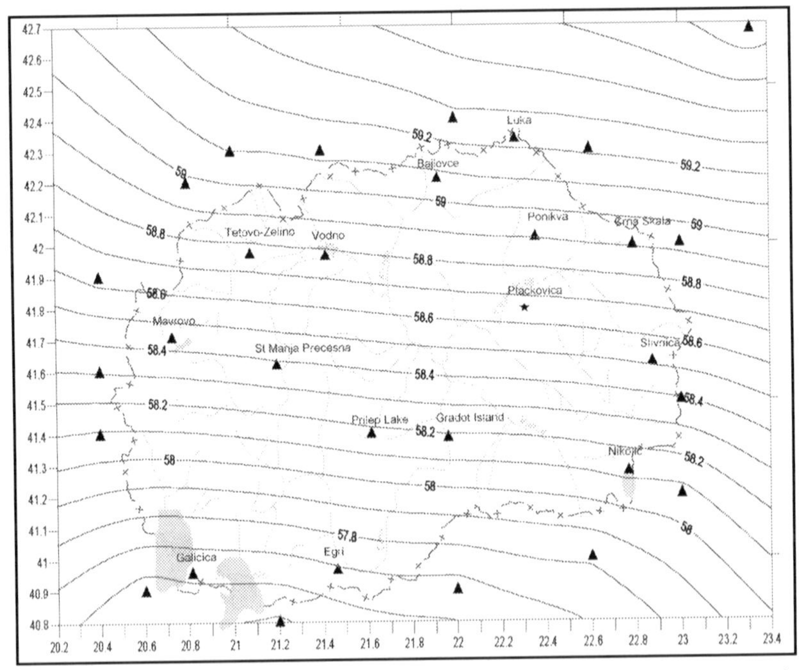


Figure 172. Map of inclination (I) from the IGRF 2000 model for epoch 2003.5, in the territory of the Republic of Macedonia.

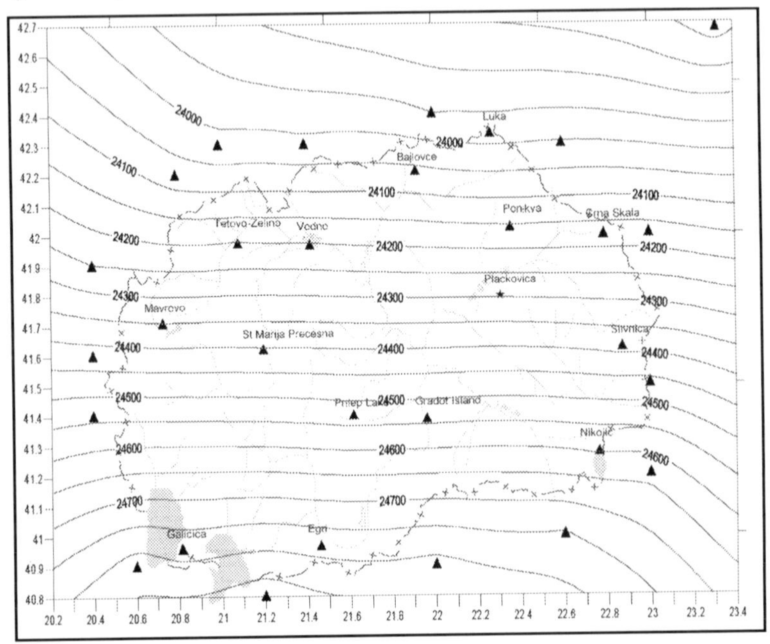


Figure 173. Map of the horizontal component (H) from the IGRF 2000 model for epoch 2003.5, in the territory of the Republic of Macedonia.

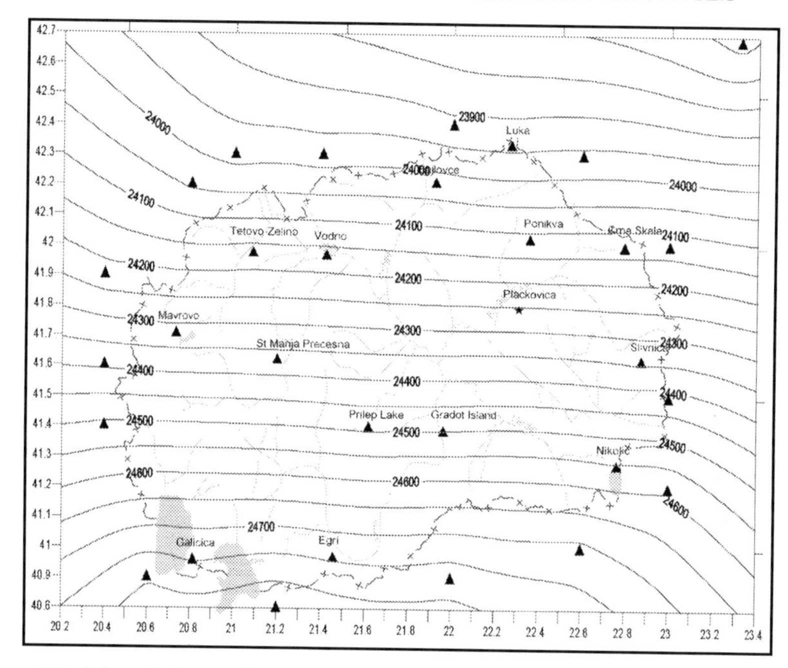


Figure 174. Map of the northern (X) component from the IGRF 2000 model for epoch 2003.5, in the territory of the Republic of Macedonia.

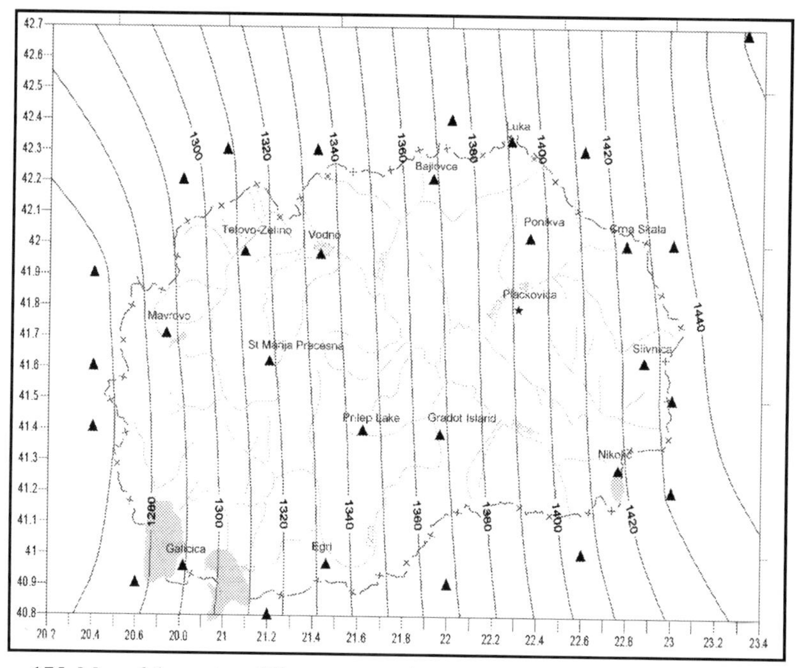


Figure 175. Map of the eastern (Y) component from the IGRF 2000 model for epoch 2003.5, on the territory of the Republic of Macedonia.

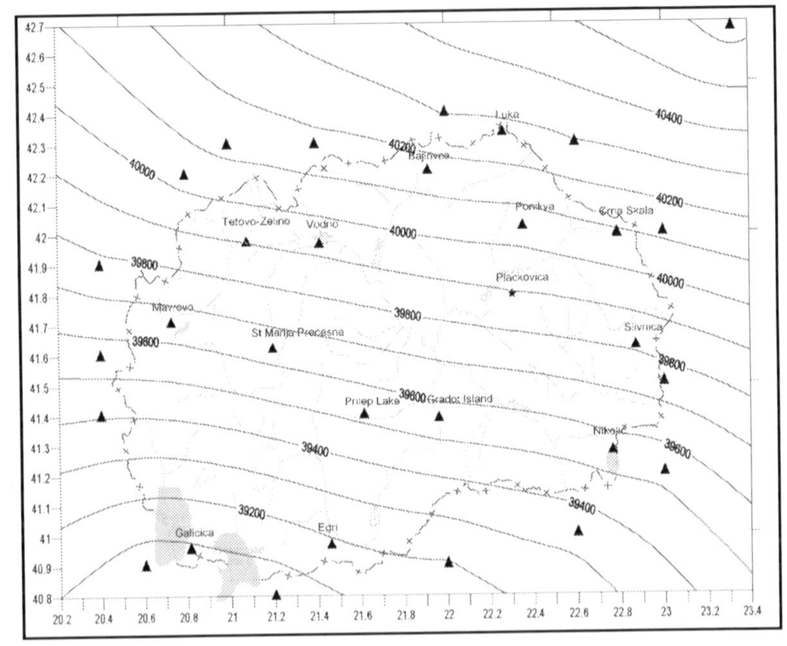


Figure 176. Map of the vertical (Z) component from the IGRF 2000 model for epoch 2003.5, in the territory of the Republic of Macedonia.

## 6. Correlation of components of the geomagnetic field in the Republic of Macedonia between model 2003.5 and IGRF model 2003.5

By comparing the data of the normal field deduced from field measurements and the data from the IGRF model for epoch 2003.5, one can come to the conclusion that the differences between the datasets lie in the intervals given in Table 38:

Table 38.

| |
|---|
| $0.0137^{o} < \Delta D < 0.3473^{o}$ |
| $0.0003^{o} < \Delta I < 0.1729^{o}$ |
| $1.32\ nT < \Delta F < 140.34\ nT$ |
| $5.17\ nT < \Delta H < 169.93\ nT$ |
| $2.81\ nT < \Delta X < 161.86\ nT$ |
| $8.42\ nT < \Delta Y < 151.\ 22\ nT$ |
| $2.06\ nT < \Delta Z < 185.05\ nT$ |

Maps for the differences of the components of the geomagnetic field between the compared models are shown in the following figures (Figure 177 – Figure 183):

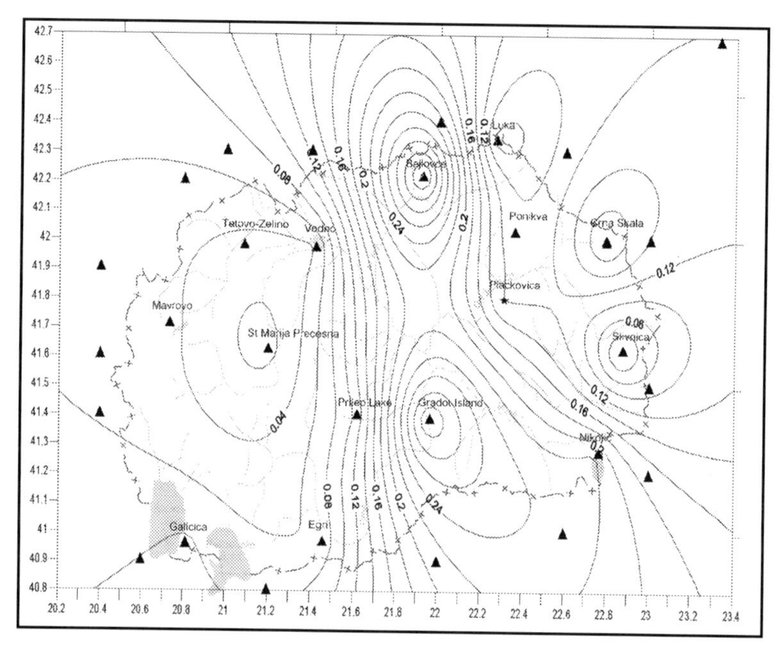


Figure 177. Map of differences between measured values of declination and the IGRF 2000 model values in the territory of the Republic of Macedonia.

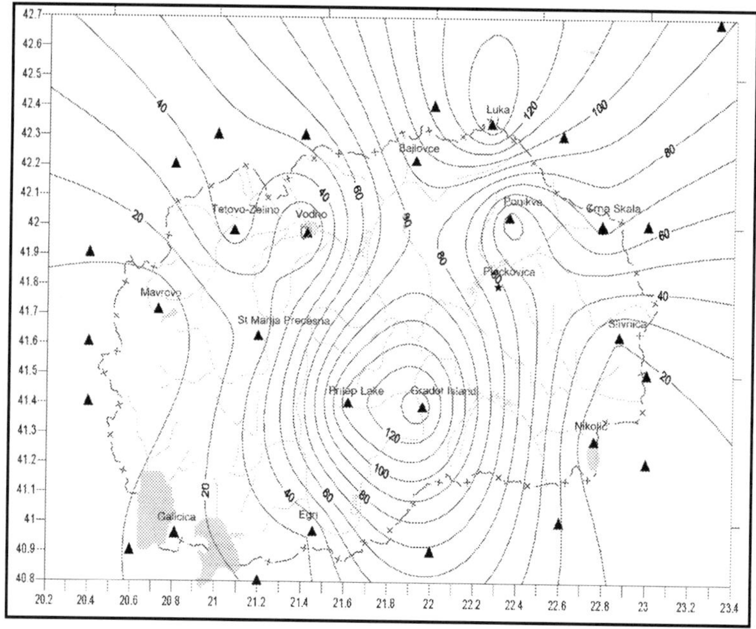


Figure 178. Map of differences between measured values of total intensity and the IGRF 2000 model values in the territory of the Republic of Macedonia.

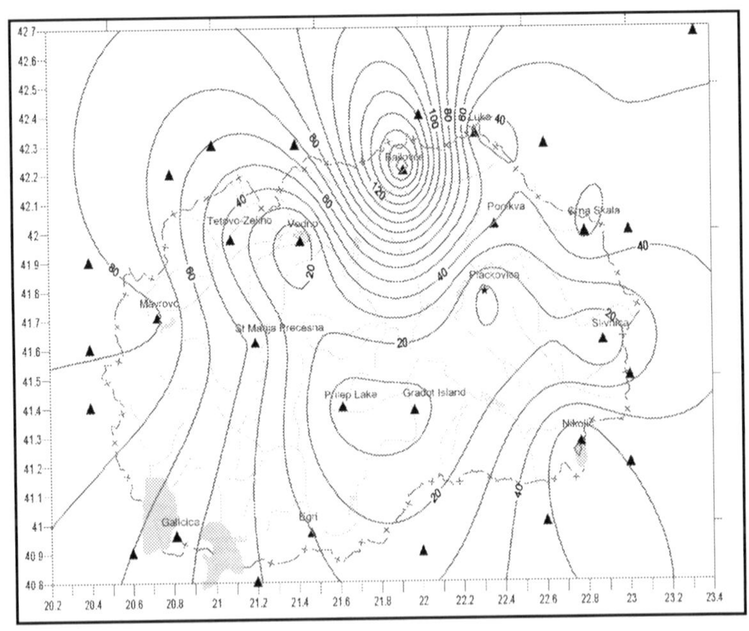


Figure 179. Map of differences between measured values of the horizontal component and the IGRF 2000 model values in the territory of the Republic of Macedonia.

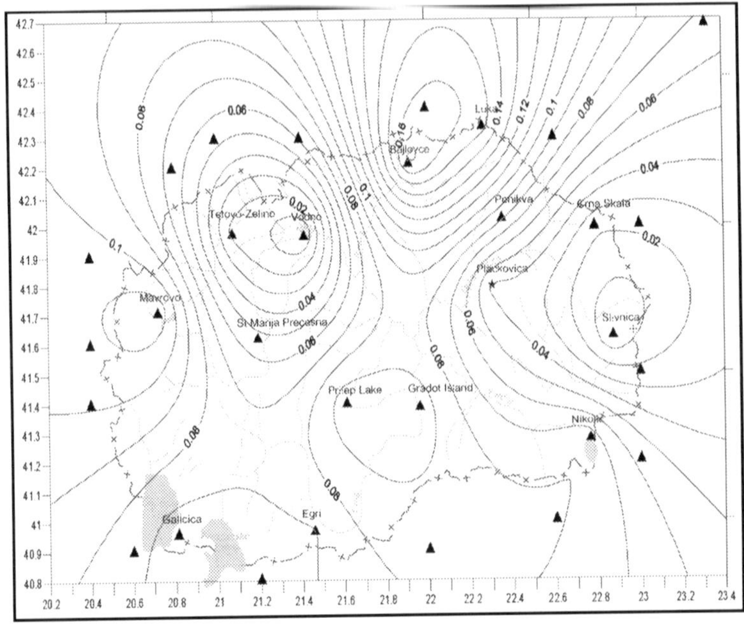


Figure 180. Map of differences between measured values of inclination and the IGRF 2000 model values in the territory of the Republic of Macedonia.

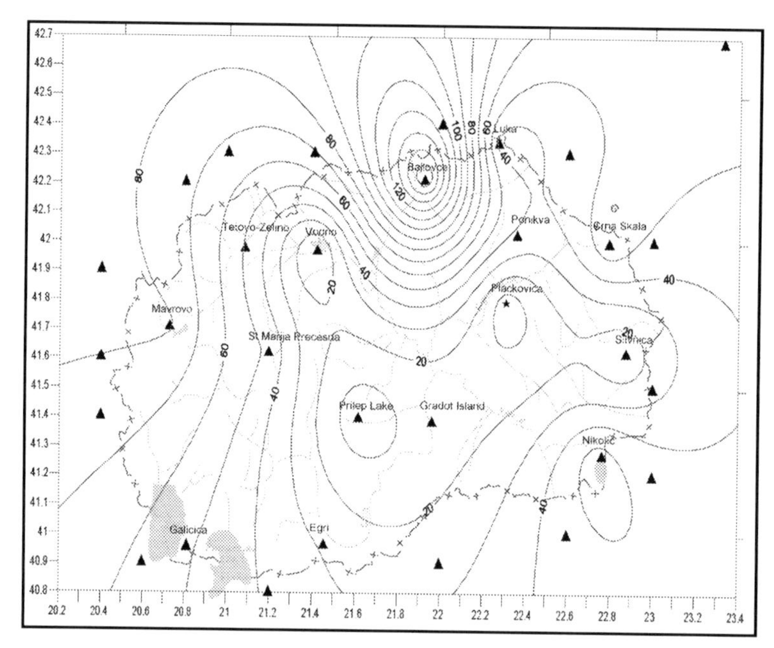


Figure 181. Map of differences between measured values of the northern (X) component and the IGRF 2000 model values in the territory of the Republic of Macedonia.

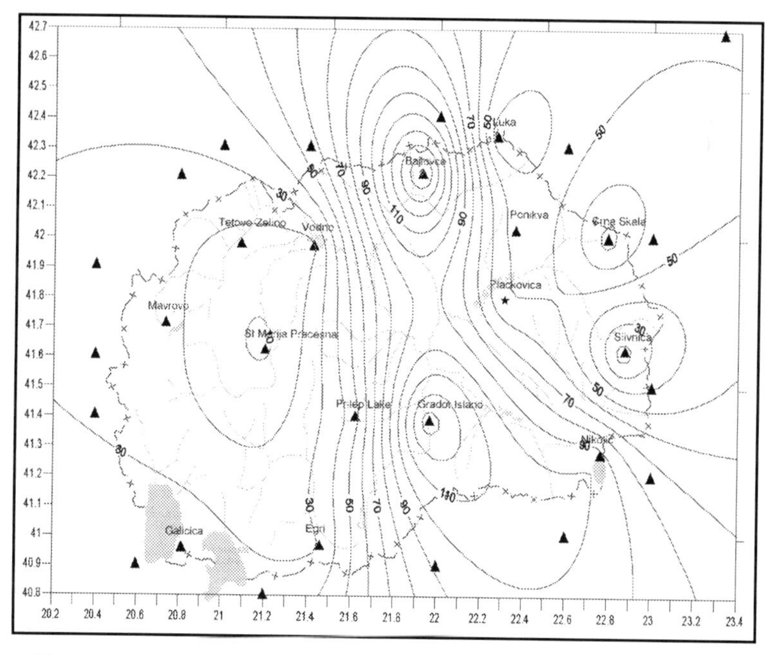


Figure 182. Map of differences between measured values of the eastern (Y) component and the IGRF 2000 model values in the territory of the Republic of Macedonia.

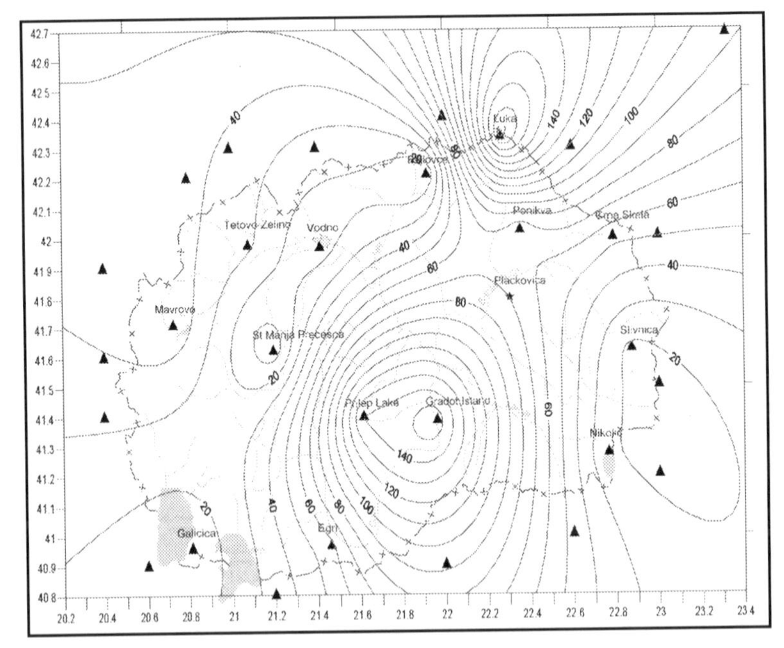


Figure 183. Map of differences between measured values of the vertical (Z) component and the IGRF 2000 model values in the territory of the Republic of Macedonia.

## 7. Interpretation

The differences in the values of the components of the geomagnetic field between the two models result mainly from:

- the relatively small number of repeat stations,
- the global nature of the IGRF model,
- the differences in altitude,
- the complex geologic composition.

The geology of the territory of the Republic of Macedonia has an impact on the geomagnetic field.

More detailed analysis and a dense net of repeat stations are necessary in order to distinguish the normal $T_N$ from the anomalous $T_A$ value in the measured results $T_M$:

$$T_M = T_N + T_A,$$

where T designates any component of the geomagnetic field.

Special investigations are necessary for the separation of the regional $T_A^R$ component, characteristic for the whole territory or large parts of the country, from local anomalies caused by local magnetic causes $T_A^L$.

$$T_A = T_A^R + T_A^L$$

We postulate that $T_A^R$ over the Republic of Macedonia is a large part of the normal model. Therefore we estimate that the differences between the IGRF and the model developed with measured data from Macedonia $T_{MAK}$ will come down to values of $T_A^R$.

$$\left|T_{IGRF} - T_{MAK}\right| = T_A^R$$

## 8. Conclusion

The analysis of the spatial distribution of the geomagnetic components obtained from our first model and of the variations between the two models with an eye on the regional tectonic setting of Macedonia is instructive. One notes in essence that the influence of the different geology of the tectonic units (Western Macedonian zone, Vardar zone, and Eastern Macedonian zone) can be seen in the geomagnetic model of Macedonia. For a more detailed and deeper understanding of the field morphology, a geomagnetic observatory and detailed investigations are necessary.

## References


- Delipetrov, T., 2003, Basics of Geophysics, Faculty of Mining and Geology, Stip, Republic of Macedonia.
- Rasson, L. J., Delipetrov, M., 2004, Magnetic Repeat Station Network Description, Dourbes, Belgium.
- Stefanovic, D., 1978, Methods of Geomagnetic Investigations, Belgrade, Serbia and Montenegro.
- USGS Geomagnetic Field Calculator; http://geomag.usgs.gov/-geomag/geomagAWT.html


**DISCUSSION**

Question (Jean Rasson): The first map you show was computed for a normal field? i.e. a second order polynomial?
Answer (Sanja Panovska): The first map was compiled on the basis of measured results observed at 15 repeat stations, reduced by the observatory data from Aquila, Penteli and Panaguirishte, without reducing according to height.

Question (Angelo De Santis): You determined the normal field for each magnetic component over your country. Have you checked geometrical and physical consistencies among the components provided by your normal field?

Answer (Sanja Panovska): Geometrical consistency among the components has been checked and it is in agreement, whereas physical consistency is within a few nT.

Question (Spomenko J. Mihajlovic): Is it better to compare your normal field values with IGRF 2000 than with the measured values?

Answer (Sanja Panovska): The consistency between IGRF2000 and the values of the normal field will be higher than the comparison between IGRF2000 and the measured values.

# UNDERSTANDING THE GEOMAGNETIC FIELD: A PRECONDITION FOR BETTER LIVING

JORDAN B. ZIVANOVIC[25]
*Faculty of Mining and Geology*
*Department of Geology and Geophysics*
SNEZANA STAVREVA-VESELINOVSKA
*Pedagogical faculty "Goce Delcev"*

**Abstract.** This paper examines the influence of the geomagnetic field on the living world and in particular, its effects on health. The paper also presents the causes of magnetic storms analyses of their impacts and calculation, and interpretation of the K-index

**Keywords:** Magnetic field; magnetogram; magnetic storms; K-index

## 1. Introduction

Certain reactions with the living world occur due to changes in the magnetic field. During the 1990's, because of the computerization of measuring instruments, access to satellite data, and the development of specially applied software, studies of the magnetic field have intensified.

## 2. Magnetic storms

Investigations of magnetic storms date from the early history of mankind. The earliest expeditions, organized to investigate and observe the geomagnetic field in sub-polar and polar zones, started in the 1920's.. At that time, magnetic storms were described as masterpieces of unknown forces hidden in space and were named after well-known women.

[25] To whom correspondence should be addressed at: Faculty of Mining and Geology, Department of Geology and Geophysics, Goce Delcev Str., 2000 Štip, Republic of Macedonia. Email: zjordan@rgf.ukim.edu.mk



*J.L. Rasson and T. Delipetrov (eds.), Geomagnetics for Aeronautical Safety,* 347–353.

Depending on how "beautiful" the magnetic storm was, scientists named the storms St Helena, Queen Elizabeth, St Mary, etc.

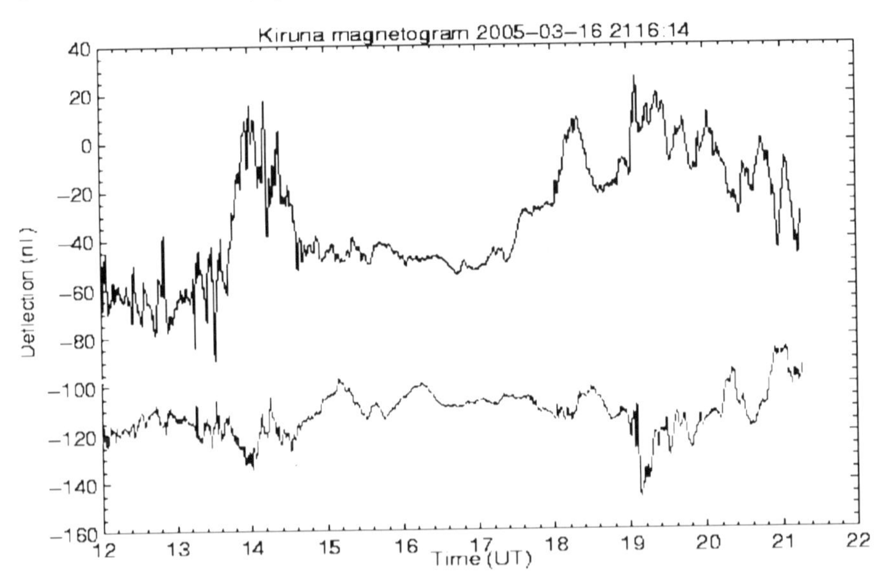


Figure 184. Magnetogram in real time.

During the 1958 International Geophysical Year, instruments in observatories across the world recorded magnetic field variations and pulsations of 0.2 to 30 minute. The study of the daily variation of the Earth's geomagnetic field indicated that the variations were the result of changes in the solar magnetic field, the magnetic fields of sunspots, and changes in the speed and density of solar wind.

Now, with satellite-borne instruments it is possible to carry out sizeable surveys of solar changes and observe the activity and its influence on the earth.

Magnetic storms are related to regular solar explosions. Explosions release strong plasma flow, elementary particles, and electromagnetic radiation. Electromagnetic radiation reaches the earth in 8 minutes, cosmic radiation in several hours, and solar wind in 24 hours. Short wave and cosmic radiation disappear in the atmosphere, but plasma flow is blocked by the Earth's magnetic field. Thus, the Earth's magnetic shield receives the shock, a process that results in disturbance of the Earth's magnetic field or its oscillation (Figure 184 - Figure 186).

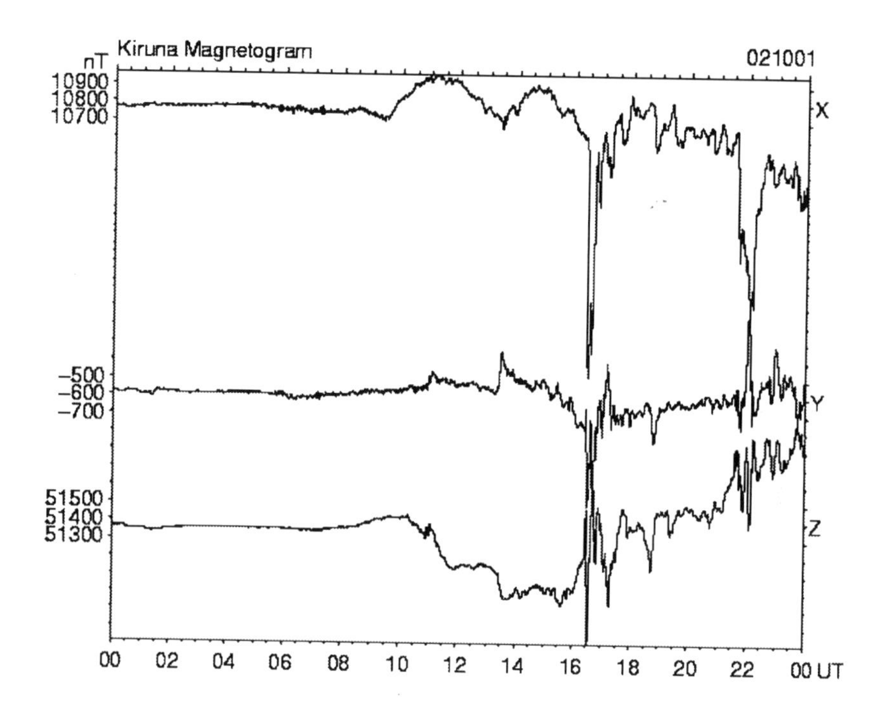


Figure 185. Magnetic storm recorded on October 2, 2001.

A magnetic storm can be mathematically defined as follows: If, over the time interval observed [$t_0$ , tn], which is divided into equal time intervals [ $t_i$, $t_{i+1}$] = $\Delta t$ = 3 hours, a change of the magnetic field value occurs $\Delta m_i$ = [ $m_i$, $m_{i+1}$], (where $m_i$ and $\Delta m$ are in [nT]) which is higher than some value given in advance $\Delta m$ (Table 39), at that moment a magnetic storm occurs.

The K-index, introduced by Bartels in 1949, denotes irregular changes of the geomagnetic field over a three-hour time interval. It is calculated as the average value of deviation from normal for the two horizontal components of magnetic field observed by base stations all over the world. K-index values are scaled from 0 to 9. The magnitude of change that defines the maximum value of the K-index varies for each base station (Table 39).

Magnetic storms and solar activity, also called heliophysical fields, of 10,000 nT to 50,000 nT size have not been studied sufficiently. It is well known that during the occurrence of such activity, a large number of accidents take place, there is an increase in job absences, and higher rates of human mortality are reported.

Magnetic fields may have positive or negative effects on living organisms depending on their intensity, frequency, orientation, exposure time, and their origins.

Negative effects include disturbances of the central nervous system, cardiovascular system, and immunological system.

The magnetic field has a positive effect when used in magnetotherapy (used in treatment of the nervous system, etc).

Most of the processes in the human body are based on electromagnetic activity, changes in the speed of chemical reactions, and the speed of nerve impulses. A major question is why weak magnetic fields cause the strongest effects on humans when they are surrounded by strong magnetic fields such as the Earth's magnetic field or artificial electromagnets.

Table 39. The 13 geomagnetic observatories and appropriate values of changes in maximum K-index (K = 9).

| **Observatory** | | | | | **Geographic** | | **Geomagnetic** | | |
|---|---|---|---|---|---|---|---|---|---|
| # | **Code** | **Name** | **Location** | **Active** | **Lat.** | **Long.** | **Lat.*** | **Long.*** | **K = 9** |
| 1 | LER | Lerwick | Scotland | 1932-present | 60°08' | 358°49' | 62.0° | 89.2° | 1000 nT |
| 2 | MEA | Meanook | Canada | 1932-present | 54°37' | 246°40' | 61.7° | 305.7° | 1500 nT |
| 3 | SIT | Sitka | Alaska (USA) | 1932-present | 57°03' | 224°40' | 60.4° | 279.8° | 1000 nT |
| 4 | ESK | Eskdalemuir | Scotland | 1932-present | 55°19' | 356°48' | 57.9° | 83.9° | 750 nT |
| 5 | LOV | Lovö | Sweden | 1954-2004 | 59°21' | 17°50' | 57.9° | 106.5° | 600 nT |
| | UPS | Uppsala | Sweden | 2004-present | 59°54' | 17°21' | 58.5° | 106.4° | 600 nT |
| 6 | AGN | Agincourt | Canada | 1932-1969 | 43°47' | 280°44' | 54.1° | 350.5° | 600 nT |
| 6 | OTT | Ottawa | Canada | 1969-present | 45°24' | 284°27' | 55.8° | 355.0° | 750 nT |
| 7 | RSV | Rude Skov | Denmark | 1932-1984 | 55°51' | 12°27' | 55.5° | 99.4° | 600 nT |
| | BFE | Brorfelde | Denmark | 1984-present | 55°37' | 11°40' | 55.4° | 98.6° | 600 nT |
| 8 | ABN | Abinger | England | 1932-1957 | 51°11' | 359°37' | 53.4° | 84.5° | 500 nT |
| | HAD | Hartland | England | 1957-present | 50°58' | 355°31' | 54.0° | 80.2° | 500 nT |
| 9 | WNG | Wingst | Germany | 1938-present | 53°45' | 9°04' | 54.1° | 95.1° | 500 nT |
| 10 | WIT | Witteveen | Netherland | 1932-1988 | 52°49' | 6°40' | 53.7° | 92.3° | 500 nT |

| Observatory | | | | | Geographic | | Geomagnetic | | |
|---|---|---|---|---|---|---|---|---|---|
| # | **Code** | **Name** | **Location** | **Active** | **Lat.** | **Long.** | **Lat.*** | **Long.*** | **K = 9** |
| | NGK | Niemegk | Germany | 1988-present | 52°04' | 12°41' | 51.9° | 97.7° | 500 nT |
| 11 | CLH | Cheltenham | USA | 1932-1957 | 38°42' | 283°12' | 49.1° | 353.8° | 500 nT |
| | FRD | Fredericksburg | USA | 1957-present | 38°12' | 282°38' | 48.6° | 353.1° | 500 nT |
| 12 | TOO | Toolangi | Australia | 1972-1981 | -37°32' | 145°28' | -45.6° | 223.0° | 500 nT |
| | CNB | Canberra | Australia | 1981-present | -35°18' | 149°00' | -42.9° | 226.8° | 450 nT |
| 13 | AML | Amberley | New Zealand | 1932-1978 | -43°09' | 172°43' | -46.9° | 254.1° | 500 nT |
| | EYR | Eyrewell | New Zealand | 1978-present | -43°25' | 172°21' | -47.2° | 253.8° | 500 nT |

*) After IGRF model 'IGRF 2000', Earth Planets Space, Vol. 52 (No. 12)

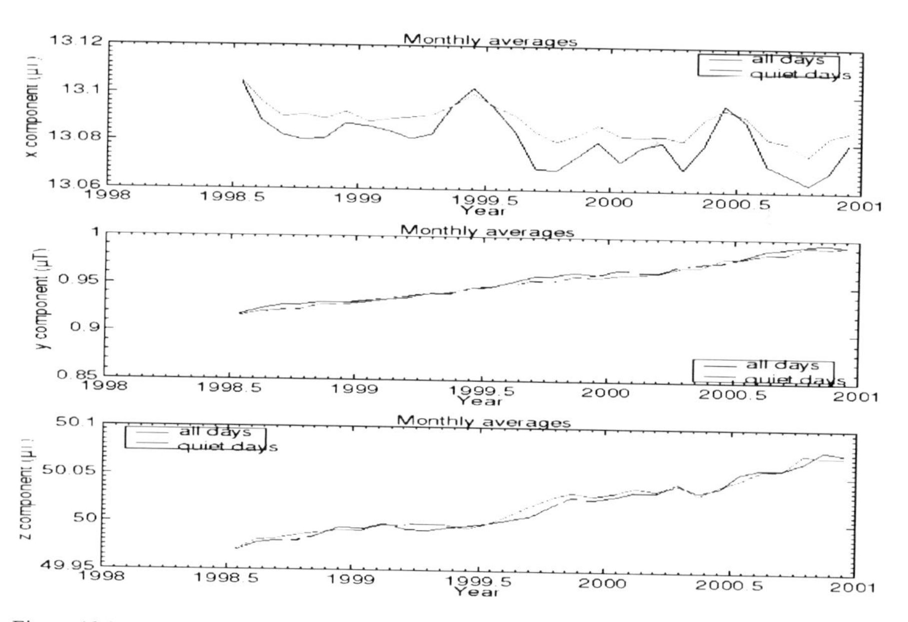


Figure 186. Magnetogram components from 1988 to 2001.

Table 40. Coordinates of Grocka, Serbia Geomagnetic Observatory.

| Observatory | | | | | Geographic | | Geomagnetic | |
|---|---|---|---|---|---|---|---|---|
| # | Code | Name | Location | Active | Lat. | Long. | Lat.* | Long.* |
| | GCK | Grocka | Serbia | 1957- present | 44°36' N | 20°46' E | 43.4° | 102.3° |

Table 41. K-index scale for the Grocka Geomagnetic Observatory.

| K-index | 0 | 1 | 2 | 3 | 4 | 5 | 6 | 7 | 8 | 9 |
|---|---|---|---|---|---|---|---|---|---|---|
| Amplitude [nT] | 0 | 3 | 7 | 15 | 27 | 48 | 80 | 140 | 230 | 350 |

According to some, a human's electromagnetic field is most active in the area of low frequencies from 0.01-100Hz. Brain activity is carried out at ultra low frequencies. Brain waves and magnetic storm waves can interfere with one another.

Russian scientists say men perceive magnetic storms only as information. The threshold of susceptibility of the human organism depends on its reaction. Some people have low thresholds and they perceive changes of the geomagnetic field that others do not feel.

Disturbed blood pressure in the cerebral system can have a negative impact on human health.

Human beings are sensitive to the Earth's magnetic field and its variations. But, with time and an enormous increase of iron they become less sensitive.

Oscillations of the geomagnetic field can be interpreted as a signal for coming danger, earthquake, tsunami, etc. The recent tsunami events in the Indian Ocean confirm this. It has been proven that a change in the Earth's magnetic field took place. On some islands inhabited by indigenous people, who avoided contact with civilization, there were no casualties; although those islands were the first to be affected by the tsunami waves. The people and the animals felt the field changes and moved to the hills in advance of the waves.

Laboratory tests have shown negative effects of rapid changes of the magnetic field. These rapid changes are the so-called pulsations and micropulsations in frequencies from 0.1 Hz to 10 Hz.

Periodic changes of the polarity of the magnetic field (a short time period geologically) have influenced the genetic system in terms of mutation and hereditary changes.

In is generally known that migrant birds navigate by the Earth's magnetic field. It is also known that pigeons maintain their direction of flight to within 0.3 degrees. Birds register changes in the magnitude of the magnetic field on the order of 1 nT and bees about 5 nT. Their nerve endings contain microscopic grains of magnetite ($Fe_3O_4$). During magnetic storms their mechanism for orientation is distorted, so that they behave in an unusual manner.

## 3. Conclusion

Measurements carried out in the Republic of Macedonia as part of the project to establish a geomagnetic observatory are not sufficient for the complex investigation on the effects of the geomagnetic field on the health of people and animals.

Permanent observation of the Earth's magnetic field would provide predictions of biological behavior which, if broadcast on the radio or other media, would be useful for the population of the Republic of Macedonia.

There is a need to finalize the project so that geomagnetic field data from Macedonia can be used, not only in a conventional manner, but be used to relate biology, medicine, ecology, and other disciplines.

## References

1. Komatina M. M.: (2001), Medicinska geologija, Tellur, Beograd
2. Mihajlovic J. S.: (1996), The morphology of geomagnetic field variations registered on geomagnetic observatory Grocka in period 1958 - 1990, Geomagnetni Institut Grocka, Beograd, Jugoslavija
3. http://www.ngdc.noaa.gov/IAGA/vmod/igrf.html
4. http://www.irf.se/mag/

## POSTFACE

At the moment of submitting the manuscript to the publisher, and while going through the many articles collected for this book, we realize that this undertaking would not have been possible without the efforts, support and help provided by many individuals, institutes and institutions.

First of all we want to extend our gratitude to NATO and Dr F. Pedrazzini, Programme Director Chemistry/Biology/Physics, for making this Advanced Research Workshop and the related book publication possible at all. We appreciated his tactful and discrete guidance in the preparation process of the Workshop and the fact that the organizers were offered an almost total freedom to act.

We wish to thank all speakers, writers and participants. We appreciated the relaxed and friendly atmosphere in which the Workshop and the article writing took place. Considering the recent history of the Former Yugoslavia and the Balkans, it was not obvious and could not be taken for granted.

We are indebted to the Faculty of Geology and Mining in Stip for providing much of the Workshop's logistics and the Director General of the Royal Meteorological Institute of Belgium Dr H Malcorps for allowing his staff to spend time, travel and participate as organizer and speakers.

We would also like to thank Betsy and Alan Berarducci for reading, correcting and polishing the English language of the papers presented in this volume. We feel that their contribution has added a lot of value to the final text.

Finally we wish to mention appreciation to the beautiful ancient city of Ohrid and its inhabitants for the wonderful hospitality.

Jean Rasson & Todor Delipetrov, Editors,<br>Dourbes - Štip,<br>February 2006.

## ADDRESSES OF AUTHORS


| | | |
|---|---|---|
| Alan M Berarducci | Geomagnetism Group, USGS Golden Box 25045, MS966, DFC Denver CO 80225 USA | berarducci@usgs.gov |
| Mario Brkic | University of Zagreb Faculty of Geodesy, Deprtment of Geomatics Kaciceva 26, HR 10000 Zagreb, Croatia | mario.brkic@geof.hr |
| Ivan Assenov Butchvarov | Geophysical Institute Acad. G. Bonchev St., Block 3, Sofia 1113, Bulgaria | Buch@geophys.bas.bg |
| Tomislav Bašić | Faculty of Geodesy, Kačićeva 26, HR 10000 Zagreb, Croatia | |
| B. Cağlayan | B.U.Kandilli Observatory and Earthquake Research Institute Cengelkoy 34680 Istanbul-Turkey | |
| Cengiz Celik | B.U.Kandilli Observatory and Earthquake Research Institute Cengelkoy 34680 Istanbul-Turkey | celikc@boun.edu.tr |
| Ilya Cholakov | Geophysical Institute Acad. G. Bonchev St., Block 3, Sofia 1113, Bulgaria | Cholakov@inter-pan.net |
| Rudi Čop | Faculty of Maritime Studies and Transport Pot pomorscakov 4, 6320 Portoroz, Slovenia | rudi.cop@fpp.edu |
| Angelo De Santis | Istituto Nazionale di Geofisica e Vulcanologia Sezioni di Roma Via di Vigna Murata, 605, I-00143 ROMA, Italia | desantisag@ingv.it |
| Marjan Delipetrov | Faculty of Mining and Geology, Department of Geology and Geophysics "Goce Delcev" 89, Stip, 2000, R. Macedonia | marjan@rgf.ukim.edu.mk |
| Todor Delipetrov | Faculty of Mining and Geology, Department of Geology and Geophysics "Goce Delcev" 89, Stip, 2000, R. Macedonia | todor.delipetrov@gmail.com |
| Guido Dominici | Istituto Nazionale di Geofisica e Vulcanologia Sezioni di Roma Via di Vigna Murata, 605, I-00143 ROMA, Italia | dominici@ingv.it |
| Bejo Duka | Fakulteti i Shkencave Natyrore Bulevardi Zog I, Tirane, ALBANIA | bejo_duka@yahoo.com |
| Dušan Fefer | University of Ljubljana, Faculty of Electrical Engineering, Ljubljana, Slovenia | |

| | | |
|---|---|---|
| Luis R. Gaya-Piqué | Istituto Nazionale di Geofisica e Vulcanologia Sezioni di Roma Via di Vigna Murata, 605, I-00143 ROMA, Italia | |
| Y. Güngörmüş | B.U.Kandilli Observatory and Earthquake Research Institute Cengelkoy 34680 Istanbul-Turkey | |
| Lazslo Hegymegi | Eotvos Lorand Geophysical Institute of Hungary Columbus utca 17-23 Budapest, H-1145, Hungary | hegymegi@elgi.hu |
| Valery Korepanov | Lviv Centre of Institute of Space Research National Academy of Sciences and National Space Agency of Ukraine 5-A Naukova St., 79000, LVIV, UKRAINE | vakor@isr.lviv.ua |
| Caslav Lazovic | Geomagnetic Institute Geomagnetic Observatory Grocka Put za Umcare 3, Grocka 11306, Belgrade SERBIA AND MONTENEGRO | |
| Danko Markovinović | Faculty of Geodesy, Kačićeva 26, HR 10000 Zagreb, Croatia | |
| Juergen Matzka | Geophysikalisches Observatorium Ludwigshöhe 8, 82256 Fuerstenfeldbruck, Germany | matzka@lmu.de |
| Spomenko J. Mihajlovic | Geomagnetic Institute Geomagnetic Observatory Grocka Put za Umcare 3, Grocka 11306, Belgrade SERBIA AND MONTENEGRO | mihas@sezampro.yu |
| Nenad Novkovski | Faculty of Natural Sciences and Mathematics, Institute of Physics Gazi Baba bb PO Box 162 1000 Skopje, R. Macedonia | nenad@iunona.pmf.ukim.edu.mk |
| Blagica Paneva | Faculty of Mining and Geology, Department of Geology and Geophysics, 2000 Stip, Macedonia | blagica@rgf.ukim.edu.mk |
| Sanja Panovska | Faculty of Mining and Geology, Department of Geology and Geophysics "Goce Delcev" 89, Stip, 2000, R. Macedonia | panovskasanja@yahoo.com |
| Lazo Pekevski | Seismological Observatory, P.O.Box 422 1000 Skopje, R. Macedonia | lpekevski@seismobsko.pmf.ukim.edu.mk |
| Dragan Popeskov | Geomagnetic Institute Geomagnetic Observatory Grocka Put za Umcare 3, Grocka 11306, Belgrade SERBIA AND MONTENEGRO | |
| Jean L. Rasson | Institut Royal Meteorologique, Centre de Physique du Globe Rue de Fagnolle, 2 B-5670 Viroinval, Dourbes, Belgium | jr@oma.be |
| Milan Rezo | Faculty of Geodesy, Kačićeva 26, HR 10000 Zagreb, Croatia | |

| | | |
|---|---|---|
| Nenad Smiljanic | Geomagnetic Institute Geomagnetic Observatory Grocka Put za Umcare 3, Grocka 11306, Belgrade SERBIA AND MONTENEGRO | smiljanic@beotel.yu |
| Snezana Stavreva-Veselinovska | Pedagogical faculty "Goce Delcev" , 2000 Stip Macedonia | sstavreva@pfst.ukim.edu.mk |
| Stanoja Stojmenov | Faculty of Natural Sciences and Mathematics, Institute of Physics Gazi Baba bb PO Box 162 1000 Skopje, R. Macedonia | stanojs@iunona.pmf.ukim.edu.mk |
| Danijel Šugar | Faculty of Geodesy, Kačićeva 26, HR 10000 Zagreb, Croatia | |
| E. Tolak | B.U.Kandilli Observatory and Earthquake Research Institute Cengelkoy 34680 Istanbul-Turkey | |
| J. Miquel Torta | Observatori de l'Ebre 43520 Roquetes (Tarragona), SPAIN | jmtorta@obsebre.es |
| Roberta Tozzi | Istituto Nazionale di Geofisica e Vulcanologia Sezioni di Roma Via di Vigna Murata, 605, I-00143 ROMA, Italia | |
| M.K. Tuncer | B.U.Kandilli Observatory and Earthquake Research Institute Cengelkoy 34680 Istanbul-Turkey | tuncer@boun.edu.tr |
| Sebastian van Loo | Institut Royal Meteorologique, Centre de Physique du Globe B-5670 Viroinval, Dourbes, Belgium | sebvl@oma.be |
| O. Yazıcı-Çakın | B.U.Kandilli Observatory and Earthquake Research Institute Cengelkoy 34680 Istanbul-Turkey | |
| Jordan Zivanovic | Faculty of Mining and Geology, Department of Geology and Geophysics "Goce Delcev" 89, Stip, 2000, R. Macedonia | zjordan@rgf.ukim.edu.mk |
| M. Zobu | B.U.Kandilli Observatory and Earthquake Research Institute Cengelkoy 34680 Istanbul-Turkey | |

# INDEX

## A



## B



## C



## D



## E



## F



## G



## H

# I



# K



# L



# M



# N



# O



# P



# Q

## R



## S



## T



## V



## W

9 781402 050244